Nanomaterials for Drug Delivery and Cancer Therapy

Nanomaterials for Drug Delivery and Cancer Therapy

Editors

Fiore P. Nicoletta
Francesca Iemma

MDPI • Basel • Beijing • Wuhan • Barcelona • Belgrade • Manchester • Tokyo • Cluj • Tianjin

Editors
Fiore P. Nicoletta
Department of Pharmacy, Health and Nutritional Sciences
University of Calabria
Rende
Italy

Francesca Iemma
Department of Pharmacy, Health and Nutritional Sciences
University of Calabria
Rende
Italy

Editorial Office
MDPI
St. Alban-Anlage 66
4052 Basel, Switzerland

This is a reprint of articles from the Special Issue published online in the open access journal *Nanomaterials* (ISSN 2079-4991) (available at: www.mdpi.com/journal/nanomaterials/special_issues/nanomaterials_drug_therapy).

For citation purposes, cite each article independently as indicated on the article page online and as indicated below:

LastName, A.A.; LastName, B.B.; LastName, C.C. Article Title. *Journal Name* **Year**, *Volume Number*, Page Range.

ISBN 978-3-0365-6703-7 (Hbk)
ISBN 978-3-0365-6702-0 (PDF)

Contents

About the Editors vii

Fiore Pasquale Nicoletta and Francesca Iemma
Nanomaterials for Drug Delivery and Cancer Therapy
Reprinted from: *Nanomaterials* **2023**, *13*, 207, doi:10.3390/nano13010207 1

Jie Tang, Larry Cai, Chuanfei Xu, Si Sun, Yuheng Liu and Joseph Rosenecker et al.
Nanotechnologies in Delivery of DNA and mRNA Vaccines to the Nasal and Pulmonary Mucosa
Reprinted from: *Nanomaterials* **2022**, *12*, 226, doi:10.3390/nano12020226 5

Rabia Arshad, Iqra Fatima, Saman Sargazi, Abbas Rahdar, Milad Karamzadeh-Jahromi and Sadanand Pandey et al.
Novel Perspectives towards RNA-Based Nano-Theranostic Approaches for Cancer Management
Reprinted from: *Nanomaterials* **2021**, *11*, 3330, doi:10.3390/nano11123330 33

Shei Li Chung, Maxine Swee-Li Yee, Ling-Wei Hii, Wei-Meng Lim, Mui Yen Ho and Poi Sim Khiew et al.
Advances in Nanomaterials Used in Co-Delivery of siRNA and Small Molecule Drugs for Cancer Treatment
Reprinted from: *Nanomaterials* **2021**, *11*, 2467, doi:10.3390/nano11102467 59

Manuela Curcio, Orazio Vittorio, Jessica Lilian Bell, Francesca Iemma, Fiore Pasquale Nicoletta and Giuseppe Cirillo
Hyaluronic Acid within Self-Assembling Nanoparticles: Endless Possibilities for Targeted Cancer Therapy
Reprinted from: *Nanomaterials* **2022**, *12*, 2851, doi:10.3390/nano12162851 79

Nicole F. Bonan, Debbie K. Ledezma, Matthew A. Tovar, Preethi B. Balakrishnan and Rohan Fernandes
Anti-Fn14-Conjugated Prussian Blue Nanoparticles as a Targeted Photothermal Therapy Agent for Glioblastoma
Reprinted from: *Nanomaterials* **2022**, *12*, 2645, doi:10.3390/nano12152645 117

Andrew Dunphy, Kamal Patel, Sarah Belperain, Aubrey Pennington, Norman H. L. Chiu and Ziyu Yin et al.
Modulation of Macrophage Polarization by Carbon Nanodots and Elucidation of Carbon Nanodot Uptake Routes in Macrophages
Reprinted from: *Nanomaterials* **2021**, *11*, 1116, doi:10.3390/nano11051116 139

Guan Zhen He and Wen Jen Lin
Peptide-Functionalized Nanoparticles-Encapsulated Cyclin-Dependent Kinases Inhibitor Seliciclib in Transferrin Receptor Overexpressed Cancer Cells
Reprinted from: *Nanomaterials* **2021**, *11*, 772, doi:10.3390/nano11030772 155

Francesca Brero, Martin Albino, Antonio Antoccia, Paolo Arosio, Matteo Avolio and Francesco Berardinelli et al.
Hadron Therapy, Magnetic Nanoparticles and Hyperthermia: A Promising Combined Tool for Pancreatic Cancer Treatment
Reprinted from: *Nanomaterials* **2020**, *10*, 1919, doi:10.3390/nano10101919 171

About the Editors

Fiore P. Nicoletta

Fiore Pasquale Nicoletta received a MSc and a PhD in Physics at the University of Calabria. He is full Professor of Physical Chemistry at the Department of Pharmacy, Health and Nutritional Sciences –University of Calabria. He has experience in the characterisation and application of the physical-chemical properties of soft matter, thermotropic and lyotropic liquid crystals, membrane models, hydrogel/porous membrane composites, responsive hydrogels and nanoparticles, electro-optical, electro-chromic, photo-chromic, and photo-electrochromic properties of liquid crystal/polymer dispersions. He is co-author of 140 papers in international journals (h-Index: 29 Scopus) and three patents.

Francesca Iemma

Francesca Iemma received a MSc and a PhD in Chemistry at the University of Calabria. She is a full Professor of Pharmaceutical Technology at the Department of Pharmacy, Health and Nutritional Sciences –University of Calabria. She has experience in the synthesis and characterisation of polymeric materials, stimuli responsive/smart hydrogels for biomedical, pharmaceutical and food applications and characterisation of physical-chemical properties of soft matter, hydrogel/membrane composites. She is co-author of 145 papers in international journals (h-Index: 40 Scopus).

nanomaterials

Editorial

Nanomaterials for Drug Delivery and Cancer Therapy

Fiore Pasquale Nicoletta * and Francesca Iemma *

Department of Pharmacy, Health and Nutritional Sciences, University of Calabria, 87036 Rende, Italy
* Correspondence: fiore.nicoletta@unical.it (F.P.N.); francesca.iemma@unical.it (F.I.)

Citation: Nicoletta, F.P.; Iemma, F. Nanomaterials for Drug Delivery and Cancer Therapy. *Nanomaterials* **2023**, *13*, 207. https://doi.org/10.3390/nano13010207

Received: 7 December 2022
Accepted: 11 December 2022
Published: 3 January 2023

In recent decades, the interest in nanomaterials has grown rapidly for their applications in many research fields, including drug delivery and cancer therapy. Polymer, metal, silica, carbon, and hybrid nanoparticles are different kinds of nanomaterials, which are currently available and utilized in a large number of applications for the treatment of disease. Nanoparticles can often be endowed with suitable functionalities to provide nanoformulations with improved performance, including superior pharmacokinetic profile and treatment efficiency, in tailored applications.

This Special Issue of *Nanomaterials*, entitled "Nanomaterials for Drug Delivery and Cancer Therapy", covers the recent advances in the use of nanoparticle systems, with an emphasis on preparation and characterization methods, functionalization chemistry, and their associated applications in drug delivery and cancer therapy. Moreover, some recent advances are presented here, with the aim to provide new ideas and to ignite a discussion among researchers working in this multidisciplinary field.

This Special Issue collects seven recent research articles and four review papers, with the latter being presented at the beginning of the book issue, since they can provide an extensive overview of the topics presented herein, with a discussion of future perspectives.

The first review article, by Jie Tang et al., provides an overview of the recent developments in nucleic acid delivery systems that target airway mucosa for vaccination purposes [1]. The article outlines the appeal of a respiratory mucosal vaccination, but at the same time stresses the challenges of several physical and biological barriers at the airway mucosal site, such as a variety of protective enzymes and mucociliary clearance. The authors discuss in detail the nanotechnologies enabling novel nucleic acid formulations for the efficient delivery of both mRNA transcribed in vitro and nucleic-acid-based vaccines.

In the second review article, Rabia Arshad et al. highlight the latest advances in cancer diagnosis and treatment by RNA nanotechnology, in particular by using multi-functionalized nanoparticles and nanobiosensors with specified ligands able to target the tissues of interest [2]. RNA-conjugated nanomaterials have demonstrated improved sensitivity and selectivity, higher therapeutic efficacy, more accurate diagnosis, lower toxicity, and more site-specific delivery than conventional techniques, resulting in better cytotoxicity management and cost-effectiveness. However, the main drawback of such technology is related to the low number of RNA-based nanoparticles that have progressed to clinical trials.

In the third review article, Shei Li Chung et al. describe the recent advancements in nanotechnology to develop novel co-delivery systems, which involve short-interfering RNA (siRNA) and small-molecule drugs for synergistic cancer therapy [3]. They discuss the problems related to the co-delivery of two distinct anti-tumor agents with different properties, showing some key examples.

The fourth review paper, by Manuela Curcio and colleagues, reports the possibilities offered by self-assembling nanoparticles based on hyaluronic acid in cancer therapy [4]. Such systems conjugate the targeting activity of hyaluronic acid towards cancer cells with the chemical versatility, ease of preparation and scalability of self-assembling nanoparticles. The authors elucidate the different hyaluronic acid derivatization strategies and preparation methods used for the fabrication of different delivery devices. After showing the biological results in in vivo and in vitro models, the pros and cons of each nanosystem are reported, opening a discussion on which approach could represent the most promising strategy for further investigation and the development of effective therapeutic protocols.

This Special Issue also includes seven research articles.

The first paper, by Nicole F. Bonan et al., presents a novel formulation of Prussian Blue nanoparticles (PBNPs) conjugated with an antibody targeting Fn14, the fibroblast growth factor-inducible 14, which is expressed in glioblastoma cell lines [5]. After having characterized the properties of the conjugate anti-Fn14-PBNPs, the authors report its ability to act as a targeted photothermal therapy agent against glioblastoma.

The second paper, by Andrew Dunphy et al., provides evidence of the interplay between macrophages and carbon nanodots, which represent a well-known target in therapeutic treatments and an important nanomaterial for biomedical applications, respectively [6]. After having found that CNDs were non-toxic in a variety of doses and that the expression of CD 206 and CD 68 could be altered by opportune treatments, they examined the potential entrance routes of CNDs into macrophages.

The third paper, by Guan Zhen He and Wen Jen Lin, is focused on the synthesis of a PLGA-PEG-maleimide copolymer, which was used for the encapsulation of Seliciclib [7]. The obtained nanoparticles were decorated with T7 peptide, which is a targeting ligand for the transferrin receptors, generally being overexpressed in cancer cells. The results show that the cellular uptake was, as expected, dependent on the overexpression of transferrin receptors, and that IC50 values were lowered for encapsulated Seliciclib.

The fourth paper, by Brero's group, aimed to evaluate a novel therapeutic protocol for the in vitro treatment of pancreatic cancer BxPC3 cells [8]. They combined hadron therapy by carbon ions with magnetic fluid hyperthermia and found that the new protocol diminished the clonogenic survival to an extent that depended on the radiation type (using photons as the control), and the decrease was amplified both by the hyperthermia protocol and the cellular uptake of magnetic nanoparticles.

The fifth paper, by Dina Farrakhova and co-workers, presents experimental evidence of a new spectroscopic approach to determine the state of a brain tumor and its microenvironment via changes in the fluorescence lifetime of Indocyanine Green (IG) [9]. In particular, IG accumulated in the tumor site with a maximum accumulation at 24 h after systemic administration and led to different values of fluorescence lifetimes for ICG and ICG aggregates, indicating promising suitability for the fluorescent diagnosis of brain tumors.

In the sixth paper, Mei-Hsiu Chen et al. synthesized a complex, Au–OMV, with gold nanoparticles (AuNPs) and outer-membrane vesicles (OMVs) [10]. Au–OMV, when combined with radiotherapy, produced radiosensitizing and immunomodulatory effects that successfully suppressed tumor growth in both subcutaneous G261 tumor-bearing and in situ (brain) tumor-bearing C57BL/6 mice. Longer survival was also noted for in situ tumor-bearing mice treated with Au–OMV and radiotherapy. The mechanisms behind the successful treatment were evaluated.

Finally, the seventh paper, by Manuela Curcio's group, investigated new dual pH/redox-responsive nanoparticles with an affinity for folate receptors, prepared by the combination of two amphiphilic dextran (DEX) derivatives [11]. The first derivative was obtained from a covalent coupling of the polysaccharide with folic acid (FA), whereas the second was derived from the reductive amination step of DEX, followed by condensation with polyethylene glycol 600. After self-assembling, nanoparticles could be destabilized in acidic pH and reducing media, and after doxorubicin loading, the proposed system was able to modulate the drug release in response to different pH and redox conditions. Then, viability and uptake experiments on healthy (MCF-10A) and metastatic cancer (MDA-MB-231) cells proved the potential applicability of the proposed system as a new drug vector in cancer therapy.

We hope that readers enjoy these selected contributions.

Author Contributions: Writing—review and editing, F.P.N. and F.I. All authors have read and agreed to the published version of the manuscript.

Conflicts of Interest: The authors declare no conflict of interest.

References

1. Tang, J.; Cai, L.; Xu, C.; Sun, S.; Liu, Y.; Rosenecker, J.; Guan, S. Nanotechnologies in Delivery of DNA and mRNA Vaccines to the Nasal and Pulmonary Mucosa. *Nanomaterials* **2022**, *12*, 226. [CrossRef] [PubMed]
2. Arshad, R.; Fatima, I.; Sargazi, S.; Rahdar, A.; Karamzadeh-Jahromi, M.; Pandey, S.; Díez-Pascual, A.M.; Bilal, M. Novel Perspectives towards RNA-Based Nano-Theranostic Approaches for Cancer Management. *Nanomaterials* **2021**, *11*, 3330. [CrossRef] [PubMed]
3. Chung, S.L.; Yee, M.S.-L.; Hii, L.-W.; Lim, W.-M.; Ho, M.Y.; Khiew, P.S.; Leong, C.-O. Advances in Nanomaterials Used in Co-Delivery of siRNA and Small Molecule Drugs for Cancer Treatment. *Nanomaterials* **2021**, *11*, 2467. [CrossRef] [PubMed]
4. Curcio, M.; Vittorio, O.; Bell, J.L.; Iemma, F.; Nicoletta, F.P.; Cirillo, G. Hyaluronic Acid within Self-Assembling Nanoparticles: Endless Possibilities for Targeted Cancer Therapy. *Nanomaterials* **2022**, *12*, 2851. [CrossRef] [PubMed]
5. Bonan, N.F.; Ledezma, D.K.; Tovar, M.A.; Balakrishnan, P.B.; Fernandes, R. Anti-Fn14-Conjugated Prussian Blue Nanoparticles as a Targeted Photothermal Therapy Agent for Glioblastoma. *Nanomaterials* **2022**, *12*, 2645. [CrossRef] [PubMed]
6. Dunphy, A.; Patel, K.; Belperain, S.; Pennington, A.; Chiu, N.H.L.; Yin, Z.; Zhu, X.; Priebe, B.; Tian, S.; Wei, J.; et al. Modulation of Macrophage Polarization by Carbon Nanodots and Elucidation of Carbon Nanodot Uptake Routes in Macrophages. *Nanomaterials* **2021**, *11*, 1116. [CrossRef] [PubMed]
7. He, G.Z.; Lin, W.J. Peptide-Functionalized Nanoparticles-Encapsulated Cyclin-Dependent Kinases Inhibitor Seliciclib in Transferrin Receptor Overexpressed Cancer Cells. *Nanomaterials* **2021**, *11*, 772. [CrossRef] [PubMed]
8. Brero, F.; Albino, M.; Antoccia, A.; Arosio, P.; Avolio, M.; Berardinelli, F.; Bettega, D.; Calzolari, P.; Ciocca, M.; Corti, M.; et al. Hadron Therapy, Magnetic Nanoparticles and Hyperthermia: A Promising Combined Tool for Pancreatic Cancer Treatment. *Nanomaterials* **2020**, *10*, 1919. [CrossRef] [PubMed]
9. Farrakhova, D.; Romanishkin, I.; Maklygina, Y.; Bezdetnaya, L.; Loschenov, V. Analysis of Fluorescence Decay Kinetics of Indocyanine Green Monomers and Aggregates in Brain Tumor Model In Vivo. *Nanomaterials* **2021**, *11*, 3185. [CrossRef] [PubMed]
10. Chen, M.-H.; Liu, T.-Y.; Chen, Y.-C.; Chen, M.-H. Combining Augmented Radiotherapy and Immunotherapy through a Nano-Gold and Bacterial Outer-Membrane Vesicle Complex for the Treatment of Glioblastoma. *Nanomaterials* **2021**, *11*, 1661. [CrossRef] [PubMed]
11. Curcio, M.; Paolì, A.; Cirillo, G.; Di Pietro, S.; Forestiero, M.; Giordano, F.; Mauro, L.; Amantea, D.; Di Bussolo, V.; Nicoletta, F.P.; et al. Combining Dextran Conjugates with Stimuli-Responsive and Folate-Targeting Activity: A New Class of Multifunctional Nanoparticles for Cancer Therapy. *Nanomaterials* **2021**, *11*, 1108. [CrossRef] [PubMed]

nanomaterials

MDPI

Review

Nanotechnologies in Delivery of DNA and mRNA Vaccines to the Nasal and Pulmonary Mucosa

Jie Tang [1,2], Larry Cai [2], Chuanfei Xu [3], Si Sun [3], Yuheng Liu [3], Joseph Rosenecker [1,*] and Shan Guan [1,3,*]

1 Department of Pediatrics, Ludwig-Maximilians University of Munich, 80337 Munich, Germany; j.tang3@uq.edu.au
2 Australian Institute for Bioengineering and Nanotechnology, The University of Queensland, Brisbane 4072, Australia; larry.cai@uq.edu.au
3 National Engineering Research Center of Immunological Products, Department of Microbiology and Biochemical Pharmacy, Third Military Medical University, Chongqing 400038, China; xu1982978434@163.com (C.X.); sunsi90@163.com (S.S.); yhliu2017@lzu.edu.cn (Y.L.)
* Correspondence: Joseph.Rosenecker@med.uni-muenchen.de (J.R.); guanshan87@163.com (S.G.); Tel.: +49-89-440057713 (J.R.); +86-23-68771645 (S.G.)

Abstract: Recent advancements in the field of in vitro transcribed mRNA (IVT-mRNA) vaccination have attracted considerable attention to such vaccination as a cutting-edge technique against infectious diseases including COVID-19 caused by SARS-CoV-2. While numerous pathogens infect the host through the respiratory mucosa, conventional parenterally administered vaccines are unable to induce protective immunity at mucosal surfaces. Mucosal immunization enables the induction of both mucosal and systemic immunity, efficiently removing pathogens from the mucosa before an infection occurs. Although respiratory mucosal vaccination is highly appealing, successful nasal or pulmonary delivery of nucleic acid-based vaccines is challenging because of several physical and biological barriers at the airway mucosal site, such as a variety of protective enzymes and mucociliary clearance, which remove exogenously inhaled substances. Hence, advanced nanotechnologies enabling delivery of DNA and IVT-mRNA to the nasal and pulmonary mucosa are urgently needed. Ideal nanocarriers for nucleic acid vaccines should be able to efficiently load and protect genetic payloads, overcome physical and biological barriers at the airway mucosal site, facilitate transfection in targeted epithelial or antigen-presenting cells, and incorporate adjuvants. In this review, we discuss recent developments in nucleic acid delivery systems that target airway mucosa for vaccination purposes.

Keywords: DNA vaccine; mRNA vaccine; mucosal immune response; intranasal delivery; pulmonary delivery; nanoparticles

Citation: Tang, J.; Cai, L.; Xu, C.; Sun, S.; Liu, Y.; Rosenecker, J.; Guan, S. Nanotechnologies in Delivery of DNA and mRNA Vaccines to the Nasal and Pulmonary Mucosa. *Nanomaterials* **2022**, *12*, 226. https://doi.org/10.3390/nano12020226

Academic Editor: Pablo Botella Asuncion

Received: 31 October 2021
Accepted: 5 January 2022
Published: 11 January 2022

Publisher's Note: MDPI stays neutral with regard to jurisdictional claims in published maps and institutional affiliations.

1. Introduction

In recent decades, nanotechnologies for the production of high-quality nucleic acids and efficient in vivo delivery systems have revolutionized the field of vaccine development [1–4]. Lauded as "third-generation vaccines", DNA and in vitro transcribed messenger RNA (IVT-mRNA) harbor huge potential to offer pioneering solutions for clinically unmet needs [5,6]. Instead of using inactivated or live, attenuated viruses synthesized through time-consuming and demanding production procedures, vaccine developers now adopt simpler, rationally designed DNA constructs or IVT-mRNA to instruct cells of the recipient to express antigenic proteins that imitate parts of the target bacteria or viruses in order to generate antigen-specific humoral and cellular immunity to eliminate the real pathogens containing said antigen when they invade the host. Since host cells are responsible for antigen production, correct folding of the protein and natural glycosylation are guaranteed [7]. However, successful transfer of these antigen-encoding genetic materials to the target site (i.e., the nucleus for DNA transcription or ribosomes in the cytoplasm for mRNA translation) within host cells is a prerequisite to achieve this goal. Apart from several exceptional cases [8–10], delivery vehicles or devices are imperative because of limited cellular uptake and the instability of naked DNA and IVT-mRNA [2,11–13]. Nucleic acid-based vaccines (NA-vaccines) utilize affordable and well-established standard manufacturing processes that can scale up rapidly in response to outbreaks of infectious diseases, avoiding the complex procedures of repeated culture and inactivation of infectious pathogens or purification of recombinant antigens [5,14]. Furthermore, these genetic vaccines offer several distinct advantages over conventional vaccines (Table 1). NA-vaccines are highly flexible, capable of encoding virtually any type of protein. The successful implementation of these vaccines would prevent or treat, at least theoretically, any disease with effective and well-characterized antigen targets. Indeed, several NA-vaccines encoding viral antigens have been tested in clinical trials to investigate their protective potency and immunogenicity [15–20]. The range of diseases that could be addressed by NA-vaccines even extends to cancer [21–23]. The emergence of severe acute respiratory syndrome coronavirus 2 (SARS-CoV-2) has turned the field's spotlight towards the prevention of coronavirus disease 2019 (COVID-19). Since then, research institutions and companies have raced to develop COVID-19 vaccines [24]. IVT-mRNA vaccines turned out to be the "biggest winner" of this vaccine competition, as they not only enabled an unimaginable speed of vaccine development but displayed extremely high protection rates against COVID-19. After the viral genome sequence was publicly released, it took just sixty-six days for Moderna's 1273-mRNA vaccine candidate to enter phase I clinical trial [25]. Both Moderna's 1273-mRNA vaccine and BioNTech's 162b2-mRNA vaccine finished phase III clinical trials with 94–95% effectiveness in protection against COVID-19 and obtained emergency use authorization within one year from the outbreak of the virus [26,27].

Table 1. Advantages of NA-vaccines compared with conventional vaccines.

Category	DNA Vaccines	RNA Vaccines
Design	Rapid design with the coding sequence of antigens A single formulation with multiple antigens is possible	Rapid design with the coding sequence of antigens A single formulation with multiple antigens is possible
Production	Rapid and reproducible production based on in vitro bacterial culture Large-scale manufacture without inactivation of infectious pathogens or purification of recombinant antigens Antigens with proper folding are produced in vivo	Rapid and reproducible production based on in vitro transcription Large-scale manufacture with "cell free" process Antigens with proper folding are produced in vivo, with the cytosol as its target, only transiently expressed
Stability	Depends on the formulation Ease of storage and transportation in most cases	Depends on the formulation Cold chain transportation is generally required
Immune responses	Both cellular and humoral immune responses	Both cellular and humoral immune responses without risk of genome integration

One key obstacle on the path of developing advanced vaccines relates to the vaccination route and mucosal immunity. Many pathogens, including SARS-CoV-2, infect the host through the respiratory mucosa, so the interaction between the virus and the immune system first occurs in the mucosa of the respiratory tract. The mucosal immune system plays a key role in the host's resistance to pathogen invasion [28,29]. However, most clinical available vaccines employ invasive parenteral routes when inoculated, such as intramuscular, subcutaneous, and intradermal injections [30]. The modality of vaccine administration could have an enormous impact on the efficacy of a vaccine, as it affects the accessibility of immune cells for priming and the extent of consequential effects (systemic and local immune responses) after the vaccination. In addition to a few reports using retinoic acid as an adjuvant to induce immunoglobulin A (IgA) secretion at mucosal cites [31,32], vaccines inoculated via the conventional intramuscular route exclusively induce a systemic immune response characterized mainly by serum immunoglobulin G (IgG) antibodies but barely induce a mucosal immune response. The ability of serum IgG to eliminate virions at the mucosal site might be very limited [33]. Mucosal routes of administration (specifically, nasal or pulmonary delivery) are still considered by the research community to be the ideal and most straightforward approach to inducing potent immune responses against respiratory infections. This is particularly important for prophylactic vaccines to prevent respiratory diseases, such as SARS-CoV-2, RSV, and influenza [34–36]. NA-vaccines administered to the mucosal layers, such as the nasal and pulmonary mucosa, could travel to the draining mucosal lymph nodes (LNs) via lymphatic systems under the airway epithelium [37–39]. Importantly, the activated mucosal immune system causes the production of pathogen-specific secretory immunoglobulin A (sIgA), which efficiently blocks and neutralizes invading pathogens in mucus at an early stage of infection and acts in the front line of defense against infection [40,41]. Immune cells stimulated by antigens in the respiratory mucosa are also able to produce corresponding specific immune responses to protect against infectious diseases in other mucosal parts of the body (e.g., the gastrointestinal and reproductive tract), which are facilitated by the common mucosal immune system [42]. Moreover, delivery of vaccines to mucosal surfaces can induce systemic immunity with levels similar to those induced by counterparts administered via the conventional parenteral route [43,44]. In addition to expanding immune protection to mucosal sites, mucosal vaccination also brings other advantages. Being an inherently needle-free delivery approach, mucosal vaccine delivery addresses various problems associated with needle injection, such as the risk of needle reuse, which may lead to cross-infections; low patient compliance due to the pain caused by injection; and the necessity of medical staff for the vaccine injection [45].

Although mucosal vaccination is highly appealing, successful delivery of nucleic acids within the airway is extremely challenging because of the complex structures and the harsh microenvironment of the respiratory tract which has evolved to proficiently remove foreign particles (Figure 1B). Compared with that of parenterally administered vaccines, the number of successful mucosal vaccines is very limited. Currently, there is only one nasally-administered vaccine (i.e., Flumist®) that has obtained market approval [46], and it is based on live attenuated viruses for the reason that, currently, only virus-based delivery systems (such as attenuated viruses or adenoviruses) have been able to achieve strong respiratory mucosal immune responses [36,47,48]. However, such virus-based vaccines still have many drawbacks, such as safety issues (virulence restoration) and preexisting immunity [46]. Biomaterials-based nonviral delivery systems (Figure 1A), on the other hand, have the advantages of excellent biocompatibility, ease of production, and high flexibility in structure modification and can hence be tailor made for the mucosal delivery of DNA- and IVT-mRNA-based therapeutics [49]. Another advantage of the nonviral delivery platform is its ability to enhance vaccine efficacy by acting as an adjuvant itself or codelivering adjuvants. However, the in vivo delivery efficiency of nonviral materials is usually inferior to that of viruses, impeding the induction of sufficient mucosal immune responses. Successful nasal or pulmonary delivery of NA-vaccines is challenging because of physical and biological barriers at the airway mucosal site. Nevertheless, the field is still developing and evolving, with numerous efforts dedicated to the exploration of various materials for successfully mediating mucosal immunity. Based on our research interest in the airway delivery of genetic payloads, this review highlights recent advances in NA-vaccines inoculated via the nasal or pulmonary routes. We discuss the role of their delivery platforms and summarize crucial factors for the development of successful DNA- or IVT-mRNA-based vaccines that work efficiently in the respiratory mucosa (Figure 1). A successful pDNA or IVT-mRNA delivery system should not only confer protective capacity and efficient mucosal delivery function by overcoming the barriers but improve the translation of antigens from their payload pDNA/mRNA in specific antigen-presenting cells (APCs) or epithelial cells (Figure 1C). This review also covers several relevant topics including the structure of the airway and other barriers to mucosal vaccine delivery and the associated immune system that processes vaccine antigens upon respiratory delivery.

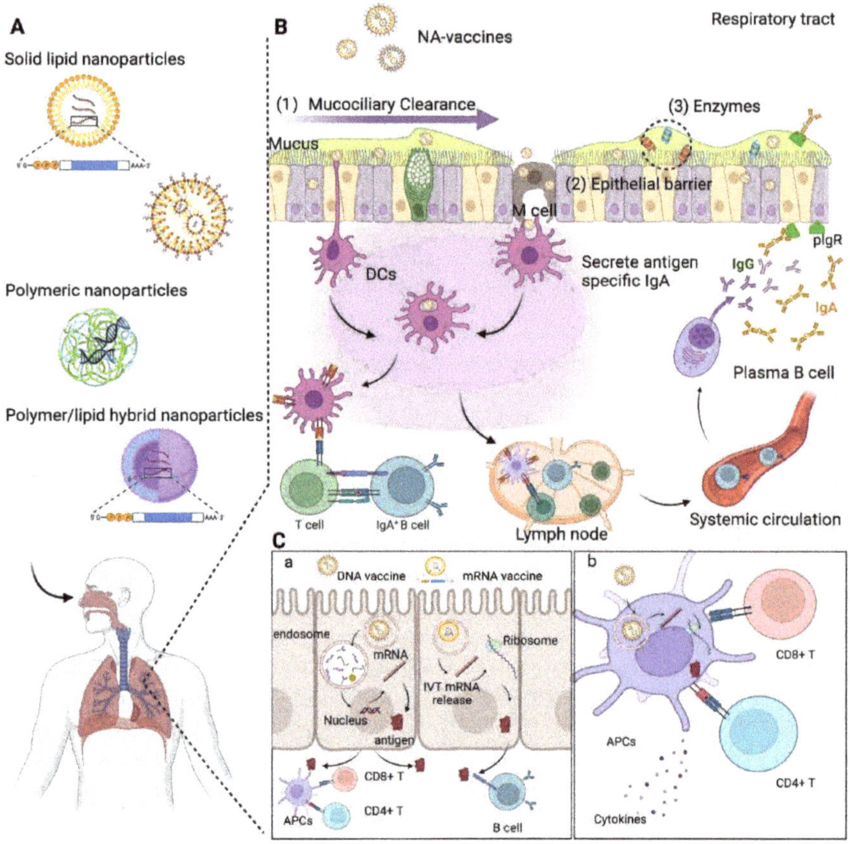

Figure 1. Overview of nucleic acid-based (NA-) vaccines administrated via the respiratory tract using nanotechnologies. (**A**) Schematic view of different nanoparticles used for intranasal and pulmonary vaccinations. (**B**) Physical and biological barriers at the airway mucosal site and mechanism of immune responses in the respiratory tract mediated by mucosal-associated lymphoid tissues (MALTs). NA-vaccines transcytose from the mucus layer into the epithelial tissues by microfold cells (M cells) or passively diffuse through epithelial cell junctions. Other NA-vaccines are captured and internalized by APCs, such as DCs, from their extension through epithelial junctions. APCs that have been transfected with genetic antigens migrate to the nearest lymph node to activate T cells and B cells. Activated B cells proliferate in the lymph node and enter the systemic circulation to the mucosal effector sites. B cells locally differentiate into antibody-secreting plasma cells to produce IgA dimers. IgA dimers are secreted via pIgR at the mucosal surface. Antigen-specific systemic IgG is also produced. (**C**) NA-vaccines are taken up by epithelial cells (**a**), and pathogen-derived antigens are then transcribed and translated from plasmid DNA or IVT mRNA and secreted into the extracellular space, where they can be taken up by professional APCs such as DCs. (**b**). APCs then present antigens to naïve T cells for activation and differentiation, promoting humoral and cell-mediated immune responses against the encoded antigen.

2. Barriers of Vaccines Inoculated via Respiratory Route

The respiratory system is critical for mammalian to maintain their life. It carries out the physiological function of exchanging gases between the organism and the environment. Unavoidably, it is subject to the presence of unwelcome foreign substances, such as pollutants and pathogens [50]. One important defense system is the mechanical–physical barrier, constituted by a layer of epithelial cells adhered on the luminal surface of the respiratory tract with tight junctions, preventing the trespassing of inhaled particles [51,52]. Among these epithelial cells, goblet cells secrete a dense, protective mucus that traps foreign particles and pathogens while also disrupting bacterial aggregation (Figure 1). The mucus, carrying trapped foreign substances, is continuously expelled via coordinated movement by the mucociliary "escalator" [53,54]. This mucociliary clearance removes large foreign materials (>5 μm) from the upper respiratory tract (URT). However, pathogens often escape this mechanical–physical barrier, reaching the pulmonary alveoli within the lower respiratory tract (LRT), which may lead to pulmonary inflammation [55]. To prevent such infections, another important line of defense is available. The airway and alveolar fluids contain a variety of soluble substances such as antimicrobial proteins (e.g., lysozymes, lactoferrins, and defensins), microbial opsonins (e.g., complement proteins and surfactant-associated proteins), and enzymes (e.g., proteases and nucleases), building up a robust antiinfection environment within the luminal surface along the entire airway [56].

Because of its important roles in protecting the host from inhaled pathogens, the mechanical–physical barrier poses huge obstacles to vaccines inoculated via the respiratory tract. Since vaccines are also perceived to be exogenous pathogens, intranasally or intratracheally administered vaccines are prone to being expelled via "mucosal clearance". This, combined with nonspecific binding with positively-charged proteins and enzymatic degradation within the mucus, makes it difficult for NA-vaccines to reach the airway epithelium to activate the mucosal immune system, resulting in limited immune response. To overcome these barriers, a safe and efficient delivery system is imperative for NA-vaccines to be administered via the respiratory tract.

3. Mucosal Immune Systems of Respiratory Tract against Infection

3.1. The Upper Respiratory Tract

The highly vascularized URT is the primary route of ingress of inhaled pathogens. A dense network of mucosal-associated lymphoid tissues (MALTs) is in the mucosal tissues to help induce pathogen-specific immune responses, reducing occurrences of infections. Peyer's patches in the small intestine, bronchus-associated lymphoid tissues (BALTs), larynx-associated lymphatic tissues (LALTs), and nasopharynx-associated lymphoid tissues (NALTs) in the rodent nasal cavity are all subsets of MALTs [57,58]. The NALT, commonly regarded to be equivalent to the Waldeyer's ring in humans [59], also plays a potentially unique role in the initiation of mucosal immune responses based on well-organized local lymphoid structures [60]. Through antigens presented by dendritic cells (DCs) from the parenchyma of nonlymphoid organs and distal sites, the NALT has an important role in generating T helper 1 and T helper 2 cells as well as IgA-secreting B cells in lymph nodes (LNs) [61].

While mucosal DCs capture antigens directly from the luminal side of the airway tract by extending DC dendrites through the tight junctions between the epithelial cells [62], M cells, a group of specialized cells located in the epithelium-overlying follicles of the MALT, are responsible for efficiently transporting antigens from the lumen to the underlying mucosal lymphoid tissues in a process termed "transcytosis" [63]. On their basolateral surface, the membranes of M cells are deeply invaginated, forming pocket structures holding DCs and/or lymphocytes. M cells are perfect antigen-transporting devices, and their reduced lysosome function enables antigen particulates entered from the airway lumen to be effectively delivered to DCs for further processing with minimal modification [63]. Apart from antigen transport, M-cells also aid immune response induction to the transported antigen by releasing a costimulatory signal for T- and B-cell proliferation [64]. Thus, M-cell targeting strategies in the development of mucosal vaccines have grown in popularity as means of improving efficacy in priming mucosal and systemic immune responses.

When captured antigens are transported to draining lymphoid nodes, they are processed into peptides following DC maturation for presentation to antigen-specific $CD8^+$ or $CD4^+$ cells in the form of major histocompatibility complex (MHC) class I or class II, respectively. Facilitated by costimulatory molecules and cytokines, the priming of $CD8^+$ or $CD4^+$ cells initiates programmed cell proliferation and differentiation within draining lymph nodes, accelerating the formation of corresponding antigen-specific effector T cells [65,66]. With the help of cytokines such as IFN-γ, IL-2, and IL-10 secreted by $CD4^+$ T helper cells, $CD8^+$ effector cells migrate from lymphoid tissues to the site of infection, inducing lysis in infected cells via Fas–FasL interactions or exocytosis of granule-associated proteases containing perforin and granzymes [67,68]. Moreover, a potent and long-term antibody response is built with help from T follicular helper (T_{FH}) cells, which provide relevant cytokines and costimulatory molecules for the affinity maturation of B cells, despite the presence of a T cell-independent mechanism for antibody production [69–71]. Following isotype switching and the selection of B cells in the germinal center, antigen-specific dimeric immunoglobulin A (dIgA) is generated [72]. Subsequently, transcytosis of dIgA is mediated by polymeric immunoglobulin receptors (pIgRs) expressed on the basolateral surface of the epithelium [73]. Endoproteolytic cleavage of pIgR on the luminal side results in the release of secretory IgA (sIgA) into the lumen [74]. Unlike serum IgA, sIgA is resistant against proteases, thereby being able to protect mucosal surfaces against pathogens despite the protease-rich environment on the nasal mucosa surface [74]. After complete clearance of infectious pathogens, the majority of effector T and B cells undergo apoptosis, leaving behind a small population of memory cells [75,76].

A study on severe acute respiratory syndrome coronavirus (SARS-CoV) showed that memory T cells remain for a long time in patients who have recovered from SARS-CoV infection, conferring a long-term immune protection effect [77]. Memory T cells can be roughly divided into two classes. After clearance of infected pathogens, central memory T cells (T_{CM}) typically remain in circulation within lymphoid organs, while effector memory T cells (T_{EM}) generally circulate through the red pulp of the spleen and nonlymphoid tissues [78,79]. Both T_{CM} and T_{EM} possess strong capabilities to proliferate and differentiate into effector T cells after encountering a reinfection. T_{EM} in the airway can rapidly remove and act on respiratory pathogens at the early stage of the reinfection [80,81]. However, T_{CM} has to undergo multiple steps, such as activation and differentiation, to initiate significantly delayed effector responses upon encountering pathogens compared to those of T_{EM} [82]. Another population of activated memory T cells does not return to circulation and is therefore known as resident memory T cells (T_{RM}). Compared with T_{EM}, T_{RM} provides a first line of defense in nonlymphoid tissues against infectious pathogens and limits further spread of infection in the host [83]. Lung T_{RM} seems more effective in promoting a long-term mucosal immune response against airway-transmitted pathogens [84,85]. Respiratory route inoculation has also been suggested to be necessary for generating of T_{RM} cells in the lung [86]. As shown in a previous study, Sendai virus-specific $CD4^+$ T_{RM} persists in lung tissues and the airway for several months after infection in C57BL/6 mice, mediating a substantial degree of protection against secondary virus infection [87]. Thus, T_{RM} levels constitute another important indicator of mucosal vaccine efficacy. Given that the mucosal surface is the primary route of ingress of most pathogens, mucosal vaccines providing longer immunity against inhaled pathogens via eliciting robust effector memory populations in mucosal tissues is of great promise.

3.2. The Lower Respiratory Tract

Most pathogens that are smaller than 1 μm tend to be able to bypass the mucosal layer on the URT luminal surface, reaching the terminal regions (bronchioles and alveoli) of the LRT. In the lung, the dominant cell population, alveolar macrophages (AMs), accounts for approximately 95% of airspace leukocytes, engulfing most of the antigens [56,88].

Macrophages play an important role in the first line of defense, while CD11c$^+$F4/80$^-$ DCs seem to be the major APC population in the lung responsible for presenting antigens to pulmonary T cells after infection in mice [89], demonstrating that DCs play an important role in adaptive immune response and in determining the magnitude of pulmonary vaccines' immune responses.

In alveoli, the recruitment and differentiation of CD11b$^+$ DCs can be facilitated by producing type I immune mediators in alveolar epithelial cells (AECs) and immune cells [90]. Pulmonary surfactant (PS)-biomimetic liposomes encapsulating 2′,3′-cyclic guanosine monophosphate-adenosine monophosphate (cGAMP) enter AMs by means of lung specific surfactant protein-A- and -D-mediated endocytosis, after which cGAMP is released into the cytosol and fluxes from AMs into AECs by way of gap junctions [90]. cGAMP, an agonist of the stimulator of interferon genes (STING), stimulates the production of type I interferons in both AECs and AMs, which in turn promotes rapid DC recruitment and differentiation. Through this mechanism, influenza-specific humoral and CD8$^+$ T cell immune responses were vigorously augmented in mice after intranasal immunization with PS-cGAMP-adjuvanted H1N1 vaccines [91]. Taken together, these findings suggest that the respiratory delivery of cGAMP as adjuvant might be an alternative strategy for developing pulmonary vaccines.

The antigen-presenting ability of different types of DCs in the respiratory system is also suggested to be different. Although there are relatively larger populations of DCs located in the lung parenchyma and alveolar spaces of the LRT, airway mucosal myeloid DCs are more endocytic and efficient than their aforementioned counterparts for presenting peptide antigens to naive CD4$^+$ T cells [92]. Similarly to NALTs, well-developed bronchus-associated lymphoid tissues (BALTs) have been found at branching sites of the bronchial tree in rabbit and feline lungs, while this type of lymphoid structure in human and mice is called inducible bronchus-associated lymphoid tissue (iBALT). The iBALT is induced only by inflammatory stimulation or infection and represents an inducible secondary lymphoid tissue for respiratory immune responses [93]. However, the biological outcomes of the immune response mediated by the iBALT could be beneficial or harmful. In some cases, the persistent exposure of antigens to iBALT during allergic reactions, chronic inflammation, or autoimmune disease may lead to exacerbation of inflammation [94]. Although the local immune mechanism of iBALT is poorly understood, the design of pulmonary vaccines aiming to trigger the formation of iBALT with subsequent protective immunity should be considered. For instance, pulmonary administration of a protein cage nanoparticle promoted the development of iBALT in the lung, resulting in enhanced viral clearance, accelerated induction of viral-specific antibody production, and significantly decreased morbidity and lung damage [95].

In conclusion, nasal and pulmonary administration are the most effective ways to elicit a substantial local immune response in addition to a systemic immune response. Mucosal immunisation stands out for its rapid and comprehensive activation of various immune subsystems in addition to its relatively fewer side effects.

4. DNA Vaccines

The first proof of concept for in vivo protein expression with nucleic acids was reported in 1990 by injecting DNA or RNA molecules into mouse skeletal muscle for the expression of chloramphenicol acetyltransferase, luciferase, and galactosidase [96]. It was thereafter demonstrated that the production of cytotoxic T lymphocytes for influenza could be induced by injecting plasmid DNA (pDNA) encoding influenza A nucleoproteins into the quadriceps of BALB/c mice [97]. These pioneer studies confirmed sufficient immunogenicity of DNA vaccines in animal models, providing evidence of this immunization platform's promising ramifications. DNA vaccines are generally constructed by inserting gene fragments encoding immunogenic antigens into a bacterial plasmid vector, forming pDNA. After pDNA is delivered into the host cell nucleus, antigenic proteins are subsequently expressed. Generally, APCs are the primary targets to be transfected with the genetic material. Following effective presentation in APCs, foreign antigenic proteins initiate specific immune responses [98]. DNA vaccines offer several advantages over conventional vaccines (e.g., live-attenuated, inactivated, or subunit vaccines), as summarized in Table 1. Also, DNA vaccines are generally stable at room temperature (though this may vary between formulations), avoiding the need for an uninterrupted cold chain during storage and transport [99]. Large-scale manufacturing of DNA vaccines primarily involves synthesis of relevant nucleic acids followed by standard cloning into plasmid vectors, avoiding the time- and labor-intensive culturing procedures required by traditional subunit and virus-based vaccines [100]. In contrast to subunit vaccines, DNA vaccines have been demonstrated to induce more potent cytotoxic T cell responses without severe side effects [101]. Although the potential possibility of genome integration remains the primary theoretical safety concern with DNA vaccines, this scenario has not been realized across large numbers of studies and reports [101]. Numerous clinical investigations have also demonstrated DNA vaccines to be largely safe in humans [3,16,102]. All these advantages make DNA vaccines an ideal candidate for rapid responses in the event of epidemic and pandemic outbreaks. Indeed, the first DNA vaccine to be approved for human use was a COVID-19 DNA vaccine (ZyCoV-D) developed in India. It was found to be 67% protective in clinical trials [103], providing evidence that DNA vaccines can be effective in controlling the pandemic [104].

In past decades, DNA vaccines have been studied for the prevention and treatment of a variety of diseases, such as infectious diseases, cancer, autoimmune diseases, and allergies. However, the immunogenicity of DNA vaccines in humans is not as sufficient as that in mouse studies to elicit significant clinical benefits [3]. The poor transport of pDNA into the nucleus of host cells results in low antigen synthesis, limiting the protective immunological responses in the recipient. Several approaches have been explored in recent years to address this issue, including the optimization of codon sequences/transcriptional elements, the incorporation of adjuvants, and enhancing delivery technologies. In this section, we discuss several well-studied delivery vehicles used for DNA vaccines that are administered via nasal or pulmonary routes (Table 2).

4.1. Delivery of DNA Vaccines via Respiratory Routes

Intranasal and pulmonary administration of DNA vaccines have attracted widespread attention recently because of various enticing properties. Early reports on inhaled DNA vaccines combined plasmids encoding ovalbumin, hepatitis B surface antigen, and HLA-A*0201-restricted T cell epitopes of *Mycobacterium tuberculosis (M. tuberculosis)*, which resulted in enhanced immunity as indicated by antibodies and cytokine production [105,106]. These pioneering investigations suggested that mucosal immune responses can be more effectively elicited when DNA vaccines are delivered directly to mucosal sites. However, there are still many obstacles that need to be overcome to more effectively utilize mucosal DNA vaccines. Vaccines must penetrate the mucus layer, translocate into target cells, and avoid extracellular and intracellular degradation in order to be effective (Figure 1). For example, delivery via the nasal cavity exposes DNA vaccines to being trapped by the nasal mucus, resulting in enzymatic breakdown. The viscosity and pore size of the mucus layer markedly affect the effective diffusivity of particles on the airway surfaces. Mucociliary clearance from cilia cells also significantly determines the fate of entrapped DNA vaccines. It continuously pushes mucus outwards, expelling mucus from the nasal channel and limiting residence time at the mucosal surface. The dilution effect in bulk mucosal fluids can also impede successful deposition onto the epithelium.

As a result, a safe and effective DNA delivery mechanism must be designed to overcome these obstacles. A suitable delivery system should target mucosal APCs for antigens processing, resulting in selective B and T cell activation. The ultimate goals of DNA delivery systems are to promote uptake of DNA into target tissues and cells, protect DNA from enzymatic breakdown, extend residence time at the target site, boost antigen expression, and optimize immune response, all without sacrificing safety. Section 4.2 discusses in detail several DNA vaccine delivery technologies that have been evaluated for respiratory administration.

4.2. Delivery Systems for DNA Vaccines via Respiratory Routes

The most prevalent technologies in the delivery of DNA constructs, such as electroporation, particle bombardment, and jet injectors, have exhibited improved transfection efficiency and huge potential in clinical trials when compared to traditional intramuscular injection of naked pDNA [107,108]. However, these attractive methods are inapplicable for the respiratory route, necessitating the use of a potent delivery strategy to overcome the barriers in the respiratory system. Advancements in nanotechnologies and material science have proven to be beneficial in the effective delivery of pDNA by the synthesis of DNA nanoparticles with diverse structures. Nanoparticles can better pass cell membranes via endocytosis while preventing premature degradation of pDNA and subsequently promote intracellular trafficking into the nucleus following endosomal escape, increasing immunogenicity in animal models and humans. Most importantly, sustained and controlled release of pDNA from nanocarriers at the delivery site recruits APCs, resulting in an improved antigen-specific immune response. The immunogenicity of DNA by vaccines can be further enhanced employing additional adjuvants. Currently, cationic lipids and polymers are two of the most employed nanomaterials for DNA vaccines.

4.2.1. Liposomes and Niosomes

Liposomes and some other vesicular systems are widely used as delivery systems for DNA vaccines. Liposomes generally consist of aqueous cores surrounded by phospholipid bilayers. For antigen delivery, two approved liposomal vaccine formulations used to be available: Inflexal® V (influenza vaccine) and Epaxal® (hepatitis A vaccine). Both of these formulations use virosome-based technology in which viral proteins are bonded to the surface of a liposome carrier similarly to how viral particles are bound [109]. The goal is to imitate a safe viral-like particle capable of eliciting significant protective immune responses. Although both examples have been discontinued, comparable technology could be used to increase the immunogenicity of DNA vaccines. To achieve this, pDNA could be either electrostatically complexed on the surface of cationic liposomes or encapsulated in the aqueous core by a dehydration–rehydration procedure. In general, cationic liposomes have shown higher in vitro transfection efficiency, while nonionic or anionic counterparts have shown enhanced antibody responses in animal models [110,111]. Surface modifications with antigenic components or targeting ligands further enhance immune responses of liposome-based vaccines [111]. It has been shown that liposomes coated with glycol chitosan are mucus adhesive and immune system stimulating [112]. Surface-modified cationic liposomes such as phosphatidylcholine (PC), dioleoyl phosphatidylethanolamine (DOPE), and cholesterol (Chol) elicited stronger humoral, mucosal, and cell-mediated immune responses post intranasal administration in mice against hepatitis than uncoated counterparts [111]. Liposomes containing noncoding pDNA have also been shown to exhibit adjuvant-like behavior, inducing elevated antibody levels and T cell immunity in mice and nonhuman primates [113].

Liposomes have been used as DNA vaccine carriers for intranasal delivery in several investigations to induce efficient immune responses against respiratory pathogens. To induce immune protection against *M. tuberculosis*, D'Souza et al. adopted pDNA encoding antigen 85A formulated with a (+/−)-N-(3-aminopropyl)-N,N-dimethyl-2,3-bis (dodecyloxy)-1-propanaminium bromide (GAP-DLRIE):DOPE liposome [114]. After intranasal immunization in mice, a positive splenic Th1-type cytokine response was induced in GAP-DLRIE:DOPE formulations, but this response was still weaker than that obtained from intramuscular administration with the same dosage. However, the combination of intranasal and intramuscular injections elicited even stronger Th1 type immune responses in the lungs [114]. Another study, performed by Rosada et al., described how a single intranasal immunization with liposome-based formulations of pDNA encoding HSP65 against *M. tuberculosis* led to a remarkable reduction in the amount of bacilli in lungs of mice [115]. The authors employed egg phosphatidylcholine (EPC), DOPE, and 1,2-dioleoyl-3-trimethylammonium-propane (DOTAP) to formulate the delivery system. This formulation also increased the production of IFN-γ and lung parenchyma protection to a level similar to that in mice vaccinated intramuscularly four times the dosage of naked pDNA encoding HSP65 [115]. In addition, intranasal immunization with liposome-based DNA vaccine provided complete protection against influenza after a viral challenge assay [116]. Mice immunized intranasally with liposome-encapsulated pDNA encoding hemagglutinin (HA) protein, but not naked plasmid, were found to produce strong serum IgA/IgG responses and increased IgA titers in bronchoalveolar lavage fluid (BALF) [117]. T cell-proliferative responses were also successfully induced in both intranasal and intramuscular administration [117]. These studies demonstrated the ability of liposomes in the delivery of DNA vaccines inoculated via the intranasal route to confer significant immune protection against respiratory infections in animal models. However, widespread adoption of liposome-based vaccines remains stunted by their relatively lower physical and chemical stability in aqueous dispersions during long-term storage [118]. Accordingly, numerous methods to improve the stability of liposome formulations during storage have been investigated, including freeze-drying, spray-drying, supercritical fluid technology, and lyophilization [119–121].

Niosomes, which are nonionic surfactant-based vesicles, have been developed as alternative delivery systems to liposomes because of their advantages such as cost-effective manufacturing, large-scale producibility, and stability [122,123]. Because of their structural similarities to liposomes, niosomes were also applied as vehicles for pDNA, small interference RNAs (siRNAs), and aptamers in target cells [124]. Cationic niosomes, containing cationic lipids, made an effective vector for pDNA delivery and achieved ~95% transfection efficiency in vitro [125]. Later, the same research team reported successful transfection of human tyrosinase gene (pMEL34) and the stability of developed cationic niosomes in transdermal delivery [126]. Perrie et al. reported that niosomes carried with H3N2 influenza virus resulted in enhanced immune response after subcutaneous administration in mice [127]. Mannolysated niosomes encapsulated with pDNA encoding HBsAg were reported to provoke protective immunity against hepatitis B as both a DNA vaccine carrier and adjuvant for oral immunization [128]. However, there have been no reports utilizing niosomes as a mucosal delivery platform in the respiratory tract as far as we know. Their efficacy for the intranasal and pulmonary delivery of DNA vaccine needs further investigation.

4.2.2. Polymers

One of the most appealing characteristics of polymer-based DNA delivery technologies is their flexibility in structure design and modification. Electrostatic interactions allow cationic polymers to form complexes (polyplexes) with DNA vaccines. Polymer synthesis is also relatively inexpensive and simple to scale up. To maximize cellular uptake and transfection effectiveness, the size and surface characteristics of polymeric particles can be adjusted by employing different polymers and methods of preparation [129,130]. It has been found that alveolar macrophages are particularly effective in absorbing particles with diameters ranging from 300 to 600 nm, so the particle size should be less than 3 μm (preferably under 500 nm) for DC-targeted absorption in the respiratory tract [131]. Aside from particle size, particle charge also influences cellular absorption in APCs in the respiratory tract. Where both DCs and macrophages are substantially present, preferential uptake of DNA vaccines into DCs is desirable. DCs can produce large quantities of peptide–MHC II complexes, which are then presented on the cellular surface to initiate T cell activation and differentiation [132]. It has been found that macrophages have higher phagocytic activity than DCs, but also that absorption in DCs can be increased by imparting positive charge to the particle [133]. Polymeric particles can also be modified with functional groups or ligands to improve the cellular uptake of DCs. DCs can preferentially uptake ligand-modified nanoparticles via receptor-mediated endocytosis using C-type lectin receptors or mannose receptors [134].

Polyethylenimine

Polyethylenimine (PEI) is one of the most well-studied polymers with high transfection efficiency and has been extensively applied for mediating in vitro and in vivo transfection of DNA molecules [135]. Compared to lipid-based formulations, DNA complexed with PEI has shown improved stability and higher levels of pulmonary transfection, even after nebulization [136,137]. For intranasal immunization, PEI/pDNA complexes encoding SARS-CoV spike proteins induced higher antigen-specific Th2 dominant IgG and IgA antibodies in BALF than naked plasmid counterparts [138]. Cellular immune responses were also detected in a PEI/pDNA treated group, with increased B cells and higher numbers of IFN-γ-, TNF-α-, and IL-2-producing T cells in the lungs [138]. A H5N1 intranasal vaccine with DNA encoding HA formulated with PEI induced potent mucosal and systemic immune responses and elicited both full protection against the parental strain and partial cross-protection against a distinct highly pathogenic strain [139]. Pulmonary immunization of DNA vaccines formulated with PEI also induced robust systemic and $CD8^+$ T-cell responses in the gut and vaginal mucosa [140]. Furthermore, mice inoculated via the intratracheal (i.t.) route elicited higher levels of interleukin-2 than those inoculated by intramuscular immunization in lung-associated antigen-specific $CD4^+$ T cells [140]. These robust T cell responses, which were induced by i.t. but not intramuscular administration, protected mice from a lethal recombinant vaccinia virus challenge [140]. A similar study also reported that robust pulmonary $CD8^+$ T cell populations effectively mediated protective immunity against influenza respiratory challenges after pulmonary immunization with PEI/pDNA [141].

Although PEI appears to be a promising delivery vector for airway inoculated DNA vaccines, one remaining major limitation is its toxicity due to its highly positively-charged and nondegradable nature. Immunization with PEI vaccine was found to provoke the activation of genes with apoptosis, stress responses, and oncogenesis [142]. As a result, biodegradable PEI derivatives with low-toxic profiles have been developed for DNA delivery. A less toxic form of PEI called deacylated PEI (dPEI) with potent transfection efficiency was applied in delivering pDNA encoding HA [143]. Essentially, dPEI is a completely hydrolyzed linear PEI with 11% more free protonatable nitrogen atoms than conventional PEI. Following intranasal administration, dPEI-complexed pDNA vaccine formulations were capable of generating strong systemic and mucosal humoral responses, activating cellular responses, and mediating a higher degree of protection in a challenge study against influenza [143]. Other strategies using covalent modification or electrostatic neutralization of PEI's cationic group to reduce zeta potential have also been investigated. Poly-lactic-co-glycolic acid (PLGA), a synthetic biodegradable copolymer, has been approved by the Food and Drug Administration (FDA) for human use in delivering therapeutic agents such as proteins and nucleic acids [144]. The negative charge and hydrophobic nature of PLGA could be used to neutralize the positive charge of PEI for safe nucleic acid delivery. Bivas-Benita et al. developed PLGA nanoparticles bearing PEI on their surfaces. Internalization of the DNA-loaded PLGA–PEI nanoparticles was also studied in the human airway submucosal epithelial cell line, Calu-3 [145]. The results suggested that DNA could be detected in the endolysosomal compartment after 6 h incubation with Calu-3 cells and that and the optimal cell viability was achieved when the weight ratio of PEI to DNA was between 1:1 and 0.5:1 [145]. A similar study reported the formulation of PLGA–PEI microparticles, in which 10% PEI (w/w) efficiently adsorbed DNA and protected DNA from enzymatic degradation [146]. Intramuscular immunization of mice with such PLGA–PEI formulations loaded with pDNA encoding immunodominant antigens of *Listeria monocytogenes* demonstrated that the formulation had an adjuvant effect [146]. These studies indicated that PEI has a favorable profile to be a nonviral gene carrier for DNA vaccines delivered through the respiratory tract, but also that further optimization is still necessary to realize their full potential.

Chitosan

Chitosan is a biodegradable and biocompatible polysaccharide derived from chitin and has been frequently employed as a DNA delivery vector because of its biodegradability and biocompatibility [147]. Furthermore, chitosan and its derivatives possess substantial mucoadhesive properties, making them ideal for intranasal administration [148,149]. Chitosan has also been reported to be immune stimulating by enhancing macrophage accumulation and activation, increasing cytokines' resilience against infections, and promoting cytotoxic T cell response [150,151]. To assess its potential in mediating mucosal immunization, chitosan was used to complex with pDNAs encoding nine different antigens (NS1, NS2, M, SH, F, M2, N, G, and P) from respiratory syncytial virus (RSV) [151]. A single intranasal administration of chitosan–pDNA resulted in a significant reduction of viral titers and viral antigen load in the lungs after an acute RSV infection [151]. In addition, significantly elevated levels of serum RSV-specific IgG antibodies, nasal IgA antibodies, cytotoxic T lymphocytes, and IFN-γ production in the lung and splenocytes were detected in comparison with controls [151]. However, when pDNA encoding the M2 proteins of RSV antigens was formulated with chitosan, virus-specific CTL responses in BALB/c mice were induced only at a level that was comparable to those induced via intradermal immunization [152]. Nonetheless, aerosolized pDNA–chitosan nanoparticles induced higher levels of IFN-γ through pulmonary administration than counterparts immunized by intratracheal and intramuscular administration [106].

In order to achieve targeted delivery of antigen to DCs, biotinylated chitosan nanoparticles loaded with pDNA encoding the nucleocapsid (N) protein of SARS-CoV were developed by Raghuwanshi et al. [153]. Chitosan was modified with bifunctional fusion protein (bfFp) consisting of truncated core-streptavidin fused with anti-DEC-205 single chain antibody (scFv) [153]. The core-streptavid in the arm of bfFp bonded with biotinylated nanoparticles, while anti-DEC-205 scFv imparted targeting specificity to the DCs' DEC-205 receptors. Intranasal administration of such targeted formulations led to the detection of an enhanced number of N protein-specific systemic IgG and nasal IgA antibodies [153]. In another study, mannosylated chitosan (MCS) formulated with DNA vaccine encoding a multi-T-epitope was employed to facilitate airway delivery and antigen targeting to the APCs in the alveoli [154]. Following intranasal immunization, HSP65-specific sIgA in the BALF was significantly elevated. A modest antigen-specific Th1 (IFN-γ, TNF-α, and IL-2) response and a potent polyfunctional $CD4^+$ T response were induced for enhancing mucosal immune protection against *M. tuberculosis* in the spleen and lung, respectively [154].

Thiolated chitosan derivatives have been found to improve the transfection efficiency of chitosan for intranasal delivery. In a study performed by Bernkop-Schnürch et al., thiol-bearing moieties were introduced on the polymeric backbone of chitosan in order to prepare thiolated chitosan that could interact with mucus glycoproteins via the formation of disulfide bonds [155]. The results indicated that thiolated chitosan improved mucosa adhesiveness to the mucus layer 6- to 100-fold compared to the unmodified counterpart, resulting in enhanced mucus permeation. Simultaneously, increased penetration at the mucosal surface was also observed in chitosan-coated PLGA nanoparticles encapsulating macromolecules [156,157]. Based on this fact, an emulsion–diffusion–evaporation technique was employed to prepare cationic nanospheres composed of biodegradable and biocompatible copolyester PLGA, with a PVA–chitosan blend stabilizing the PLGA nanospheres [158]. Despite the charge on the nanospheres being sufficient to bind the negatively charged DNA, the immunity of DNA vaccines complexed by this formulation remains to be illuminated. In one study, chitosan-coated PLGA was employed to deliver pDNA encoding foot-and-mouth disease (FMDV) capsid protein and bovine IL-6 to protect mice against FMDV infections [159]. This chitosan/PLGA/pDNA vaccine formulation provided enhanced protective immunity against FMDV post-intranasal immunization [159].

In a recent publication, chitosan was adopted to coat star-shaped gold-nanoparticles to yield a gold-nanostar chitosan (AuNS–chitosan) nanoformulation for intranasal delivery of a DNA vector expressing S protein of SARS-CoV-2 [160]. Six-time repeated dosing of AuNS–chitosan DNA vaccine induced high levels of S protein-specific IgG, IgM, and IgA antibodies in serum up to 8 weeks in both BALB/c and C57BL/6 mice as assessed using an ELISA assay [160]. IgG and IgA sustained their levels until week 8, then rose back to a high level with a single 7th dose in week 14. The serum neutralization IC_{50} (serum dilution factor) for infectivity inhibition concentrations were determined to be 1:83.8, 1:47.5, and 1:150 (collected in week 18 of the study) for pseudoviruses engineered with S proteins from SARS-CoV-2-Wuhan, SARS-CoV-2-beta mutant, and SARS-CoV-2-D614G mutant variants, respectively [160]. With the help of immunophenotyping and histological analyses, the authors further revealed chronological events involved in the recognition of S antigen by resident DCs and alveolar macrophages in the lungs of DNA vaccine transfected mice [160]. These APCs further primed the draining lymph nodes and spleen for peak antigen-specific cellular and humoral immune responses. Although this proof-of-concept study suggested the capabilities of the AuNS–chitosan DNA vaccine to elicit potent mucosal immune responses, further development with a reduced dosing regimen and more key evidence (e.g., sIgA levels in BALF samples, details in the subsets of effector T cells and memory T cells, the protection efficiency of SARS-CoV-2 challenge studies, etc.) will be required to validate its potential for clinical translation.

Table 2. Summary of DNA vaccines inoculated via respiratory tract.

Disease	Nanoparticle	Coding Antigens	Experimental Animal	Administration	Immune Response [1]	Ref.
Hepatitis	PC/DOPE/Chol	S protein	mice	i.n.	HIR(+)/MIR(+++)/CIR(+)	[111]
Tuberculosis	GAP-DLRIE:DOPE	85A	mice	i.n.	Th1 CIR(+)	[114]
Tuberculosis	EPC/DOPE/DOTAP	HSP65	mice	i.n.	Th1 CIR(++++)	[115]
Tuberculosis	MCS	HSP65	mice	i.n.	MIR(+++)/CIR(++)	[154]
Tuberculosis	Chitosan	Multiantigens	HLA-A2	i.t.	CIR(++)	[106]
Influenza	DODAC/DOPE/PEG	HA	mice	i.n.	HIR(++)/MIR(+)	[116,117]
Influenza	PEI	HA	mice	i.n.	HIR(+++)/MIR(++)	[139]
Influenza	dPEI	HA	mice	i.n.	HIR(++++)/MIR(++++)/CIR(+)	[143]
SARS-CoV	PEI	S protein	mice	i.n.	HIR(+++)/MIR(+++)/CIR(++)	[138]
SARS-CoV	Chitosan	N	mice	i.n.	HIR(++++)/MIR(++++)	[153]
HIV	PEI	HXBc2 gp120	mice	i.t.	CIR(++)	[140,141]
RSV	Chitosan	Multiantigens	mice	i.n.	HIR(++++)/MIR(++++)/CIR(+)	[151]
RSV	Chitosan	M2	mice	i.n.	CIR(+)	[152]
COVID-19	Chitosan–gold	S-protein	Mice	i.n.	MIR(N.A.)/HIR(++)/CIR(+)	[160]

[1] Responses are geometric means of postvaccination increases in specific antibodies versus control in vaccine recipients: ++++, >10-fold; +++, 5- to 10-fold; ++, 2.5- to 5-fold; +, 1.5- to 2.5-fold. RSV: respiratory syncytial virus; HA: hemagglutinin protein; HIV: human immunodeficiency virus; HIR: humoral immune responses; MIR: mucosal immune responses; CIR: cellular immune responses; SARS-CoV: the severe acute respiratory syndrome coronavirus; i.n.: intranasal administration; i.t.: intrathecal administration; N.A.: not available.

5. IVT-mRNA Vaccines

In 2019, the outbreak of COVID-19 caused by SARS-CoV-2 spread throughout the world, developing into a global pandemic. Application of safe and effective vaccines is expected to be the most efficient medical approach in controlling and stopping the pandemic of COVID-19. BNT162b2 (BioNTech/Pfizer, Mainz, Germany/New York, NY, USA) and mRNA-1273 (Moderna, Cambridge, MA, USA) are both lipid nanoparticle-formulated, nucleoside-modified IVT-mRNA vaccines that received approval for emergency use by the FDA with extremely high protection rates against COVID-19 [25–27,161]. The area of IVT-mRNA vaccines is rapidly evolving; a considerable amount of clinical evidence has been gathered over the last several years, widely establishing IVT-mRNA vaccines as a highly promising medical strategy [162]. The design and synthesis of IVT-mRNA and associated delivery technologies are key to the success of IVT-mRNA vaccines (Figure 2). IVT-mRNA encoding specific proteins of interest (POI) is transcribed in vitro according to a linearized DNA template. Once delivered into host cells, IVT-mRNA utilizes the translation machinery of the host to produce corresponding POI in cytoplasm without the need to enter the nucleus to be functional [5]. If the POI is an appropriate antigen, the resulting POI induces antigen-specific immune responses following effective antigen presentation by APCs. Currently, two major classes of IVT-mRNA have been investigated broadly as vaccines: conventional nonreplicating mRNA and self-amplifying mRNA (saRNA). Both share elements of a eukaryotic mRNA that are essential for translation and stability: i.e., a cap structure (m7Gp3N), 5′- and 3′- untranslated regions (UTRs), an open reading frame (ORF), and a poly(A) tail. Compared with conventional nonreplicating mRNA, saRNA has the potential to produce more antigen protein with the viral replication machinery. The design and development of saRNA vaccines with validated immunogenicity and efficacy has been recently reviewed elsewhere [163]. For IVT-mRNA-based therapeutics, robust delivery systems are generally necessary for the successful transport and uptake of IVT-mRNA into target organs. The instability, innate immunogenicity, and inefficient in vivo delivery to organs beyond the liver and spleen are major problems that need to be solved in the future applications of IVT-mRNA. As a result, a wide range of in vitro and in vivo delivery systems have been designed. Some vaccine formulations contain adjuvants proven to potentiate subunit vaccines [164–166], while others generate strong responses despite the absence of known adjuvants. Furthermore, highly effective RNA carriers with low toxicity have been produced, allowing for sustained antigen expression in vivo in some instances [11,12,167]. Apart from that, rational design of RNA sequences has also greatly improved the potential of IVT-mRNA. In this section, we briefly summarize the recent advances in conventional IVT-mRNA vaccines regarding their structure optimization and specifically focus on the delivery system of IVT-mRNA vaccines inoculated via the respiratory tract route (Table 3).

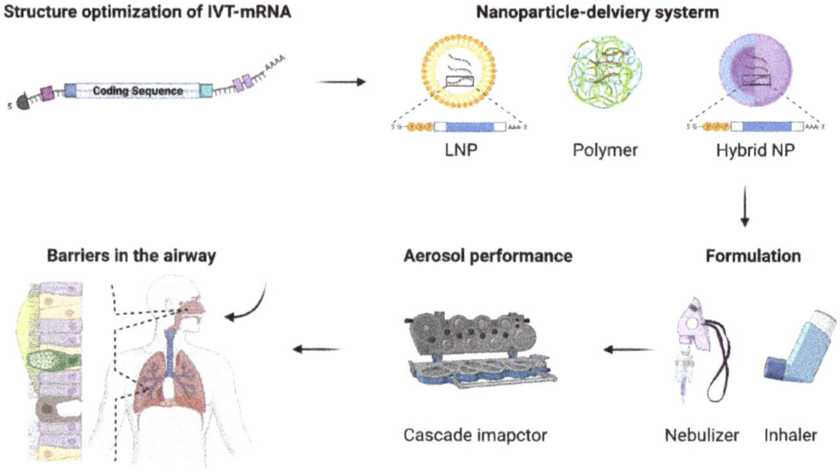

Figure 2. Schematic illustration of key factors associated with the technology development of IVT-mRNA vaccines via the respiratory route. IVT-mRNA structural elements, including elongation of the poly(A) tail, the 5′ cap, the structure of UTRs, and the ORF with optional incorporation of modified nucleotides, are optimized to enhance the stability and reduce the innate immunogenicity. Nanoparticle-based carriers are designed to facilitate the delivery of IVT-mRNA across the barriers in the airway. Liquid aerosol or dry powderformulations are then developed with the identification of a suitable inhalation device (nebulizer or powder inhaler) for clinical applications. The aerosol performance, IVT-mRNA stability after aerosolization, immunogenicity, and vaccine efficacy of the inhaled formulation should be thoroughly characterized and evaluated.

5.1. Structural Optimization of IVT-mRNA

Because of mRNA's susceptibility to nuclease degradation and its high innate immunogenicity, optimization of IVT-mRNA structural elements is crucial to enhance the expression of POI. Over the past decades, a great deal of effort has been given to modifying the structural elements of IVT-mRNA in order to improve its stability and translation efficiency. As extensively discussed in previous reviews [1,5], a functional 5′-cap structure required for the initiation of translation can be added in vitro via either post- or cotranscriptional processes. Vaccinia virus capping enzyme (VCE) is the most widely used commercially available reagent for posttranscriptional addition. First, IVT-mRNA can be modified with a Cap 0 (m7GpppN) structure using VCE. Then, using 2′ *O*-methyltransferase, Cap 0 can be switched to Cap 1 (m7GpppmN), which enhances the translation efficiency of IVT-mRNA [161,162]. However, large-scale manufacturing of IVT-mRNA may consume a huge amount of enzymes using this approach, which would increase the production costs and limit the approach's widespread application [5]. Alternatively, the 5′-cap can be added simultaneously during the in vitro transcription process of IVT-mRNA with cap analogs (e.g., m7GpppG). However, this method can lead to inefficient capping and reversed cap orientation (i.e., Gpppm7G) [163,168]. To overcome this problem, anti-reverse cap analogues (ARCAs; m27,3′-OGpppG) can be introduced into the in vitro transcription and enhance IVT-mRNA translation efficacy [169,170]. For cotranscriptional mRNA capping, an ARCAs to GTP ratio of 4:1 in a single reaction is optimal for maximizing both RNA yield and the proportion of capped transcripts with commercial transcription kits, yielding up to 100% capping efficiency [171]. Recently, CleanCap®, a method developed by TriLink Biotechnology, was reported to generate IVT-mRNA with high translational efficiency, simplifying the production of 5′-capped IVT-mRNA. Based on these advantages, the approach of CleanCap® has been used for 5′ capping of BNT162b2 and mRNA-1273 vaccines [161,172].

In synergy with a 5′-cap, the poly(A) tail regulates translation efficiency and stability of IVT-mRNA. Incorporation of poly(A) can be introduced by recombinant polymerase or by direct transcription from a DNA template with a defined length of poly(T) nucleotides [173]. Although the incorporation of modified nucleotides into the poly(A) tail can be introduced with recombinant poly(A) polymerase to inhibit deadenylation mediated by poly(A)-specific nucleases, a mixture of mRNAs with inconsistent lengths of RNA poly(A) tail is the limitation for this approach, especially for clinical applications [174]. Therefore, the optimal length of the poly(A) tail, around 100–150 nucleotides, is preferably added from the encoding DNA template [175]. However, poly(T) nucleotides in the plasmid template are often lost during the passaging of *E. coli*. Thus, adding a precise length of poly(A) to IVT-mRNA is still a challenge for IVT-mRNA production.

Optimization of the 5′- and 3′- UTRs is also important to enhance the expression and stability of IVT-mRNA [176]. UTRs contain regulatory sequence elements usually harbored by viral or eukaryotic genes, which can be designed via de novo methods or deep-learning methods [5,177]. Previous studies have shown that optimization on AU-rich elements and G–C ratios increase the stability of IVT-mRNA and duration of protein expression [178,179]. Furthermore, Steve Pascolo et al. added an eIF4G-recruiting aptamer sequence to the 5′ UTR of IVT-mRNA in order to increase the expression of IVT-mRNA [180]. Recently, a sequence of IVT-mRNAs was systematically engineered through deep-learning methods, significantly enhancing the expression of SARS-CoV-2 antigens [177]. However, codon optimization replacing rare codons with frequently used synonymous codons may not be necessary. This approach will affect the generation of potent cryptic T cell epitopes for some IVT-mRNA encoded vaccines [181,182], thus affecting the proper folding of some proteins that need a slow translation process because of rare codons [183].

IVT-mRNA is inherently immunogenic, it activates pattern recognition receptors (PRRs) such as Toll-like receptors (TLR-3, TLR-7, and TLR-8) in the endosomal compartments of host cells [184,185]. In some cases, activation of innate immunity by IVT-mRNA is potentially advantageous for vaccines, as it promotes recruitment of DCs to the administration site and subsequent maturation of DCs to elicit potent adaptive immunity. However, this feature of IVT-mRNA may also lead to improper or excessive activation, resulting in the inhibition of antigen expression and negative immune responses [186]. Currently, modified nucleotides (e.g., pseudouridine and 5-methylcytidine) can be incorporated into IVT-mRNA during transcription, and chromatographic purification methods have been used to remove contaminating double-stranded RNA after transcription [187–189]. This approach has been shown to effectively avoid undesired immune stimulation, subsequently increasing production of POI [190]. This nucleotide modification technology has been successfully employed in the production of the BNT162b2 (BioNTech and Pfizer) and mRNA-1273 (Moderna) COVID-19 mRNA vaccines.

5.2. Delivery Systems for IVT-mRNA Vaccine Inoculated via the Respiratory Route

In addition to structural optimization and high-quality manufacturing of IVT-mRNA, another major technical barrier in the clinical translation process of IVT-mRNA therapeutics is the development of safe and efficient delivery systems. Whether a potent protective immunity can be established by an IVT-mRNA vaccine is largely attributed to the choice of a safe and efficient delivery system. Basically, a safe IVT-mRNA delivery system should at least condense IVT-mRNA, so that the fragile molecule can be protected from degradation in extracellular space, and facilitate endosomal escape following cellular uptake in cytosol. The successful delivery of IVT-mRNA to the lungs at low dosages would reveal a meaningful region for therapeutics or vaccines designed for pulmonary application. This endeavor, however, remains difficult, because nanocarriers that carry IVT-mRNA to the lungs behave differently than counterparts using the parenteral route. For example, the potency of IVT-mRNA may be significantly compromised by the nebulization process. Despite the fact that modern nebulizers are designed to gently aerosolize medicine, nanoparticles are nevertheless subjected to shear force that could compromise the structures of vehicle–mRNA complexes. Biological aspects of the airway pose other barriers against nebulized IVT-mRNA administration. The cells, proteins, biomolecules, and physical barriers with which nanoparticles interact when delivered via nebulization differ from those with which they interact in bloodstream [191–193]. Airway cells are also highly heterogeneous, which may further influence delivery properties [194]. Nevertheless, there have already been several reports on the successful delivery of IVT-mRNA to the airways.

5.2.1. Lipids

Currently, the most successful and frequently used delivery vehicle for the in vivo application of IVT-mRNA is lipid nanoparticles (LNPs) which consist of distinct components with variable proportions [195]. For most LNPs, the ionizable lipid is a key element responsible for packaging the negatively charged IVT-mRNA and enabling the endosomal escape of IVT-mRNA molecules into the cytoplasm. A recent publication from Norbert Pardi et al. suggested that LNP formulations have favorable immunostimulatory profiles that promote the induction of strong responses from T_{FH} cells, germinal center B cells, long-lived plasma cells, and memory B cells that are associated with durable and protective antibodies in mice. The authors further revealed that the adjuvant activity of the LNP relied on the ionizable lipid component and IL-6 cytokine induction [196]. The resulting magnitude and duration of IVT-mRNA expression were favorable for inducing potent immune responses, as evidenced by a study performed by Tam et al. The authors found that exponentially increasing dosing profiles led to prolonged antigen retention in lymph nodes and increased numbers of T_{FH} cells and germinal center B-cells, as well as eliciting >10-fold increases in antibody production relative to single bolus vaccination postprime [197]. Apart from ionizable lipids, phospholipids play a structural role in the formation and disruption of the lipid bilayer. Neutral lipids, such as cholesterol, are usually used as a stabilizing element for increasing transfection efficiency. Lipid-anchored polyethylene glycol (PEG) is a biocompatible and inert polymer, sterically stabilizing the LNPs and reducing nonspecific binding to proteins for increased circulation time in vivo.

LNPs were originally developed as highly efficient carriers of short-interfering RNAs (siRNA) to hepatocytes via intravenous administration [198]. Intramuscular, intradermal, and intratracheal administration of LNP-encapsulated mRNA have been found to produce higher and more prolonged protein levels at the site of inoculation than those produced by systemic delivery [195]. However, LNP needs to be specifically designed and optimized for the efficient pulmonary delivery of IVT-mRNA [199]. In one study, cationic LNPs and mannose-conjugated LNP (Man-LNP) were separately applied to encapsulate IVT-mRNA encoding the HA gene of the H1N1 influenza A virus. Following intranasal administration, both formulations induced HA-specific responses with high levels of IgG2a subtype and enhanced the production of both Th1-associated IFN-γ and Th2-associated IL-4 [34]. However, Man-LNP induced significantly higher hemagglutinin inhibition titers than cationic LNP-based counterparts did after boosting immunization [34]. Furthermore, LNP represents a powerful technology for the respiratory delivery of therapeutic genes. LNP-carrying chemically modified IVT-mRNA encoding cystic fibrosis transmembrane conductance regulator (CFTR) was successfully applied for the treatment of cystic fibrosis following nasal administration in CFTR knockout mice. The chloride response in two consecutive doses of CFTR-mRNA/LNP-treated mice recovered up to 55% of the levels observed in wild-type mice and lasted up to 2 weeks posttransfection [200]. Motivated by the huge unmet needs for lung IVT-mRNA delivery vehicles, as well as by the lack of established LNP design principles, Lokugamage et al. recently reported an in vivo cluster-based iterative screening approach to identify LNP chemical traits that promote IVT-mRNA lung delivery [199]. They discovered that a low PEG molar ratio increased the performance of LNPs with neutral helper lipids, whereas a high PEG molar ratio improved the performance of cationic helper lipids, uncovering and optimizing LNPs for low-dose IVT-mRNA delivery.

The optimized LNP nebulized delivery of an IVT-mRNA generating a broadly neutralizing antibody against hemagglutinin protected mice against a fatal challenge of the H1N1 subtype of the influenza A virus and delivered IVT-mRNA more effectively than LNPs previously tuned for systemic delivery. Several limitations of this study should be kept in mind: (1) the findings focused on LNPs containing 7C1, but more research is needed to quantify the link between LNP composition and nebulized delivery; (2) more studies are needed to quantify the link between payload and delivery efficiency, as the protein expression kinetics observed may vary depending on the IVT-mRNA payload; (3) these findings, from mice models, may or may not translate predictably to larger animals, such as nonhuman primates; (4) LNP size and zeta potential, as well as other altered physiologies associated with diseased pulmonary tissues, may impede pulmonary LNP delivery.

5.2.2. PEI

PEI has also been proven to be functional as a potent mucosal adjuvant for intranasal administration [201]. However, how to maximize its adjuvant effect and how to minimize its toxicity remain open challenges. Optimizing the chemical structure of PEI seems to be a practical solution to this problem. Cationic cyclodextrin–PEI conjugate was developed by Li et al. and complexed with anionic IVT-mRNA encoding ovalbumin (OVA) through electrostatic interactions [39]. This formulation showed the ability to stimulate DC maturation and migration after intranasal inoculation in mice, further enhancing both humoral and cellular immune responses [39]. In addition, a cyclodextrin–PEI-based formulation was further evaluated for its ability to penetrate the airway epithelial barrier and its paracellular delivery efficiency using IVT-mRNA encoding HIV gp120 [202]. With prolonged residence in the nasal cavity and excellent intracellular delivery, potent systemic and mucosal anti-HIV immune responses were induced after intranasal dosing [202].

5.2.3. Other Nonviral Vectors

Chitosan was reported to condense saRNA encoding hemagglutinin and nucleoprotein of influenza virus, and it expressed antigen in DCs after SC injection [203]. Recently, intranasal delivery of chitosan nanoparticles encapsulating mRNA encoding influenza HA2 and M2e antigens was reported to elicit efficient protective immune responses against avian influenza in chickens [167]. The aforementioned AuNS–chitosan nanoformulation also showed robust delivery of IVT-mRNA encoding firefly luciferase (fLuc-mRNA) in the lungs of mice upon intranasal delivery as measured by bioluminescence imaging [160]. Using a similar fLuc system, Yingshan Qiu et al. reported that $PEG_{12}KL4$ (a monodisperse linear PEG of 12-mers attached to synthetic cationic KL4 peptide) formed nanosized complexes with fLuc-mRNA and that the intratracheal administration of $PEG_{12}KL4$/fLuc-mRNA complexes resulted in luciferase expression in the deep lung region of mice 24 h posttransfection that was superior to expression induced by naked IVT-mRNA and Lipofectamine 2000 based lipoplexes [204]. However, both of these studies using luciferase reporter systems did not evaluate the antigen-specific immune responses induced by the IVT-mRNA formulations [160,204], making it difficult to affirm whether these delivery systems are suitable for the application of IVT-mRNA vaccines.

Meanwhile, lipid/polymer hybrid complexes have also been employed for efficient pulmonary delivery of IVT-mRNA. Compared with single lipid or polymer delivery systems, a stable IVT-mRNA vaccine particle could be formulated using positively charged protamine to concentrate IVT-mRNA followed by encapsulation with DOTAP/Chol/DSPE-PEG [205]. Such liposome/protamine hybrid complexes exhibit stronger capacities to stimulate DCs' maturation, which further induces potent antitumor cellular immune responses after intranasal administration with IVT-mRNA encoding cytokeratin [205].

Some commercial transfection reagents have also been employed for IVT-mRNA vaccine delivery via the respiratory tract. Intranasal administration of mRNA encoding HSP65 protein dissolved in Ringer's lactate solution prompted the specific production of IL-10 and TNF-α in the spleens of mice, and protected the mice from subsequent challenge with *M. tuberculosis* [206]. In both prophylactic and therapeutic immunization models, IVT-mRNA delivered in nanoparticles prepared with Stemfect^{TM} induced tumor immunity correlated with splenic antigen-specific $CD8^+$ T cells, while naked counterparts failed to elicit antigen-specific immune responses [207].

Table 3. Summary of RNA vaccines inoculated via the respiratory tract.

Disease	Nanoparticle	Coding Antigens	Model Tested	Administration	Immune Response [1]	Ref.
Influenza	LNP	HA	mice	i.n.	HIR(+)/CIR(++)	[34]
Influenza	Chitosan	HA and M2	chicken	i.n.	HIR(++)/MIR(++)/CIR(+)	[167]
HIV	Cyclodextrin–PEI conjugate	gp120	mice	i.n.	HIR(+)/CIR(+)	[202]
Model antigen	Cyclodextrin–PEI conjugate	OVA	mice	i.n.	HIR(+)/MIR(+)CIR(+)	[39]
Aggressive Lewis lung cancer model	Cationic liposome/protamine	cytokeratin 19	mice	i.n.	CIR(+)	[205]
Tuberculosis	Ringer's lactate solution	HSP65	mice	i.n.	CIR(++)	[206]
E.G7-OVA tumor	Stemfect mRNA transfection reagent	OVA	mice	i.n.	CIR(++++)	[207]

[1] Responses are geometric means of postvaccination increase in specific antibodies versus control in vaccine recipients: ++++, >10-fold; +++, 5- to 10-fold; ++, 2.5- to 5-fold; +, 1.5- to 2.5-fold. HA: hemagglutinin protein; M2: matrix protein 2; OVA: ovalbumin; HIV: human immunodeficiency virus; HIR: humoral immune responses; MIR: mucosal immune response; CIR: cellular immune responses; SARS-CoV: severe acute respiratory syndrome coronavirus; i.n.: intranasal administration.

6. Conclusions and Perspective

The mucosal immune system provides a first line of defense in the host's resistance to pathogen invasion and controls further infections at peripheral tissues. For the delivery of NA-vaccines, mucosal routes could be as good as or even better than parenteral counterparts. However, the complexity of the pulmonary environment poses huge obstacles for vaccines in inducing mucosal immunity compared with the parenteral route. This is especially difficult for NA-vaccines when considering their distinct characteristics, such as their high susceptibility to degradation by nuclease, negative charge, and high molecular weight. To overcome the mechanical–physical barriers of the respiratory tract, developing safe and efficient delivery systems is crucial for the successful clinical translation of NA-vaccines, as proven by the LNP delivery system in COVID-19 mRNA vaccines. Because of their similarity to microorganisms in both size and structure, nanodelivery vehicles play multifunctional roles in NA-vaccines by acting as a cargo carriers and potentially also as adjuvants [196]. Such nanosized systems can augment mucosal immune responses via simulating the natural infection process. For DNA vaccines, efficient delivery systems can improve immune responses by enhancing pDNA delivery into the nuclei of the host cells, which increases the expression of antigens. For mRNA vaccines, efficient carriers protect mRNA from premature degradation while optimizing in vivo expression of antigens, inducing a potent protective immunity. Although evidence has shown that NA-vaccines inoculated via the respiratory tract could elicit strong and long-term mucosal, humoral, and cell-mediated immune responses in animal models, there is still no NA-vaccine approved for mucosal vaccination purposes. As a result, innovative mucosal delivery technologies that could mediate successful clinical translations are urgently needed. Based on previous studies and our experiences, an ideal delivery system could provide the NA-vaccine with further unique features: (1) an electrically neutral structure with a hydrophilic shell, to bestow an efficient mucus-penetrating ability and to mediate rapid deposition onto epithelial surfaces; (2) APCs or M-cell targeting, to enhance the uptake of nucleic acids; (3) structures with protonated ability and/or membrane disruptive properties, which may facilitate efficient endosomal escape, avoiding enzymatic degradation.

Another interesting but scarcely discussed topic is the determination of the more advantageous route (between intranasal or intratracheal inoculation) for NA-vaccines. As shown in Tables 2 and 3, most studies have chosen the nasal route because of its convenient and noninvasive features. Nevertheless, there are still certain concerns regarding nasal administration. For example, negative perception for nasal vaccines was generated from reported cases of Bell's palsy after intranasal dosing of inactivated influenza vaccines [208,209]. Thus, neurotoxicity tests are necessary for nasal NA-vaccines in order to confirm their safety profiles. On the other hand, the intratracheal route also suffers from limitations. For example, it is not possible to apply intratracheal instillation in humans because of poor compliance and its invasive characteristics, the latter of which may cause unnecessary damage to the airway. Alternatively, one of the most appropriate approaches of delivering NA-vaccines to human airway would be nebulized formulations. Not only do nebulized formulations tend to be more evenly distributed throughout the respiratory tract, but the inhalation of aerosolized vaccine is also highly acceptable and tolerable for the recipient [210]. Unfortunately, nebulization of nucleic acid-based formulations tends to be inefficient [211]. A previous study found that as little as 10% of the nucleic acid payload in a nebulization device chamber could be successfully emitted [212]. Advanced nebulization strategies and optimized formulations have significantly improved the situation; now, even fragile IVT-mRNA has been successfully utilized in aerosolized formulations for in vitro and in vivo investigations [213,214]. As a result, nebulized formulations appear to be a feasible and attractive approach for the pulmonary delivery of NA-vaccines.

In summary, the field of respiratory mucosal vaccine has been bolstered by advances in novel nucleic acid formulations based on nanotechnologies. Thus, we hope that successful clinical nasal or pulmonary administration of NA-vaccines will be on the horizon, revolutionizing the vaccine development field as IVT-mRNA vaccines did during the COVID-19 pandemic.

Author Contributions: J.T., L.C., C.X., S.S. and S.G. wrote the manuscript; L.C., Y.L., J.R. and S.G. critically revised all versions of the article. All authors have read and agreed to the published version of the manuscript.

Funding: This work was supported by the National Natural Science Foundation of China (NSFC, Grant Nos. 82173764 and 82041045), the Chongqing Talents: Exceptional Young Talents Project (CQYC202005027), and the Natural Science Foundation of Chongqing (cstc2021jcyj-msxmX0136).

Acknowledgments: S.G. would like to acknowledge the financial support from Chongqing Talents: Exceptional Young Talents Project. The funders had no role in the literature search, the writing of the manuscript, or the decision to publish. The authors are grateful to Xiaoyan Ding (Third Military

Medical University, China) for her contributions to the manuscript regarding the topic of IVT-mRNA structure modification.

Conflicts of Interest: The authors declare no conflict of interest.

Abbreviations

AEC: alveolar epithelial cells; AM, alveola macrophages; APC, antigen-presenting cell; ARCA, anti-reverse cap analogue; AuNS, gold nanostar; BALB, Bagg Albino; BALF, bronchoalveolar lavage fluid; BALT, bronchus-associated lymphoid tissue; CFTR, cystic fibrosis transmembrane conductance regulator; cGAMP, 2′,3′-cyclic guanosine monophosphate–adenosine monophosphate; CIR, cellular immune responses; COVID, coronavirus disease; CTL, cytotoxic T lymphocytes; DC, dendritic cell; DNA, deoxyribonucleic acid; DODAC, N-dimethylammonium chloride; DOPE, dioleoyl phosphatidylethanolamine; DOTAP, 1,2-dioleoyl-3-trimethylammonium-propane; DSPE, 1,2-Distearoyl-sn-glycero-3-phosphorylethanolamine; EPC, egg phosphatidylcholine; FDA, Food and Drug Administration; fLuc, firefly luciferase; FMDV, foot-and-mouth disease; GAP-DLRIE, (+/−)-N-(3-aminopropyl)-N, N-dimethyl-2,3-bis (dodecyloxy)-1-propanaminium bromide; HA, hyaluronic acid; HA2, influenza virus hemagglutinin 2; HIR, humoral immune response; HIV, human immunodeficiency virus; IC50, serum dilution factor; IgA, immunoglobulin A; IgG, immunoglobulin G; IVT, in vitro transcribed; LALT, larynx-associated lymphatic tissues; LN, lymph nodes; LNP, lipid nanoparticles; LRT, lower respiratory tract; MALT, mucosal-associated lymphoid tissues; MCS, mannosylated chitosan; MHC, major histocompatibility complex; MIR, mucosal immune response; mRNA, messenger ribonucleic acid; NALT, nasopharynx-associated lymphoid tissues; NA-vaccines, nucleic acid-based vaccines; ORF, open reading frame; OVA, ovalbumin; PC, phosphatidylcholine; PEG, polyethylene glycol; PEI, polyethylenimine; PLGA, poly-lactic-co-glycolic acid; POI, protein(s) of interest; PRR, pattern recognition receptors; PS, pulmonary surfactant; PVA, polyvinyl alcohol; RNA, ribonucleic acid; RSV, respiratory syncytial virus; SARS-CoV, severe acute respiratory syndrome coronavirus; SC, subcutaneous; sIgA, secreted immunoglobulin A; STING, stimulator of interferon genes; T_{CM}, central memory T cells; T_{EM}, effector memory T cells; T_{FH}, follicular helper T cells; TLR, toll-like receptors; TNF, tumor necrosis factor; TRM, resident memory T cells; VCE, virus capping enzyme.

References

1. Pardi, N.; Hogan, M.J.; Porter, F.W.; Weissman, D. mRNA vaccines-a new era in vaccinology. *Nat. Rev. Drug Discov.* **2018**, *17*, 261–279. [CrossRef]
2. Hajj, K.A.; Whitehead, K.A. Tools for translation: Non-viral materials for therapeutic mRNA delivery. *Nat. Rev. Mater.* **2017**, *2*, 17056. [CrossRef]
3. Kutzler, M.A.; Weiner, D.B. DNA vaccines: Ready for prime time? *Nat. Rev. Genet.* **2008**, *9*, 776–788. [CrossRef]
4. Sadeghi, I.; Byrne, J.; Shakur, R.; Langer, R. Engineered drug delivery devices to address Global Health challenges. *J. Control. Release* **2021**, *331*, 503–514. [CrossRef] [PubMed]
5. Sahin, U.; Karikó, K.; Türeci, Ö. mRNA-based therapeutics—Developing a new class of drugs. *Nat. Rev. Drug Discov.* **2014**, *13*, 759–780. [CrossRef] [PubMed]
6. Rice, J.; Ottensmeier, C.H.; Stevenson, F.K. DNA vaccines: Precision tools for activating effective immunity against cancer. *Nat. Rev. Cancer* **2008**, *8*, 108–120. [CrossRef] [PubMed]
7. Lederer, K.; Castaño, D.; Atria, D.G.; Oguin, T.H.; Wang, S.; Manzoni, T.B.; Muramatsu, H.; Hogan, M.J.; Amanat, F.; Cherubin, P.; et al. SARS-CoV-2 mRNA vaccines foster potent antigen-specific germinal center responses associated with neutralizing antibody generation. *Immunity* **2020**, *53*, 1281–1295. [CrossRef] [PubMed]
8. Anttila, V.; Saraste, A.; Knuuti, J.; Jaakkola, P.; Hedman, M.; Svedlund, S.; Lagerström-Fermér, M.; Kjaer, M.; Jeppsson, A.; Gan, L.-M. Synthetic mRNA Encoding VEGF-A in Patients Undergoing Coronary Artery Bypass Grafting: Design of a Phase 2a Clinical Trial. *Mol. Ther. Methods Clin. Dev.* **2020**, *18*, 464–472. [CrossRef]
9. Yu, J.; Tostanoski, L.H.; Peter, L.; Mercado, N.B.; McMahan, K.; Mahrokhian, S.H.; Nkolola, J.P.; Liu, J.; Li, Z.; Chandrashekar, A.; et al. DNA vaccine protection against SARS-CoV-2 in rhesus macaques. *Science* **2020**, *369*, 806–811. [CrossRef]
10. Weide, B.; Carralot, J.-P.; Reese, A.; Scheel, B.; Eigentler, T.K.; Hoerr, I.; Rammensee, H.-G.; Garbe, C.; Pascolo, S. Results of the first phase I/II clinical vaccination trial with direct injection of mRNA. *J. Immunother.* **2008**, *31*, 180–188. [CrossRef]
11. Guan, S.; Munder, A.; Hedtfeld, S.; Braubach, P.; Glage, S.; Zhang, L.; Lienenklaus, S.; Schultze, A.; Hasenpusch, G.; Garrels, W.; et al. Self-assembled peptide–poloxamine nanoparticles enable in vitro and in vivo genome restoration for cystic fibrosis. *Nat. Nanotechnol.* **2019**, *14*, 287–297. [CrossRef] [PubMed]

12. Guan, S.; Rosenecker, J. Nanotechnologies in delivery of mRNA therapeutics using nonviral vector-based delivery systems. *Gene Ther.* **2017**, *24*, 133–143. [CrossRef]
13. Lim, M.; Badruddoza, A.Z.M.; Firdous, J.; Azad, M.; Mannan, A.; Al-Hilal, T.A.; Cho, C.-S.; Islam, M.A. Engineered Nanodelivery Systems to Improve DNA Vaccine Technologies. *Pharmaceutics* **2020**, *12*, 30. [CrossRef]
14. Bolhassani, A.; Yazdi, S.R. DNA immunization as an efficient strategy for vaccination. *Avicenna J. Med. Biotechnol.* **2009**, *1*, 71–88.
15. Smith, L.R.; Wloch, M.K.; Ye, M.; Reyes, L.R.; Boutsaboualoy, S.; Dunne, C.E.; Chaplin, J.A.; Rusalov, D.; Rolland, A.P.; Fisher, C.L.; et al. Phase 1 clinical trials of the safety and immunogenicity of adjuvanted plasmid DNA vaccines encoding influenza A virus H5 hemagglutinin. *Vaccine* **2010**, *28*, 2565–2572. [CrossRef] [PubMed]
16. Tebas, P.; Yang, S.; Boyer, J.D.; Reuschel, E.L.; Patel, A.; Christensen-Quick, A.; Andrade, V.M.; Morrow, M.P.; Kraynyak, K.; Agnes, J.; et al. Safety and immunogenicity of INO-4800 DNA vaccine against SARS-CoV-2: A preliminary report of an open-label, Phase 1 clinical trial. *EClinicalMedicine* **2020**, *31*, 100689. [CrossRef]
17. Jacobson, J.M.; Routy, J.-P.; Welles, S.; DeBenedette, M.; Tcherepanova, I.; Angel, J.B.; Asmuth, D.M.; Stein, D.K.; Baril, J.-G.; McKellar, M.; et al. Dendritic Cell Immunotherapy for HIV-1 Infection Using Autologous HIV-1 RNA: A Randomized, Double-Blind, Placebo-Controlled Clinical Trial. *J. Acquir. Immune Defic. Syndr.* **2016**, *72*, 31–38. [CrossRef]
18. Alberer, M.; Gnad-Vogt, U.; Hong, H.S.; Mehr, K.T.; Backert, L.; Finak, G.; Gottardo, R.; Bica, M.A.; Garofano, A.; Koch, S.D.; et al. Safety and immunogenicity of a mRNA rabies vaccine in healthy adults: An open-label, non-randomised, prospective, first-in-human phase 1 clinical trial. *Lancet* **2017**, *390*, 1511–1520. [CrossRef]
19. Richner, J.M.; Himansu, S.; Dowd, K.A.; Butler, S.L.; Salazar, V.; Fox, J.M.; Julander, J.G.; Tang, W.W.; Shresta, S.; Pierson, T.C.; et al. Modified mRNA Vaccines Protect against Zika Virus Infection. *Cell* **2017**, *168*, 1114–1125. [CrossRef] [PubMed]
20. Bahl, K.; Senn, J.J.; Yuzhakov, O.; Bulychev, A.; Brito, L.A.; Hassett, K.J.; Laska, M.E.; Smith, M.; Almarsson, O.; Thompson, J. Preclinical and Clinical Demonstration of Immunogenicity by mRNA Vaccines against H10N8 and H7N9 Influenza Viruses. *Mol. Ther. J. Am. Soc. Gene Ther.* **2017**, *25*, 1316–1327. [CrossRef]
21. Aarntzen, E.H.J.G.; Schreibelt, G.; Bol, K.; Lesterhuis, W.J.; Croockewit, A.J.; de Wilt, J.H.W.; van Rossum, M.M.; Blokx, W.A.M.; Jacobs, J.F.M.; Duiveman-de Boer, T.; et al. Vaccination with mRNA-electroporated dendritic cells induces robust tumor antigen-specific CD4+ and CD8+ T cells responses in stage III and IV melanoma patients. *Clin. Cancer Res. Off. J. Am. Assoc. Cancer Res.* **2012**, *18*, 5460–5470. [CrossRef]
22. Weide, B.; Pascolo, S.; Scheel, B.; Derhovanessian, E.; Pflugfelder, A.; Eigentler, T.K.; Pawelec, G.; Hoerr, I.; Rammensee, H.-G.; Garbe, C. Direct injection of protamine-protected mRNA: Results of a phase 1/2 vaccination trial in metastatic melanoma patients. *J. Immunother.* **2009**, *32*, 498–507. [CrossRef] [PubMed]
23. Kranz, L.M.; Diken, M.; Haas, H.; Kreiter, S.; Loquai, C.; Reuter, K.C.; Meng, M.; Fritz, D.; Vascotto, F.; Hefesha, H.; et al. Systemic RNA delivery to dendritic cells exploits antiviral defence for cancer immunotherapy. *Nature* **2016**, *534*, 396–401. [CrossRef]
24. Dong, Y.; Dai, T.; Wei, Y.; Zhang, L.; Zheng, M.; Zhou, F. A systematic review of SARS-CoV-2 vaccine candidates. *Signal Transduct. Target. Ther.* **2020**, *5*, 237. [CrossRef]
25. Jackson, L.A.; Anderson, E.J.; Rouphael, N.G.; Roberts, P.C.; Makhene, M.; Coler, R.N.; McCullough, M.P.; Chappell, J.D.; Denison, M.R.; Stevens, L.J.; et al. An mRNA Vaccine against SARS-CoV-2—Preliminary Report. *N. Engl. J. Med.* **2020**, *383*, 1920–1931. [CrossRef]
26. Baden, L.R.; El Sahly, H.M.; Essink, B.; Kotloff, K.; Frey, S.; Novak, R.; Diemert, D.; Spector, S.A.; Rouphael, N.; Creech, C.B.; et al. Efficacy and Safety of the mRNA-1273 SARS-CoV-2 Vaccine. *N. Engl. J. Med.* **2021**, *384*, 403–416. [CrossRef]
27. Walsh, E.E.; Frenck, R.W.J.; Falsey, A.R.; Kitchin, N.; Absalon, J.; Gurtman, A.; Lockhart, S.; Neuzil, K.; Mulligan, M.J.; Bailey, R.; et al. Safety and Immunogenicity of Two RNA-Based COVID-19 Vaccine Candidates. *N. Engl. J. Med.* **2020**, *383*, 2439–2450. [CrossRef]
28. Russell, M.W.; Ogra, P.L. Chapter 1—Historical perspectives on mucosal vaccines. In *Mucosal Vaccines*, 2nd ed.; Kiyono, H., Pascual, D.W.B.T.-M.V., Eds.; Academic Press: Cambridge, MA, USA, 2020; pp. 3–17. ISBN 978-0-12-811924-2.
29. Iwasaki, A.; Foxman, E.F.; Molony, R.D. Early local immune defences in the respiratory tract. *Nat. Rev. Immunol.* **2017**, *17*, 7–20. [CrossRef] [PubMed]
30. Zhang, L.; Wang, W.; Wang, S. Effect of vaccine administration modality on immunogenicity and efficacy. *Expert Rev. Vaccines* **2015**, *14*, 1509–1523. [CrossRef] [PubMed]
31. Riccomi, A.; Piccaro, G.; Christensen, D.; Palma, C.; Andersen, P.; Vendetti, S. Parenteral Vaccination with a Tuberculosis Subunit Vaccine in Presence of Retinoic Acid Provides Early but Transient Protection to M. Tuberculosis Infection. *Front. Immunol.* **2019**, *10*, 934. [CrossRef] [PubMed]
32. Christensen, D.; Bøllehuus Hansen, L.; Leboux, R.; Jiskoot, W.; Christensen, J.P.; Andersen, P.; Dietrich, J. A Liposome-Based Adjuvant Containing Two Delivery Systems with the Ability to Induce Mucosal Immunoglobulin a Following a Parenteral Immunization. *ACS Nano* **2019**, *13*, 1116–1126. [CrossRef]
33. Moreno-Fierros, L.; García-Silva, I.; Rosales-Mendoza, S. Development of SARS-CoV-2 vaccines: Should we focus on mucosal immunity? *Expert Opin. Biol. Ther.* **2020**, *20*, 831–836. [CrossRef] [PubMed]
34. Zhuang, X.; Qi, Y.; Wang, M.; Yu, N.; Nan, F.; Zhang, H.; Tian, M.; Li, C.; Lu, H.; Jin, N. mRNA Vaccines Encoding the HA Protein of Influenza a H1N1 Virus Delivered by Cationic Lipid Nanoparticles Induce Protective Immune Responses in Mice. *Vaccines* **2020**, *8*, 123. [CrossRef]

35. Tiwari, P.M.; Vanover, D.; Lindsay, K.E.; Bawage, S.S.; Kirschman, J.L.; Bhosle, S.; Lifland, A.W.; Zurla, C.; Santangelo, P.J. Engineered mRNA-expressed antibodies prevent respiratory syncytial virus infection. *Nat. Commun.* **2018**, *9*, 3999. [CrossRef]
36. Hassan, A.O.; Kafai, N.M.; Dmitriev, I.P.; Fox, J.M.; Brittany, K.; Harvey, I.B.; Chen, R.E.; Winkler, E.S.; Wessel, A.W.; Brett, J.; et al. A single-dose intranasal ChAd vaccine protects upper and lower respiratory tracts against SARS-CoV-2. *Cell* **2020**, *183*, 169–184. [CrossRef]
37. Trevaskis, N.L.; Kaminskas, L.M.; Porter, C.J.H. From sewer to saviour—targeting the lymphatic system to promote drug exposure and activity. *Nat. Rev. Drug Discov.* **2015**, *14*, 781–803. [CrossRef] [PubMed]
38. Lindsay, K.E.; Vanover, D.; Thoresen, M.; King, H.; Xiao, P.; Badial, P.; Araínga, M.; Park, S.B.; Tiwari, P.M.; Peck, H.E.; et al. Aerosol Delivery of Synthetic mRNA to Vaginal Mucosa Leads to Durable Expression of Broadly Neutralizing Antibodies against HIV. *Mol. Ther.* **2020**, *28*, 805–819. [CrossRef]
39. Li, M.; Li, Y.; Peng, K.; Wang, Y.; Gong, T.; Zhang, Z.; He, Q.; Sun, X. Engineering intranasal mRNA vaccines to enhance lymph node trafficking and immune responses. *Acta Biomater.* **2017**, *64*, 237–248. [CrossRef]
40. Li, Y.; Jin, L.; Chen, T. The Effects of Secretory IgA in the Mucosal Immune System. *Biomed. Res. Int.* **2020**, *2020*, 2032057. [CrossRef] [PubMed]
41. Shakya, A.K.; Chowdhury, M.Y.E.; Tao, W.; Gill, H.S. Mucosal vaccine delivery: Current state and a pediatric perspective. *J. Control. Release* **2016**, *240*, 394–413. [CrossRef]
42. Fujkuyama, Y.; Tokuhara, D.; Kataoka, K.; Gilbert, R.S.; McGhee, J.R.; Yuki, Y.; Kiyono, H.; Fujihashi, K. Novel vaccine development strategies for inducing mucosal immunity. *Expert Rev. Vaccines* **2012**, *11*, 367–379. [CrossRef] [PubMed]
43. Griffin, D.E. Current progress in pulmonary delivery of measles vaccine. *Expert Rev. Vaccines* **2014**, *13*, 751–759. [CrossRef]
44. Hoft, D.F.; Lottenbach, K.R.; Blazevic, A.; Turan, A.; Blevins, T.P.; Pacatte, T.P.; Yu, Y.; Mitchell, M.C.; Hoft, S.G.; Belshe, R.B. Comparisons of the Humoral and Cellular Immune Responses Induced by Live Attenuated Influenza Vaccine and Inactivated Influenza Vaccine in Adults. *Clin. Vaccine Immunol.* **2017**, *24*, 414–416. [CrossRef]
45. Lobaina Mato, Y. Nasal route for vaccine and drug delivery: Features and current opportunities. *Int. J. Pharm.* **2019**, *572*, 118813. [CrossRef] [PubMed]
46. Miquel-Clopés, A.; Bentley, E.G.; Stewart, J.P.; Carding, S.R. Mucosal vaccines and technology. *Clin. Exp. Immunol.* **2019**, *196*, 205–214. [CrossRef] [PubMed]
47. Van Doremalen, N.; Purushotham, J.; Schulz, J.; Holbrook, M.; Bushmaker, T.; Carmody, A.; Port, J.; Yinda, K.C.; Okumura, A.; Saturday, G.; et al. Intranasal ChAdOx1 nCoV-19/AZD1222 vaccination reduces shedding of SARS-CoV-2 D614G in rhesus macaques. *bioRxiv Prepr. Serv. Biol.* **2021**. [CrossRef]
48. Loes, A.N.; Gentles, L.E.; Greaney, A.J.; Crawford, K.H.D.; Bloom, J.D. Attenuated Influenza Virions Expressing the SARS-CoV-2 Receptor-Binding Domain Induce Neutralizing Antibodies in Mice. *Viruses* **2020**, *12*, 987. [CrossRef]
49. Yin, H.; Kanasty, R.L.; Eltoukhy, A.A.; Vegas, A.J.; Dorkin, J.R.; Anderson, D.G. Non-viral vectors for gene-based therapy. *Nat. Rev. Genet.* **2014**, *15*, 541–555. [CrossRef]
50. Winslow, C.E.A. A new method of enumerating bacteria in air. *Science* **1908**, *28*, 28–31. [CrossRef]
51. Marasini, N.; Haque, S.; Kaminskas, L.M. Polymer-drug conjugates as inhalable drug delivery systems: A review. *Curr. Opin. Colloid Interface Sci.* **2017**, *31*, 18–29. [CrossRef]
52. Tsukita, S.; Yamazaki, Y.; Katsuno, T.; Tamura, A.; Tsukita, S. Tight junction-based epithelial microenvironment and cell proliferation. *Oncogene* **2008**, *27*, 6930–6938. [CrossRef]
53. Kaliner, M.; Shelhamer, J.H.; Borson, B.; Nadel, J.; Patow, C.; Marom, Z. Human respiratory mucus. *Am. Rev. Respir. Dis.* **1986**, *134*, 612–621. [CrossRef]
54. Wanner, A. Mucociliary clearance in the trachea. *Clin. Chest Med.* **1986**, *7*, 247–258. [CrossRef]
55. Sato, S.; Kiyono, H. The mucosal immune system of the respiratory tract. *Curr. Opin. Virol.* **2012**, *2*, 225–232. [CrossRef]
56. Martin, T.R.; Frevert, C.W. Innate immunity in the lungs. *Proc. Am. Thorac. Soc.* **2005**, *2*, 403–411. [CrossRef] [PubMed]
57. Kiyono, H.; Fukuyama, S. Nalt-versus Peyer's-patch-mediated mucosal immunity. *Nat. Rev. Immunol.* **2004**, *4*, 699–710. [CrossRef]
58. Pabst, R. Lymphatic tissue of the nose (NALT) and larynx (LALT) in species comparison: Human, rat, mouse. *Pneumologie* **2010**, *64*, 445–446. [CrossRef] [PubMed]
59. Kuper, C.F.; Koornstra, P.J.; Hameleers, D.M.; Biewenga, J.; Spit, B.J.; Duijvestijn, A.M.; van Breda Vriesman, P.J.; Sminia, T. The role of nasopharyngeal lymphoid tissue. *Immunol. Today* **1992**, *13*, 219–224. [CrossRef]
60. Zuercher, A.W.; Coffin, S.E.; Thurnheer, M.C.; Fundova, P.; Cebra, J.J. Nasal-Associated Lymphoid Tissue Is a Mucosal Inductive Site for Virus-Specific Humoral and Cellular Immune Responses. *J. Immunol.* **2002**, *168*, 1796–1803. [CrossRef]
61. Allan, R.S.; Waithman, J.; Bedoui, S.; Jones, C.M.; Villadangos, J.A.; Zhan, Y.; Lew, A.M.; Shortman, K.; Heath, W.R.; Carbone, F.R. Migratory Dendritic Cells Transfer Antigen to a Lymph Node-Resident Dendritic Cell Population for Efficient CTL Priming. *Immunity* **2006**, *25*, 153–162. [CrossRef]
62. Sung, S.-S.J.; Fu, S.M.; Rose, C.E.; Gaskin, F.; Ju, S.-T.; Beaty, S.R. A Major Lung CD103 (α E)-β 7 Integrin-Positive Epithelial Dendritic Cell Population Expressing Langerin and Tight Junction Proteins. *J. Immunol.* **2006**, *176*, 2161–2172. [CrossRef] [PubMed]
63. Park, H.-S.; Francis, K.P.; Yu, J.; Cleary, P.P. Membranous Cells in Nasal-Associated Lymphoid Tissue: A Portal of Entry for the Respiratory Mucosal Pathogen Group A Streptococcus. *J. Immunol.* **2003**, *171*, 2532–2537. [CrossRef] [PubMed]
64. Pappo, J.; Mahlman, R.T. Follicle epithelial M cells are a source of interleukin-1 in Peyer's patches. *Immunology* **1993**, *78*, 505–507.

65. Lawrence, C.W.; Braciale, T.J. Activation, Differentiation, and Migration of Naive Virus-Specific CD8 + T Cells during Pulmonary Influenza Virus Infection. *J. Immunol.* **2004**, *173*, 1209–1218. [CrossRef]
66. Román, E.; Miller, E.; Harmsen, A.; Wiley, J.; Von Andrian, U.H.; Huston, G.; Swain, S.L. CD4 effector T cell subsets in the response to influenza: Heterogeneity, migration, and function. *J. Exp. Med.* **2002**, *196*, 957–968. [CrossRef]
67. Hou, S.; Doherty, P.C. Clearance of Sendai virus by CD8+ T cells requires direct targeting to virus-infected epithelium. *Eur. J. Immunol.* **1995**, *25*, 111–116. [CrossRef] [PubMed]
68. Topham, D.J.; Tripp, R.A.; Doherty, P.C. CD8+ T cells clear influenza virus by perforin or Fas-dependent processes. *J. Immunol.* **1997**, *159*, 5197–5200.
69. Allie, S.R.; Randall, T.D. Pulmonary immunity to viruses. *Clin. Sci.* **2017**, *131*, 1737–1762. [CrossRef]
70. De Silva, N.S.; Klein, U. Dynamics of B cells in germinal centres. *Nat. Rev. Immunol.* **2015**, *15*, 137–148. [CrossRef] [PubMed]
71. Lee, B.O.; Rangel-Moreno, J.; Moyron-Quiroz, J.E.; Hartson, L.; Makris, M.; Sprague, F.; Lund, F.E.; Randall, T.D. CD4 T Cell-Independent Antibody Response Promotes Resolution of Primary Influenza Infection and Helps to Prevent Reinfection. *J. Immunol.* **2005**, *175*, 5827–5838. [CrossRef] [PubMed]
72. Van Riet, E.; Ainai, A.; Suzuki, T.; Hasegawa, H. Mucosal IgA responses in influenza virus infections; thoughts for vaccine design. *Vaccine* **2012**, *30*, 5893–5900. [CrossRef]
73. Kaetzel, C.S. The polymeric immunoglobulin receptor: Bridging innate and adaptive immune responses at mucosal surfaces. *Immunol. Rev.* **2005**, *206*, 83–99. [CrossRef]
74. Johansen, F.E.; Kaetzel, C.S. Regulation of the polymeric immunoglobulin receptor and IgA transport: New advances in environmental factors that stimulate pIgR expression and its role in mucosal immunity. *Mucosal Immunol.* **2011**, *4*, 598–602. [CrossRef]
75. Häcker, G.; Bauer, A.; Villunger, A. Apoptosis in activated T cells: What are the triggers, and what the signal transducers? *Cell Cycle* **2006**, *5*, 2421–2424. [CrossRef] [PubMed]
76. Kurosaki, T.; Kometani, K.; Ise, W. Memory B cells. *Nat. Rev. Immunol.* **2015**, *15*, 149–159. [CrossRef] [PubMed]
77. Li, C.K.; Wu, H.; Yan, H.; Ma, S.; Wang, L.; Zhang, M.; Tang, X.; Temperton, N.J.; Weiss, R.A.; Brenchley, J.M.; et al. T cell responses to whole SARS coronavirus in humans. *J. Immunol.* **2008**, *181*, 5490–5500. [CrossRef] [PubMed]
78. Sallusto, F.; Geginat, J.; Lanzavecchia, A. Central memory and effector memory T cell subsets: Function, generation, and maintenance. *Annu. Rev. Immunol.* **2004**, *22*, 745–763. [CrossRef]
79. Roberts, A.D.; Ely, K.H.; Woodland, D.L. Differential contributions of central and effector memory T cells to recall responses. *J. Exp. Med.* **2005**, *202*, 123–133. [CrossRef]
80. Harari, A.; Bellutti Enders, F.; Cellerai, C.; Bart, P.-A.; Pantaleo, G. Distinct profiles of cytotoxic granules in memory CD8 T cells correlate with function, differentiation stage, and antigen exposure. *J. Virol.* **2009**, *83*, 2862–2871. [CrossRef] [PubMed]
81. Hansen, S.G.; Vieville, C.; Whizin, N.; Coyne-Johnson, L.; Siess, D.C.; Drummond, D.D.; Legasse, A.W.; Axthelm, M.K.; Oswald, K.; Trubey, C.M.; et al. Effector memory T cell responses are associated with protection of rhesus monkeys from mucosal simian immunodeficiency virus challenge. *Nat. Med.* **2009**, *15*, 293–299. [CrossRef] [PubMed]
82. Robinson, H.L.; Amara, R.R. T cell vaccines for microbial infections. *Nat. Med.* **2005**, *11*, S25–S32. [CrossRef]
83. Schenkel, J.M.; Masopust, D. Tissue-resident memory T cells. *Immunity* **2014**, *41*, 886–897. [CrossRef]
84. Teijaro, J.R.; Turner, D.; Pham, Q.; Wherry, E.J.; Lefrançois, L.; Farber, D.L. Cutting edge: Tissue-retentive lung memory CD4 T cells mediate optimal protection to respiratory virus infection. *J. Immunol.* **2011**, *187*, 5510–5514. [CrossRef]
85. Turner, D.L.; Farber, D.L. Mucosal resident memory CD4 T cells in protection and immunopathology. *Front. Immunol.* **2014**, *5*, 331. [CrossRef] [PubMed]
86. Morabito, K.M.; Ruckwardt, T.R.; Redwood, A.J.; Moin, S.M.; Price, D.A.; Graham, B.S. Intranasal administration of RSV antigen-expressing MCMV elicits robust tissue-resident effector and effector memory CD8+ T cells in the lung. *Mucosal Immunol.* **2017**, *10*, 545–554. [CrossRef]
87. Hogan, R.J.; Zhong, W.; Usherwood, E.J.; Cookenham, T.; Roberts, A.D.; Woodland, D.L. Protection from respiratory virus infections can be mediated by antigen-specific CD4(+) T cells that persist in the lungs. *J. Exp. Med.* **2001**, *193*, 981–986. [CrossRef] [PubMed]
88. Blank, F.; Stumbles, P.A.; Seydoux, E.; Holt, P.G.; Fink, A.; Rothen-Rutishauser, B.; Strickland, D.H.; Von Garnier, C. Size-dependent uptake of particles by pulmonary antigen-presenting cell populations and trafficking to regional lymph nodes. *Am. J. Respir. Cell Mol. Biol.* **2013**, *49*, 67–77. [CrossRef]
89. Sun, X.; Jones, H.P.; Dobbs, N.; Bodhankar, S.; Simecka, J.W. Dendritic Cells Are the Major Antigen Presenting Cells in Inflammatory Lesions of Murine Mycoplasma Respiratory Disease. *PLoS ONE* **2013**, *8*, e55984. [CrossRef]
90. Wang, J.; Li, P.; Yu, Y.; Fu, Y.; Jiang, H.; Lu, M.; Sun, Z.; Jiang, S.; Lu, L.; Wu, M.X. Pulmonary surfactant-biomimetic nanoparticles potentiate heterosubtypic influenza immunity. *Science* **2020**, *367*, 6480. [CrossRef]
91. She, L.; Barrera, G.D.; Yan, L.; Alanazi, H.H.; Brooks, E.G.; Dube, P.H.; Sun, Y.; Zan, H.; Chupp, D.P.; Zhang, N.; et al. STING activation in alveolar macrophages and group 2 innate lymphoid cells suppresses IL-33-driven type 2 immunopathology. *JCI Insight* **2021**, *6*, e143509. [CrossRef] [PubMed]
92. Von Garnier, C.; Filgueira, L.; Wikstrom, M.; Smith, M.; Thomas, J.A.; Strickland, D.H.; Holt, P.G.; Stumbles, P.A. Anatomical Location Determines the Distribution and Function of Dendritic Cells and Other APCs in the Respiratory Tract. *J. Immunol.* **2005**, *175*, 1609–1618. [CrossRef] [PubMed]

93. Moyron-Quiroz, J.E.; Rangel-Moreno, J.; Kusser, K.; Hartson, L.; Sprague, F.; Goodrich, S.; Woodland, D.L.; Lund, F.E.; Randall, T.D. Role of inducible bronchus associated lymphoid tissue (iBALT) in respiratory immunity. *Nat. Med.* **2004**, *10*, 927–934. [CrossRef] [PubMed]
94. Marin, N.D.; Dunlap, M.D.; Kaushal, D.; Khader, S.A. Friend or Foe: The Protective and Pathological Roles of Inducible Bronchus-Associated Lymphoid Tissue in Pulmonary Diseases. *J. Immunol.* **2019**, *202*, 2519–2526. [CrossRef]
95. Wiley, J.A.; Richert, L.E.; Swain, S.D.; Harmsen, A.; Barnard, D.L.; Randall, T.D.; Jutila, M.; Douglas, T.; Broomell, C.; Young, M.; et al. Inducible bronchus-associated lymphoid tissue elicited by a protein cage nanoparticle enhances protection in mice against diverse respiratory viruses. *PLoS ONE* **2009**, *4*, e7142. [CrossRef] [PubMed]
96. Wolff, J.A.; Malone, R.W.; Williams, P.; Chong, W.; Acsadi, G.; Jani, A.; Felgner, P.L. Direct gene transfer into mouse muscle in vivo. *Science* **1990**, *247*, 1465–1468. [CrossRef]
97. Ulmer, J.B.; Donnelly, J.J.; Parker, S.E.; Rhodes, G.H.; Felgner, P.L.; Dwarki, V.J.; Gromkowski, S.H.; Deck, R.R.; DeWitt, C.M.; Friedman, A.; et al. Heterologous protection against influenza by injection of DNA encoding a viral protein. *Science* **1993**, *259*, 1745–1749. [CrossRef] [PubMed]
98. Silveira, M.M.; Moreira, G.M.S.G.; Mendonça, M. DNA vaccines against COVID-19: Perspectives and challenges. *Life Sci.* **2021**, *267*, 118919. [CrossRef] [PubMed]
99. Ingolotti, M.; Kawalekar, O.; Shedlock, D.J.; Muthumani, K.; Weiner, D.B. DNA vaccines for targeting bacterial infections. *Expert Rev. Vaccines* **2010**, *9*, 747–763. [CrossRef]
100. Xu, Y.; Yuen, P.-W.; Lam, J.K.-W. Intranasal DNA Vaccine for Protection against Respiratory Infectious Diseases: The Delivery Perspectives. *Pharmaceutics* **2014**, *6*, 378–415. [CrossRef] [PubMed]
101. Mallapaty, S.; Callaway, E. What scientists do and don't know about the Oxford-AstraZeneca COVID vaccine. *Nature* **2021**, *592*, 15–17. [CrossRef]
102. Modjarrad, K.; Roberts, C.C.; Mills, K.T.; Castellano, A.R.; Paolino, K.; Muthumani, K.; Reuschel, E.L.; Robb, M.L.; Racine, T.; Oh, M.-D.; et al. Safety and immunogenicity of an anti-Middle East respiratory syndrome coronavirus DNA vaccine: A phase 1, open-label, single-arm, dose-escalation trial. *Lancet. Infect. Dis.* **2019**, *19*, 1013–1022. [CrossRef]
103. Sheridan, C. First COVID-19 DNA vaccine approved, others in hot pursuit. *Nat. Biotechnol.* **2021**, *39*, 1479–1482. [CrossRef] [PubMed]
104. Mallapaty, S. India's DNA COVID vaccine is a world first—More are coming. *Nature* **2021**, *597*, 161–162. [CrossRef]
105. Lombry, C.; Marteleur, A.; Arras, M.; Lison, D.; Louahed, J.; Renauld, J.-C.; Préat, V.; Vanbever, R. Local and systemic immune responses to intratracheal instillation of antigen and DNA vaccines in mice. *Pharm. Res.* **2004**, *21*, 127–135. [CrossRef]
106. Bivas-Benita, M.; van Meijgaarden, K.E.; Franken, K.L.M.C.; Junginger, H.E.; Borchard, G.; Ottenhoff, T.H.M.; Geluk, A. Pulmonary delivery of chitosan-DNA nanoparticles enhances the immunogenicity of a DNA vaccine encoding HLA-A*0201-restricted T-cell epitopes of Mycobacterium tuberculosis. *Vaccine* **2004**, *22*, 1609–1615. [CrossRef] [PubMed]
107. Low, L.; Mander, A.; McCann, K.; Dearnaley, D.; Tjelle, T.; Mathiesen, I.; Stevenson, F.; Ottensmeier, C.H. DNA vaccination with electroporation induces increased antibody responses in patients with prostate cancer. *Hum. Gene Ther.* **2009**, *20*, 1269–1278. [CrossRef]
108. Rottinghaus, S.T.; Poland, G.A.; Jacobson, R.M.; Barr, L.J.; Roy, M.J. Hepatitis B DNA vaccine induces protective antibody responses in human non-responders to conventional vaccination. *Vaccine* **2003**, *21*, 4604–4608. [CrossRef]
109. Brito, L.A.; Malyala, P.; O'Hagan, D.T. Vaccine adjuvant formulations: A pharmaceutical perspective. *Semin. Immunol.* **2013**, *25*, 130–145. [CrossRef]
110. Legendre, J.Y.; Szoka, F.C.J. Delivery of plasmid DNA into mammalian cell lines using pH-sensitive liposomes: Comparison with cationic liposomes. *Pharm. Res.* **1992**, *9*, 1235–1242. [CrossRef] [PubMed]
111. Khatri, K.; Goyal, A.K.; Gupta, P.N.; Mishra, N.; Mehta, A.; Vyas, S.P. Surface modified liposomes for nasal delivery of DNA vaccine. *Vaccine* **2008**, *26*, 2225–2233. [CrossRef]
112. Gogev, S.; de Fays, K.; Versali, M.-F.; Gautier, S.; Thiry, E. Glycol chitosan improves the efficacy of intranasally administrated replication defective human adenovirus type 5 expressing glycoprotein D of bovine herpesvirus 1. *Vaccine* **2004**, *22*, 1946–1953. [CrossRef] [PubMed]
113. Lay, M.; Callejo, B.; Chang, S.; Hong, D.K.; Lewis, D.B.; Carroll, T.D.; Matzinger, S.; Fritts, L.; Miller, C.J.; Warner, J.F.; et al. Cationic lipid/DNA complexes (JVRS-100) combined with influenza vaccine (Fluzone) increases antibody response, cellular immunity, and antigenically drifted protection. *Vaccine* **2009**, *27*, 3811–3820. [CrossRef] [PubMed]
114. D'Souza, S.; Rosseels, V.; Denis, O.; Tanghe, A.; De Smet, N.; Jurion, F.; Palfliet, K.; Castiglioni, N.; Vanonckelen, A.; Wheeler, C.; et al. Improved tuberculosis DNA vaccines by formulation in cationic lipids. *Infect. Immun.* **2002**, *70*, 3681–3688. [CrossRef] [PubMed]
115. Rosada, R.S.; de la Torre, L.G.; Frantz, F.G.; Trombone, A.P.F.; Zárate-Bladés, C.R.; Fonseca, D.M.; Souza, P.R.M.; Brandão, I.T.; Masson, A.P.; Soares, E.G.; et al. Protection against tuberculosis by a single intranasal administration of DNA-hsp65 vaccine complexed with cationic liposomes. *BMC Immunol.* **2008**, *9*, 38. [CrossRef] [PubMed]
116. Wong, J.P.; Zabielski, M.A.; Schmaltz, F.L.; Brownlee, G.G.; Bussey, L.A.; Marshall, K.; Borralho, T.; Nagata, L.P. DNA vaccination against respiratory influenza virus infection. *Vaccine* **2001**, *19*, 2461–2467. [CrossRef]

117. Wang, D.; Christopher, M.E.; Nagata, L.P.; Zabielski, M.A.; Li, H.; Wong, J.P.; Samuel, J. Intranasal immunization with liposome-encapsulated plasmid DNA encoding influenza virus hemagglutinin elicits mucosal, cellular and humoral immune responses. *J. Clin. Virol. Off. Publ. Pan Am. Soc. Clin. Virol.* **2004**, *31* (Suppl. 1), 99–106. [CrossRef]
118. Chen, C.; Han, D.; Cai, C.; Tang, X. An overview of liposome lyophilization and its future potential. *J. Control. Release* **2010**, *142*, 299–311. [CrossRef]
119. Mishima, K. Biodegradable particle formation for drug and gene delivery using supercritical fluid and dense gas. *Adv. Drug Deliv. Rev.* **2008**, *60*, 411–432. [CrossRef]
120. Lo, Y.; Tsai, J.; Kuo, J. Liposomes and disaccharides as carriers in spray-dried powder formulations of superoxide dismutase. *J. Control. Release* **2004**, *94*, 259–272. [CrossRef]
121. Kadimi, U.S.; Balasubramanian, D.R.; Ganni, U.R.; Balaraman, M.; Govindarajulu, V. In vitro studies on liposomal amphotericin B obtained by supercritical carbon dioxide-mediated process. *Nanomedicine* **2007**, *3*, 273–280. [CrossRef] [PubMed]
122. Baldino, L.; Reverchon, E. Niosomes formation using a continuous supercritical CO_2 assisted process. *J. CO2 Util.* **2021**, *52*, 101669. [CrossRef]
123. Durak, S.; Esmaeili Rad, M.; Alp Yetisgin, A.; Eda Sutova, H.; Kutlu, O.; Cetinel, S.; Zarrabi, A. Niosomal Drug Delivery Systems for Ocular Disease-Recent Advances and Future Prospects. *Nanomaterials* **2020**, *10*, 1191. [CrossRef] [PubMed]
124. Grijalvo, S.; Puras, G.; Zárate, J.; Sainz-Ramos, M.; Qtaish, N.A.L.; López, T.; Mashal, M.; Attia, N.; Díaz, D.; Pons, R.; et al. Cationic Niosomes as Non-Viral Vehicles for Nucleic Acids: Challenges and Opportunities in Gene Delivery. *Pharmaceutics* **2019**, *11*, 50. [CrossRef] [PubMed]
125. Rose, J.K.; Buonocore, L.; Whitt, M.A. A new cationic liposome reagent mediating nearly quantitative transfection of animal cells. *Biotechniques* **1991**, *10*, 520–525.
126. Manosroi, J.; Khositsuntiwong, N.; Manosroi, W.; Götz, F.; Werner, R.G.; Manosroi, A. Enhancement of transdermal absorption, gene expression and stability of tyrosinase plasmid (pMEL34)-loaded elastic cationic niosomes: Potential application in vitiligo treatment. *J. Pharm. Sci.* **2010**, *99*, 3533–3541. [CrossRef]
127. Perrie, Y.; Barralet, J.E.; McNeil, S.; Vangala, A. Surfactant vesicle-mediated delivery of DNA vaccines via the subcutaneous route. *Int. J. Pharm.* **2004**, *284*, 31–41. [CrossRef]
128. Jain, S.; Singh, P.; Mishra, V.; Vyas, S.P. Mannosylated niosomes as adjuvant-carrier system for oral genetic immunization against hepatitis B. *Immunol. Lett.* **2005**, *101*, 41–49. [CrossRef] [PubMed]
129. Singh, M.; Chakrapani, A.; O'Hagan, D. Nanoparticles and microparticles as vaccine-delivery systems. *Expert Rev. Vaccines* **2007**, *6*, 797–808. [CrossRef]
130. Fifis, T.; Gamvrellis, A.; Crimeen-Irwin, B.; Pietersz, G.A.; Li, J.; Mottram, P.L.; McKenzie, I.F.C.; Plebanski, M. Size-dependent immunogenicity: Therapeutic and protective properties of nano-vaccines against tumors. *J. Immunol.* **2004**, *173*, 3148–3154. [CrossRef]
131. Gratton, S.E.A.; Ropp, P.A.; Pohlhaus, P.D.; Luft, J.C.; Madden, V.J.; Napier, M.E.; DeSimone, J.M. The effect of particle design on cellular internalization pathways. *Proc. Natl. Acad. Sci. USA* **2008**, *105*, 11613–11618. [CrossRef] [PubMed]
132. Blank, F.; Wehrli, M.; Lehmann, A.; Baum, O.; Gehr, P.; von Garnier, C.; Rothen-Rutishauser, B.M. Macrophages and dendritic cells express tight junction proteins and exchange particles in an in vitro model of the human airway wall. *Immunobiology* **2011**, *216*, 86–95. [CrossRef] [PubMed]
133. Thiele, L.; Rothen-Rutishauser, B.; Jilek, S.; Wunderli-Allenspach, H.; Merkle, H.P.; Walter, E. Evaluation of particle uptake in human blood monocyte-derived cells in vitro. Does phagocytosis activity of dendritic cells measure up with macrophages? *J. Control. Release* **2001**, *76*, 59–71. [CrossRef]
134. De Temmerman, M.-L.; Rejman, J.; Demeester, J.; Irvine, D.J.; Gander, B.; De Smedt, S.C. Particulate vaccines: On the quest for optimal delivery and immune response. *Drug Discov. Today* **2011**, *16*, 569–582. [CrossRef] [PubMed]
135. Boussif, O.; LezoualC'H, F.; Zanta, M.A.; Mergny, M.D.; Scherman, D.; Demeneix, B.; Behr, J.P. A versatile vector for gene and oligonucleotide transfer into cells in culture and in vivo: Polyethylenimine. *Proc. Natl. Acad. Sci. USA* **1995**, *92*, 7297–7301. [CrossRef]
136. Densmore, C.L.; Orson, F.M.; Xu, B.; Kinsey, B.M.; Waldrep, J.C.; Hua, P.; Bhogal, B.; Knight, V. Aerosol Delivery of Robust Polyethyleneimine-DNA Complexes for Gene Therapy and Genetic Immunization. *Mol. Ther.* **2000**, *1*, 180–188. [CrossRef]
137. Gautam, A.; Densmore, C.L.; Xu, B.; Waldrep, J.C. Enhanced gene expression in mouse lung after PEI-DNA aerosol delivery. *Mol. Ther.* **2000**, *2*, 63–70. [CrossRef]
138. Shim, B.-S.; Park, S.-M.; Quan, J.-S.; Jere, D.; Chu, H.; Song, M.K.; Kim, D.W.; Jang, Y.-S.; Yang, M.-S.; Han, S.H.; et al. Intranasal immunization with plasmid DNA encoding spike protein of SARS-coronavirus/polyethylenimine nanoparticles elicits antigen-specific humoral and cellular immune responses. *BMC Immunol.* **2010**, *11*, 65. [CrossRef]
139. Torrieri-Dramard, L.; Lambrecht, B.; Ferreira, H.L.; Van den Berg, T.; Klatzmann, D.; Bellier, B. Intranasal DNA vaccination induces potent mucosal and systemic immune responses and cross-protective immunity against influenza viruses. *Mol. Ther.* **2011**, *19*, 602–611. [CrossRef]
140. Bivas-Benita, M.; Bar, L.; Gillard, G.O.; Kaufman, D.R.; Simmons, N.L.; Hovav, A.-H.; Letvin, N.L. Efficient Generation of Mucosal and Systemic Antigen-Specific CD8+ T-Cell Responses following Pulmonary DNA Immunization. *J. Virol.* **2010**, *84*, 5764–5774. [CrossRef]

141. Bivas-Benita, M.; Gillard, G.O.; Bar, L.; White, K.A.; Webby, R.J.; Hovav, A.H.; Letvin, N.L. Airway CD8+ T cells induced by pulmonary DNA immunization mediate protective anti-viral immunity. *Mucosal Immunol.* **2013**, *6*, 156–166. [CrossRef]
142. Regnström, K.; Ragnarsson, E.G.E.; Köping-Höggård, M.; Torstensson, E.; Nyblom, H.; Artursson, P. PEI—A potent, but not harmless, mucosal immuno-stimulator of mixed T-helper cell response and FasL-mediated cell death in mice. *Gene Ther.* **2003**, *10*, 1575–1583. [CrossRef] [PubMed]
143. Mann, J.F.S.; McKay, P.F.; Arokiasamy, S.; Patel, R.K.; Klein, K.; Shattock, R.J. Pulmonary delivery of DNA vaccine constructs using deacylated PEI elicits immune responses and protects against viral challenge infection. *J. Control. Release* **2013**, *170*, 452–459. [CrossRef]
144. Panyam, J.; Labhasetwar, V. Biodegradable nanoparticles for drug and gene delivery to cells and tissue. *Adv. Drug Deliv. Rev.* **2003**, *55*, 329–347. [CrossRef]
145. Bivas-Benita, M.; Romeijn, S.; Junginger, H.E.; Borchard, G. PLGA-PEI nanoparticles for gene delivery to pulmonary epithelium. *Eur. J. Pharm. Biopharm. Off. J. Arb. fur Pharm. Verfahr. e.V* **2004**, *58*, 1–6. [CrossRef]
146. Oster, C.G.; Kim, N.; Grode, L.; Barbu-Tudoran, L.; Schaper, A.K.; Kaufmann, S.H.E.; Kissel, T. Cationic microparticles consisting of poly(lactide-co-glycolide) and polyethylenimine as carriers systems for parental DNA vaccination. *J. Control. Release* **2005**, *104*, 359–377. [CrossRef]
147. Kang, M.L.; Cho, C.S.; Yoo, H.S. Application of chitosan microspheres for nasal delivery of vaccines. *Biotechnol. Adv.* **2009**, *27*, 857–865. [CrossRef]
148. Casettari, L.; Vllasaliu, D.; Lam, J.K.W.; Soliman, M.; Illum, L. Biomedical applications of amino acid-modified chitosans: A review. *Biomaterials* **2012**, *33*, 7565–7583. [CrossRef] [PubMed]
149. Issa, M.M.; Köping-Höggård, M.; Artursson, P. Chitosan and the mucosal delivery of biotechnology drugs. *Drug Discov. Today. Technol.* **2005**, *2*, 1–6. [CrossRef]
150. Seferian, P.G.; Martinez, M.L. Immune stimulating activity of two new chitosan containing adjuvant formulations. *Vaccine* **2000**, *19*, 661–668. [CrossRef]
151. Kumar, M.; Behera, A.K.; Lockey, R.F.; Zhang, J.; Bhullar, G.; De La Cruz, C.P.; Chen, L.-C.; Leong, K.W.; Huang, S.-K.; Mohapatra, S.S. Intranasal gene transfer by chitosan-DNA nanospheres protects BALB/c mice against acute respiratory syncytial virus infection. *Hum. Gene Ther.* **2002**, *13*, 1415–1425. [CrossRef] [PubMed]
152. Iqbal, M.; Lin, W.; Jabbal-Gill, I.; Davis, S.S.; Steward, M.W.; Illum, L. Nasal delivery of chitosan-DNA plasmid expressing epitopes of respiratory syncytial virus (RSV) induces protective CTL responses in BALB/c mice. *Vaccine* **2003**, *21*, 1478–1485. [CrossRef]
153. Raghuwanshi, D.; Mishra, V.; Das, D.; Kaur, K.; Suresh, M.R. Dendritic cell targeted chitosan nanoparticles for nasal DNA immunization against SARS-CoV nucleocapsid protein. *Mol. Pharm.* **2012**, *9*, 946–956. [CrossRef]
154. Wu, M.; Zhao, H.; Li, M.; Yue, Y.; Xiong, S.; Xu, W. Intranasal vaccination with mannosylated chitosan formulated DNA vaccine enables robust IgA and cellular response induction in the lungs of mice and improves protection against pulmonary mycobacterial challenge. *Front. Cell. Infect. Microbiol.* **2017**, *7*, 445. [CrossRef]
155. Bernkop-Schnürch, A.; Hornof, M.; Guggi, D. Thiolated chitosans. *Eur. J. Pharm. Biopharm. Off. J. Arb. fur Pharm. Verfahrenstechnik e.V* **2004**, *57*, 9–17. [CrossRef]
156. Nafee, N.; Taetz, S.; Schneider, M.; Schaefer, U.F.; Lehr, C.-M. Chitosan-coated PLGA nanoparticles for DNA/RNA delivery: Effect of the formulation parameters on complexation and transfection of antisense oligonucleotides. *Nanomedicine* **2007**, *3*, 173–183. [CrossRef] [PubMed]
157. Janes, K.A.; Calvo, P.; Alonso, M.J. Polysaccharide colloidal particles as delivery systems for macromolecules. *Adv. Drug Deliv. Rev.* **2001**, *47*, 83–97. [CrossRef]
158. Ravi Kumar, M.N.V.; Bakowsky, U.; Lehr, C.M. Preparation and characterization of cationic PLGA nanospheres as DNA carriers. *Biomaterials* **2004**, *25*, 1771–1777. [CrossRef] [PubMed]
159. Wang, G.; Pan, L.; Zhang, Y.; Wang, Y.; Zhang, Z.; Lü, J.; Zhou, P.; Fang, Y.; Jiang, S. Intranasal delivery of cationic PLGA nano/microparticles-loaded FMDV DNA vaccine encoding IL-6 elicited protective immunity against FMDV challenge. *PLoS ONE* **2011**, *6*, e27605. [CrossRef]
160. Kumar, U.S.; Afjei, R.; Ferrara, K.; Massoud, T.F.; Paulmurugan, R. Gold-Nanostar-Chitosan-Mediated Delivery of SARS-CoV-2 DNA Vaccine for Respiratory Mucosal Immunization: Development and Proof-of-Principle. *ACS Nano* **2021**, *15*, 17582–17601. [CrossRef]
161. Teo, S.P. Review of COVID-19 mRNA Vaccines: BNT162b2 and mRNA-1273. *J. Pharm. Pract.* **2021**, 8971900211009650. [CrossRef]
162. Chaudhary, N.; Weissman, D.; Whitehead, K.A. mRNA vaccines for infectious diseases: Principles, delivery and clinical translation. *Nat. Rev. Drug Discov.* **2021**, *20*, 1–22. [CrossRef] [PubMed]
163. Blakney, A.K.; Ip, S.; Geall, A.J. An Update on Self-Amplifying mRNA Vaccine Development. *Vaccines* **2021**, *9*, 97. [CrossRef] [PubMed]
164. Brito, L.A.; Chan, M.; Shaw, C.A.; Hekele, A.; Carsillo, T.; Schaefer, M.; Archer, J.; Seubert, A.; Otten, G.R.; Beard, C.W.; et al. A cationic nanoemulsion for the delivery of next-generation RNA vaccines. *Mol. Ther.* **2014**, *22*, 2118–2129. [CrossRef]
165. Li, Q.; Ren, J.; Liu, W.; Jiang, G.; Hu, R. CpG Oligodeoxynucleotide Developed to Activate Primate Immune Responses Promotes Antitumoral Effects in Combination with a Neoantigen-Based mRNA Cancer Vaccine. *Drug Des. Devel. Ther.* **2021**, *15*, 3953–3963. [CrossRef] [PubMed]

166. Haabeth, O.A.W.; Lohmeyer, J.J.K.; Sallets, A.; Blake, T.R.; Sagiv-Barfi, I.; Czerwinski, D.K.; McCarthy, B.; Powell, A.E.; Wender, P.A.; Waymouth, R.M.; et al. An mRNA SARS-CoV-2 Vaccine Employing Charge-Altering Releasable Transporters with a TLR-9 Agonist Induces Neutralizing Antibodies and T Cell Memory. *ACS Cent. Sci.* **2021**, *7*, 1191–1204. [CrossRef] [PubMed]
167. Hajam, I.A.; Senevirathne, A.; Hewawaduge, C.; Kim, J.; Lee, J.H. Intranasally administered protein coated chitosan nanoparticles encapsulating influenza H9N2 HA2 and M2e mRNA molecules elicit protective immunity against avian influenza viruses in chickens. *Vet. Res.* **2020**, *51*, 37. [CrossRef] [PubMed]
168. Pasquinelli, A.E.; Dahlberg, J.E.; Lund, E. Reverse 5′ caps in RNAs made in vitro by phage RNA polymerases. *RNA* **1995**, *1*, 957–967. [PubMed]
169. Stepinski, J.; Waddell, C.; Stolarski, R.; Darzynkiewicz, E.; Rhoads, R.E. Synthesis and properties of mRNAs containing the novel "anti-reverse" cap analogs 7-methyl(3′-O-methyl)GpppG and 7-methyl(3′-deoxy)GpppG. *RNA* **2001**, *7*, 1486–1495.
170. Martin, S.A.; Paoletti, E.; Moss, B. Purification of mRNA guanylyltransferase and mRNA(guanine 7) methyltransferase from vaccinia virions. *J. Biol. Chem.* **1975**, *24*, 9322–9329. [CrossRef]
171. Hornblower, B.; Robb, G.B.; Tzertzinis, G.; England, N. Minding Your Caps and Tails—Considerations for Functional mRNA Synthesis. *N. Engl. Biolabs.* Available online: https://international.neb.com/tools-and-resources/feature-articles/minding-your-caps-and-tails (accessed on 31 October 2021).
172. Henderson, J.M.; Ujita, A.; Hill, E.; Yousif-Rosales, S.; Smith, C.; Ko, N.; McReynolds, T.; Cabral, C.R.; Escamilla-Powers, J.R.; Houston, M.E. Cap 1 Messenger RNA Synthesis with Co-transcriptional CleanCap(®) Analog by In Vitro Transcription. *Curr. Protoc.* **2021**, *1*, e39. [CrossRef]
173. Holtkamp, S.; Kreiter, S.; Selmi, A.; Simon, P.; Koslowski, M.; Huber, C.; Türeci, O.; Sahin, U. Modification of antigen-encoding RNA increases stability, translational efficacy, and T-cell stimulatory capacity of dendritic cells. *Blood* **2006**, *108*, 4009–4017. [CrossRef] [PubMed]
174. Körner, C.G.; Wahle, E. Poly(A) tail shortening by a mammalian poly(A)-specific 3′-exoribonuclease. *J. Biol. Chem.* **1997**, *272*, 10448–10456. [CrossRef] [PubMed]
175. Mockey, M.; Gonçalves, C.; Dupuy, F.P.; Lemoine, F.M.; Pichon, C.; Midoux, P. mRNA transfection of dendritic cells: Synergistic effect of ARCA mRNA capping with Poly(A) chains in cis and in trans for a high protein expression level. *Biochem. Biophys. Res. Commun.* **2006**, *340*, 1062–1068. [CrossRef] [PubMed]
176. Ross, J.; Sullivan, T.D. Half-lives of beta and gamma globin messenger RNAs and of protein synthetic capacity in cultured human reticulocytes. *Blood* **1985**, *66*, 1149–1154. [CrossRef]
177. Zhang, H.; Zhang, L.; Lin, A.; Xu, C.; Li, Z.; Liu, K.; Liu, B.; Ma, X.; Zhao, F.; Yao, W.; et al. LinearDesign: Efficient Algorithms for Optimized mRNA Sequence Design. *arXiv* **2020**, arXiv:2004.10177.
178. Chen, C.Y.A.; Shyu, A. Bin AU-rich elements: Characterization and importance in mRNA degradation. *Trends Biochem. Sci.* **1995**, *20*, 465–470. [CrossRef]
179. Thess, A.; Grund, S.; Mui, B.L.; Hope, M.J.; Baumhof, P.; Fotin-Mleczek, M.; Schlake, T. Sequence-engineered mRNA Without Chemical Nucleoside Modifications Enables an Effective Protein Therapy in Large Animals. *Mol. Ther.* **2015**, *23*, 1456–1464. [CrossRef]
180. Tusup, M.; Kundig, T.; Pascolo, S. An eIF4G-recruiting aptamer increases the functionality of in vitro transcribed mRNA. *EPH Int. J. Med. Health Sci.* **2018**, *4*, 29–37.
181. Saulquin, X.; Scotet, E.; Trautmann, L.; Peyrat, M.-A.; Halary, F.; Bonneville, M.; Houssaint, E. +1 Frameshifting as a novel mechanism to generate a cryptic cytotoxic T lymphocyte epitope derived from human interleukin 10. *J. Exp. Med.* **2002**, *195*, 353–358. [CrossRef]
182. Schwab, S.R.; Li, K.C.; Kang, C.; Shastri, N. Constitutive display of cryptic translation products by MHC class I molecules. *Science* **2003**, *301*, 1367–1371. [CrossRef] [PubMed]
183. Kimchi-Sarfaty, C.; Oh, J.M.; Kim, I.-W.; Sauna, Z.E.; Calcagno, A.M.; Ambudkar, S.V.; Gottesman, M.M. A "silent" polymorphism in the MDR1 gene changes substrate specificity. *Science* **2007**, *315*, 525–528. [CrossRef]
184. Alexopoulou, L.; Holt, A.C.; Medzhitov, R.; Flavell, R.A. Recognition of double-stranded RNA and activation of NF-κB by Toll-like receptor 3. *Nature* **2001**, *413*, 732–738. [CrossRef]
185. Heil, F.; Hemmi, H.; Hochrein, H.; Ampenberger, F.; Kirschning, C.; Akira, S.; Lipford, G.; Wagner, H.; Bauer, S. Species-Specific Recognition of Single-Stranded RNA via Till-like Receptor 7 and 8. *Science* **2004**, *303*, 1526–1529. [CrossRef]
186. Iavarone, C.; O'hagan, D.T.; Yu, D.; Delahaye, N.F.; Ulmer, J.B. Mechanism of action of mRNA-based vaccines. *Expert Rev. Vaccines* **2017**, *16*, 871–881. [CrossRef] [PubMed]
187. Karikó, K.; Buckstein, M.; Ni, H.; Weissman, D. Suppression of RNA recognition by Toll-like receptors: The impact of nucleoside modification and the evolutionary origin of RNA. *Immunity* **2005**, *23*, 165–175. [CrossRef] [PubMed]
188. Karikó, K.; Muramatsu, H.; Ludwig, J.; Weissman, D. Generating the optimal mRNA for therapy: HPLC purification eliminates immune activation and improves translation of nucleoside-modified, protein-encoding mRNA. *Nucleic Acids Res.* **2011**, *39*, e142. [CrossRef]
189. Pascolo, S. Messenger RNA-based vaccines. *Expert Opin. Biol. Ther.* **2004**, *4*, 1285–1294. [CrossRef] [PubMed]
190. Karikó, K.; Muramatsu, H.; Welsh, F.A.; Ludwig, J.; Kato, H.; Akira, S.; Weissman, D. Incorporation of pseudouridine into mRNA yields superior nonimmunogenic vector with increased translational capacity and biological stability. *Mol. Ther.* **2008**, *16*, 1833–1834. [CrossRef]

191. Yin, B.; Chan, C.K.W.; Liu, S.; Hong, H.; Wong, S.H.D.; Lee, L.K.C.; Ho, L.W.C.; Zhang, L.; Leung, K.C.-F.; Choi, P.C.-L.; et al. Intrapulmonary Cellular-Level Distribution of Inhaled Nanoparticles with Defined Functional Groups and Its Correlations with Protein Corona and Inflammatory Response. *ACS Nano* **2019**, *13*, 14048–14069. [CrossRef] [PubMed]
192. Raesch, S.S.; Tenzer, S.; Storck, W.; Rurainski, A.; Selzer, D.; Ruge, C.A.; Perez-Gil, J.; Schaefer, U.F.; Lehr, C.-M. Proteomic and Lipidomic Analysis of Nanoparticle Corona upon Contact with Lung Surfactant Reveals Differences in Protein, but Not Lipid Composition. *ACS Nano* **2015**, *9*, 11872–11885. [CrossRef]
193. Monopoli, M.P.; Aberg, C.; Salvati, A.; Dawson, K.A. Biomolecular coronas provide the biological identity of nanosized materials. *Nat. Nanotechnol.* **2012**, *7*, 779–786. [CrossRef]
194. Montoro, D.T.; Haber, A.L.; Biton, M.; Vinarsky, V.; Lin, B.; Birket, S.E.; Yuan, F.; Chen, S.; Leung, H.M.; Villoria, J.; et al. A revised airway epithelial hierarchy includes CFTR-expressing ionocytes. *Nature* **2018**, *560*, 319–324. [CrossRef]
195. Pardi, N.; Tuyishime, S.; Muramatsu, H.; Kariko, K.; Mui, B.L.; Tam, Y.K.; Madden, T.D.; Hope, M.J.; Weissman, D. Expression kinetics of nucleoside-modified mRNA delivered in lipid nanoparticles to mice by various routes. *J. Control. Release* **2015**, *217*, 345–351. [CrossRef] [PubMed]
196. Alameh, M.-G.; Tombácz, I.; Bettini, E.; Lederer, K.; Sittplangkoon, C.; Wilmore, J.R.; Gaudette, B.T.; Soliman, O.Y.; Pine, M.; Hicks, P.; et al. Lipid nanoparticles enhance the efficacy of mRNA and protein subunit vaccines by inducing robust T follicular helper cell and humoral responses. *Immunity* **2021**, *54*, 2877–2892. [CrossRef] [PubMed]
197. Tam, H.H.; Melo, M.B.; Kang, M.; Pelet, J.M.; Ruda, V.M.; Foley, M.H.; Hu, J.K.; Kumari, S.; Crampton, J.; Baldeon, A.D.; et al. Sustained antigen availability during germinal center initiation enhances antibody responses to vaccination. *Proc. Natl. Acad. Sci. USA* **2016**, *113*, 6639–6648. [CrossRef] [PubMed]
198. Akinc, A.; Querbes, W.; De, S.; Qin, J.; Frank-Kamenetsky, M.; Jayaprakash, K.N.; Jayaraman, M.; Rajeev, K.G.; Cantley, W.L.; Dorkin, J.R.; et al. Targeted delivery of RNAi therapeutics with endogenous and exogenous ligand-based mechanisms. *Mol. Ther.* **2010**, *18*, 1357–1364. [CrossRef]
199. Lokugamage, M.P.; Vanover, D.; Beyersdorf, J.; Hatit, M.Z.C.; Rotolo, L.; Echeverri, E.S.; Peck, H.E.; Ni, H.; Yoon, J.-K.; Kim, Y.; et al. Optimization of lipid nanoparticles for the delivery of nebulized therapeutic mRNA to the lungs. *Nat. Biomed. Eng.* **2021**, *5*, 1059–1068. [CrossRef]
200. Robinson, E.; MacDonald, K.D.; Slaughter, K.; McKinney, M.; Patel, S.; Sun, C.; Sahay, G. Lipid Nanoparticle-Delivered Chemically Modified mRNA Restores Chloride Secretion in Cystic Fibrosis. *Mol. Ther.* **2018**, *26*, 2034–2046. [CrossRef] [PubMed]
201. Wegmann, F.; Gartlan, K.H.; Harandi, A.M.; Brinckmann, S.A.; Coccia, M.; Hillson, W.R.; Kok, W.L.; Cole, S.; Ho, L.-P.; Lambe, T.; et al. Polyethyleneimine is a potent mucosal adjuvant for viral glycoprotein antigens. *Nat. Biotechnol.* **2012**, *30*, 883–888. [CrossRef]
202. Li, M.; Zhao, M.; Fu, Y.; Li, Y.; Gong, T.; Zhang, Z.; Sun, X. Enhanced intranasal delivery of mRNA vaccine by overcoming the nasal epithelial barrier via intra- and paracellular pathways. *J. Control. Release* **2016**, *228*, 9–19. [CrossRef]
203. McCullough, K.C.; Bassi, I.; Milona, P.; Suter, R.; Thomann-Harwood, L.; Englezou, P.; Démoulins, T.; Ruggli, N. Self-replicating Replicon-RNA Delivery to Dendritic Cells by Chitosan-nanoparticles for Translation In Vitro and In Vivo. *Mol. Ther. Nucleic Acids* **2014**, *3*, e173. [CrossRef] [PubMed]
204. Qiu, Y.; Man, R.C.H.; Liao, Q.; Kung, K.L.K.; Chow, M.Y.T.; Lam, J.K.W. Effective mRNA pulmonary delivery by dry powder formulation of PEGylated synthetic KL4 peptide. *J. Control. Release* **2019**, *314*, 102–115. [CrossRef] [PubMed]
205. Mai, Y.; Guo, J.; Zhao, Y.; Ma, S.; Hou, Y.; Yang, J. Intranasal delivery of cationic liposome-protamine complex mRNA vaccine elicits effective anti-tumor immunity. *Cell. Immunol.* **2020**, *354*, 104143. [CrossRef] [PubMed]
206. Lorenzi, J.C.C.; Trombone, A.P.F.; Rocha, C.D.; Almeida, L.P.; Lousada, R.L.; Malardo, T.; Fontoura, I.C.; Rossetti, R.A.M.; Gembre, A.F.; Silva, A.M.; et al. Intranasal vaccination with messenger RNA as a new approach in gene therapy: Use against tuberculosis. *BMC Biotechnol.* **2010**, *10*, 77. [CrossRef]
207. Phua, K.K.L.; Staats, H.F.; Leong, K.W.; Nair, S.K. Intranasal mRNA nanoparticle vaccination induces prophylactic and therapeutic anti-tumor immunity. *Sci. Rep.* **2014**, *4*, 5128. [CrossRef] [PubMed]
208. Mutsch, M.; Zhou, W.; Rhodes, P.; Bopp, M.; Chen, R.T.; Linder, T.; Spyr, C.; Steffen, R. Use of the inactivated intranasal influenza vaccine and the risk of Bell's palsy in Switzerland. *N. Engl. J. Med.* **2004**, *350*, 896–903. [CrossRef]
209. Izurieta, H.S.; Haber, P.; Wise, R.P.; Iskander, J.; Pratt, D.; Mink, C.; Chang, S.; Braun, M.M.; Ball, R. Adverse events reported following live, cold-adapted, intranasal influenza vaccine. *JAMA* **2005**, *294*, 2720–2725. [CrossRef]
210. Beck, S.E.; Laube, B.L.; Barberena, C.I.; Fischer, A.C.; Adams, R.J.; Chesnut, K.; Flotte, T.R.; Guggino, W.B. Deposition and expression of aerosolized rAAV vectors in the lungs of Rhesus macaques. *Mol. Ther.* **2002**, *6*, 546–554. [CrossRef]
211. Guan, S.; Darmstädter, M.; Xu, C.; Rosenecker, J. In Vitro Investigations on Optimizing and Nebulization of IVT-mRNA Formulations for Potential Pulmonary-Based Alpha-1-Antitrypsin Deficiency Treatment. *Pharmaceutics* **2021**, *13*, 1281. [CrossRef]
212. Birchall, J.C.; Kellaway, I.W.; Gumbleton, M. Physical stability and in-vitro gene expression efficiency of nebulised lipid-peptide-DNA complexes. *Int. J. Pharm.* **2000**, *197*, 221–231. [CrossRef]
213. Patel, A.K.; Kaczmarek, J.C.; Bose, S.; Kauffman, K.J.; Mir, F.; Heartlein, M.W.; DeRosa, F.; Langer, R.; Anderson, D.G. Inhaled Nanoformulated mRNA Polyplexes for Protein Production in Lung Epithelium. *Adv. Mater.* **2019**, *31*, e1805116. [CrossRef] [PubMed]
214. Johler, S.M.; Rejman, J.; Guan, S.; Rosenecker, J. Nebulisation of IVT mRNA Complexes for Intrapulmonary Administration. *PLoS ONE* **2015**, *10*, e0137504.

nanomaterials MDPI

Review

Novel Perspectives towards RNA-Based Nano-Theranostic Approaches for Cancer Management

Rabia Arshad [1], Iqra Fatima [2], Saman Sargazi [3], Abbas Rahdar [4,*], Milad Karamzadeh-Jahromi [5], Sadanand Pandey [6], Ana M. Díez-Pascual [7,*] and Muhammad Bilal [8]

1 Faculty of Pharmacy, University of Lahore, Lahore 45320, Pakistan; rabia.arshad@pharm.uol.edu.pk
2 Department of Pharmacy, Quaid-i-Azam University, Islamabad 45320, Pakistan; iqraf332@gmail.com
3 Cellular and Molecular Research Center, Research Institute of Cellular and Molecular Sciences in Infectious Diseases, Zahedan University of Medical Sciences, Zahedan 98167-43463, Iran; sgz.biomed@gmail.com
4 Department of Physics, University of Zabol, Zabol 98613-35856, Iran
5 Department of Physics, University of Kashan, Kashan 87317-51167, Iran; milad71karamzade@yahoo.com
6 Department of Chemistry, College of Natural Science, Yeungnam University, 280 Daehak-Ro, Gyeongsan 38541, Korea; sadanand.au@gmail.com
7 Universidad de Alcalá, Facultad de Ciencias, Departamento de Química Analítica, Química Física e Ingeniería Química, Ctra. Madrid-Barcelona, Km. 33.6, 28805 Alcalá de Henares, Madrid, Spain
8 School of Life Science and Food Engineering, Huaiyin Institute of Technology, Huai'an 223003, China; bilaluaf@hotmail.com
* Correspondence: a.rahdar@uoz.ac.ir (A.R.); am.diez@uah.es (A.M.D.-P.)

Abstract: In the fight against cancer, early diagnosis is critical for effective treatment. Traditional cancer diagnostic technologies, on the other hand, have limitations that make early detection difficult. Therefore, multi-functionalized nanoparticles (NPs) and nano-biosensors have revolutionized the era of cancer diagnosis and treatment for targeted action via attaching specified and biocompatible ligands to target the tissues, which are highly over-expressed in certain types of cancers. Advancements in multi-functionalized NPs can be achieved via modifying molecular genetics to develop personalized and targeted treatments based on RNA interference. Modification in RNA therapies utilized small RNA subunits in the form of small interfering RNAs (siRNA) for overexpressing the specific genes of, most commonly, breast, colon, gastric, cervical, and hepatocellular cancer. RNA-conjugated nanomaterials appear to be the gold standard for preventing various malignant tumors through focused diagnosis and delivering to a specific tissue, resulting in cancer cells going into programmed death. The latest advances in RNA nanotechnology applications for cancer diagnosis and treatment are summarized in this review.

Keywords: RNA nanotechnology; cancer; theranostic; nano-biosensor

Citation: Arshad, R.; Fatima, I.; Sargazi, S.; Rahdar, A.; Karamzadeh-Jahromi, M.; Pandey, S.; Díez-Pascual, A.M.; Bilal, M. Novel Perspectives towards RNA-Based Nano-Theranostic Approaches for Cancer Management. *Nanomaterials* **2021**, *11*, 3330. https://doi.org/10.3390/nano11123330

Academic Editors: Fiore P. Nicoletta and Francesca Iemma

Received: 28 October 2021
Accepted: 5 December 2021
Published: 8 December 2021

1. Introduction

Despite decades of basic and clinical investigation, as well as trials of new therapeutic modalities, cancer remains a substantial cause of mortality worldwide [1]. Various resistance strategies have led to the near inefficiency of anticancer drugs. As illustrated in Figure 1, these drugs have been rendered almost ineffective. Recent approaches in nanomedicine have prompted the development of effective theranostic platforms for a myriad of biological and biomedical applications [2]. Nanomaterials (i.e., niosomes [3], polymer-based nanocapsules [4], nanoparticles (NPs) [5–8], metal nanocages [9], nanocomposites [10], nanoliposomes [11], and engineered nanohydrogels [12]), with highly controlled geometry and physic-chemical properties, have been introduced as promising tools for recognizing cancer tissues and also serve as novel drug delivery systems (DDSs) to achieve active targeting [2,13–21]. It is now believed that nanotechnology can purposefully improve the clinical outcome of cancer therapies through improving the tolerability of the efficacy of novel drugs [22] or delivering proteins, DNA, RNA, and various types of molecules to cancerous cells [23,24]. Several nanomaterial-based biosensors have also been developed for the accurate sensing of tumor markers [24,25].

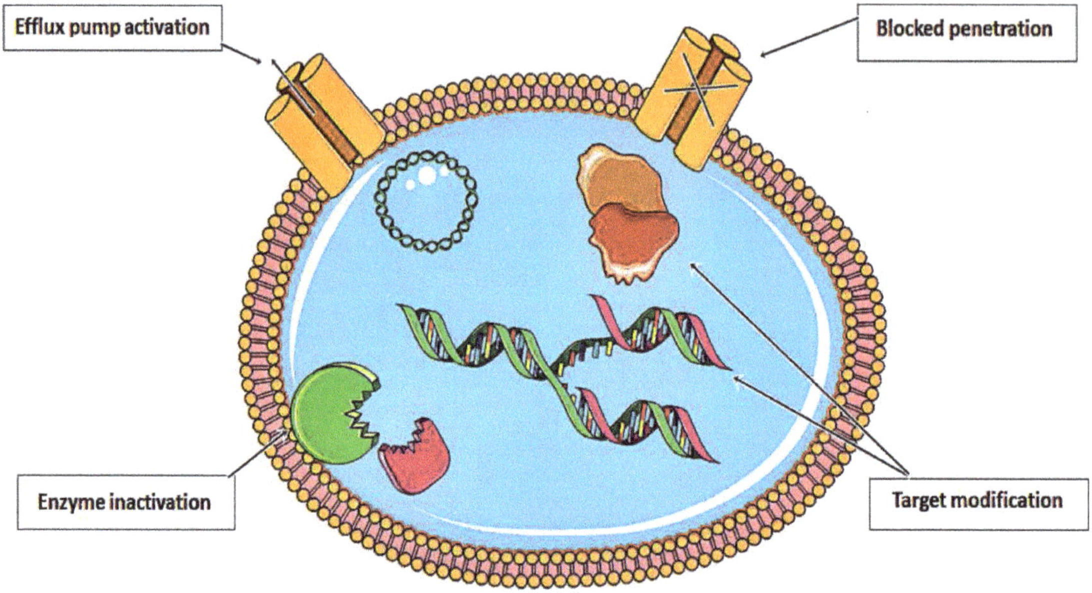

Figure 1. Resistance strategies developed by anticancer drugs that limit their therapeutic efficiency.

Similarly, biodegradable and biocompatible natural and engineered biomolecules (including proteins, peptides, polysaccharides, and nucleic acids) have been broadly examined to synthesize nanostructures [26,27]. Biomolecule-based building blocks provide particular features that make it not feasible to be reproduced in these synthetic materials. However, multifunctional approaches can be developed by exploiting biomolecule-derived elements concerning cancer targeting and therapy [27].

It is scientifically established that nucleic acids can be manipulated and designed to create many nanostructures [28]. In the past, researchers have studied the preparation and characterization of DNA nanoparticles (DNA NPs), utilizing complex processes, such as coacervation [29,30]. In this respect, charge-neural biodegradable DNA NPs were synthesized by compacting a small-sized single DNA molecule and loading it on magnetic NPs [31]. This strategy is now considered a beneficial translatable gene therapy platform for overcoming challenging biological barriers by enhancing nuclear uptake across tiny nuclear pores of dividing cells and is being widely investigated to treat various conditions (i.e., malignancies and respiratory diseases) [31–33].

Due to its increased thermodynamic stability, the RNA structure can be more flexible while folding into different structures (i.e., rigid structural motifs), and it produces diverse building blocks for numerous therapeutic applications [34], including the fabrication of nanosensors and nanodevices [28]. Furthermore, the thermal stability of RNA also allows it to produce multivalent nanostructures possessing specific stoichiometry [28,35]. In this view, introducing novel methods for the re-assembling of RNA molecules has recently spurred interest in investigating the biomedical application of RNA nanostructures.

Nanomedicine has led to the development of a variety of nanoscale therapeutics and diagnostics to treat a variety of diseases, specifically cancer [6,14,15,17,18,36–46]. This fact is broadly exploited in the field of DNA nanotechnology [47]. Generally, methodologies for DNA nanotechnology can be applied to RNA nanotechnology [48]. Despite their similarities, DNA and RNA nanotechnology differ in a few key aspects. Inter-and intramolecular interactions, as well as stem-loops, are abundant in the RNA molecules. These allow the formation of complex secondary and tertiary structures (i.e., bulges, stems, junctions, loops, etc.) and thus can be utilized to create 'dovetail' joints among the building blocks [35]. RNA can serve as a potential new therapeutic modality for cancer due to its lack of accumulation in vital organs [49].

In the past, numerous RNA-NPs have been prepared by an automated self-assembly process and studied in the context of cancer research [50,51]. A key challenge in programming RNA strands to assemble into nanostructures is to create a folding pathway to avoid kinetic traps [52]. In addition, by introducing chemical modification into nucleotides without substantial changes in RNA content, RNA-NPs will escape host RNA decay pathways. In addition, RNA nanostructures might have immunologic properties, making them valuable tools for in-vivo applications [49].

The conjugation of polymeric NPs with RNA molecules, such as RNA aptamers, cause these bio-conjugates to be easily absorbed by specific tumor cells, and therefore, can be considered as a beneficial strategy towards the controlled release of polymeric drug delivery vehicles [53]. In the growing field of RNA nanotechnology to fight cancer, RNA aptamers have attracted great attention as tools for delivering other RNA therapeutics, such as short interfering RNAs (SiRNAs), to specific organs [54]. Moreover, polyvalent RNA nanostructures have been effectively fabricated as carriers of siRNA, ribozyme, and anticancer agents to tumor sites [55].

Using in-vitro and in-vivo experimental models, Yin et al. exploited RNA-based technology to efficiently deliver anti-microRNA to cancer cells derived from breast tissue [56]. Ghimire and colleagues showed that RNA-NPs could be utilized as rubber for constraining vessel extravasation to improve the targeting of cancer cells and increase their renal excretion, thus reducing their toxic effects [57]. Recently, radiolabeled RNA NPs were developed for specific targeting and efficient tumor accumulation with desirable in-vivo biodistribution [58]. According to Kim et al., dual-targeting polymeric siRNA NPs were synthesized by multiple processes, including electrostatic deposition and poly-L-lysine condensation. Researchers found that these nanostructures are capable of efficiently delivering siRNA to tumor cells [59]. Xu et al. successfully delivered delta-5-desaturase via dihomo-γ-linolenic acid-loaded RNA-NPs for suppressing the growth of cancerous colon cells via the induction of apoptotic cell death [60]. Haque and colleagues systemically injected synergistic tetravalent RNA-NPs into the tail-vein of mice. They observed that these RNA nanostructures preserve their biological function within cancer cells without entering other tissues or organs [61].

RNA therapies provide new insights for cancer treatment. The escalating growth of RNA nanotechnology in cancer theranostics demands the preparation of an updated review. This article comprehensively discusses recent findings and highlights the promising avenues of RNA nanotechnology implemented to design stable RNA nanostructures for diagnostic and prognostic purposes. Finally, we will discuss strategies to improve and overcome many technical obstacles in this field.

2. RNA Nanotechnology for Diagnosis of Cancers

Biosensors are diagnostic systems that convert a natural response into a programmable signal [62]. The calculable signal can be electrochemical, optical, thermal, or piezoelectric. Low detection limits, accuracy, and high sensitivity make electrochemical biosensors the most reliable of all. Electrochemical biosensors have a great prospective in the real-sample analysis [63]. The combination of nanotechnology with biosensors is a hallmark for disease assessment and the planning of its cure. Nanoscience is the science of the molecular and atomic manipulation of materials. It entails creating and managing chemical, physical, and systems in sizes of 1–200 nm. Nanotechnology has many applications in the biomedical field, especially in medical imaging for disease diagnosis [64]. The exclusive physicochemical properties of NPs are used to develop biosensors of point-of-care accuracy, known as nanosensors. The performance of other electrochemical and enzymatic biosensors increases due to their small size as the distance between enzyme and electrode decreases. Some optical biosensors use noble metal NPs to improve optical properties and increase localized surface plasmon resonance (SPR); e.g., inter-particle plasmon coupling changes the color of NPs, which is used for improving the properties of biosensors grounded on the aggregation of NPs [65]. DNA oligonucleotides were the first nucleic acid-based NPs that served as the foundation of DNA origami technology, but nowadays, RNA oligonucleotide nanotechnology is an important alternative to DNA technology [49]. DNA and RNA have operational differences because of their different structural properties. RNA nanotechnology uses single-stranded oligonucleotides for designing diverse and functional RNA nanostructures. The specific structure and organization of functional groups in NPs make them an excellent tool for diagnosing and treating different diseases [66]. The main edge of RNA NPs includes therapeutic elements, regulatory moieties, and targeting ligands. The field of RNA nanotechnology is different from traditional RNA research, which targets 2D/3D structure-function relationships and intra-RNA connections, as it emphasizes mainly quaternary exchanges and inter-RNA interactions [51].

2.1. Benefits of RNA Nanotechnology in Targeting Cancer Treatment

RNA NPs are discrete entities, quite different from classical therapeutic RNA, including siRNA, anti-miRNA, miRNA, mRNA, ribosomal-RNA, and viral immune-stimulatory RNA. These traditional RNAs have a broad fundamental history of research in the RNA field. These small RNAs are caught by cells, and they stimulate RNA-sensing pattern recognition receptors (PRRs). Some of these are reported to be immunogenic.

There are a lot of benefits of RNA NPs as compared to traditional RNA, for example: (a) enhanced permeability and retention (EPR) effect, (b) the small size offers to promise pharmacokinetic and pharmacodynamic properties, (c) decreased liver accumulation, (d) Small non-coding RNAs or microRNAs serve as scaffolding and active elements in the bottom-up self-assembly of more complex nanomaterials, and (d) untraceable toxicity in-vivo [67]. Multinational characteristics of RNA nanostructures, such as targeting ligands and multi-drug loading, are useful for combination therapy [68]. RNA NPs show chemical, metabolic, and thermal stability in biological systems. The standardized volume of distribution V_d, i.e., 1.2 L/kg of pRNA NPs, indicates the presence of a valuable amount in peripheral tissues, especially in a tumor. The comparatively small amount of clearance (Cl) value indicates the insufficient filtration of the NPs from the kidneys. A specific targeted delivery can be achieved by incorporating traditional RNA into NPs by taking advantage of these properties. These NPs show specific targeted delivery, higher therapeutic efficacy, and an increased in-vivo half-life [69].

Cancer nanotechnology faces a serious challenge of non-specific accumulation of delivered NPs in healthy organs, such as lungs, liver, spleen, and kidneys [70]. The non-specific accumulation leads to poor transport of NPs to the tumor site, unwanted side effects, and toxicity. The Guo laboratory combined targeting ligands and a sequence of pRNA-3WJ-based NPs for advanced targeted delivery. After systemic delivery in tumor-bearing mice, these RNA NPs are precisely transported to tumors within almost 4 h. These NPs remain at the tumor site for more than 24 h. After more than a few hours of injection, no or minimal organ accumulation was detected. Several common cancer models were used to get consistent results, such as breast, colorectal, prostate, glioblastoma, gastric, and head and neck cancers. The targeting ligands were altered based on the overexpression of specific targets in tumor tissues [71]. RNA with negative charge limits non-specific interactions with negatively charged cell membranes, which is important for high selectivity. The ratchet-like shape and strong elasticity of pRNA-3WJ-based NPs show improved EPR effects and higher tumor penetration. Overall, these findings showed that pRNA-3WJ-based NPs could be synthesized simply, with high selectivity and low side effects for healthy tissues. This promising bio-distribution is a significant signal of pharmacological profiles of RNA NPs [51].

2.2. Nano-Biosensors as Developing Trend in Cancer Diagnostics

The nano-biosensor is an innovative unit that creates nano-conjugated biological systems, which function as signaling mediators to detect the specific contents of medical, biochemical, or physical agents. The related data is transferred in the form of signals using thermometric, piezoelectric, optical, magnetic, electrochemical, and micromechanical methods. The signals produced by these methods depend on the bio-recognition of a cancer cells-related surface or intracellular biomarkers through bio-ligands or antibodies. In a new era of research, nano-biosensors will increasingly be classified based on bio-recognition and signal transduction elements. RNA-based nano-biosensors are displayed in Figure 2.

In the basic structure of a sensor, there are two main components: (i) the target analyte that can be a nucleic acid, antibody, drug, protein, or cell-surface molecule; and (ii) the transducer used for altering a signal into a form of energy. It can be detected electrochemically by detecting the energy it produces (voltage and current), (absorption and luminescence), and mechanically (resonance). The exclusive and tunable physicochemical properties of nanostructures, such as enhanced electric conductivity, greater area to volume ratio, high reactivity, unique magnetic properties, and powerful scattering and absorption, make them a fascinating tool for bio-sensing. The nanomaterials' ability to interact effectively with biologically important analytes and convert those interactions into considerably enhanced signals has enabled a new class of early diagnostic procedures. Gold NPs (AuNPs) and QDs are excellent signal transducers because the existence of a specific analyte regulates the signal generated by the material's optical properties. Nano and microsensors reduce the size of the receptor to improve their responsiveness. Increased signal-to-noise (S/N) ratio is responsible for this enhanced sensitivity. However, a running device's success depends on its ability to detect a tremendously low concentration in a reasonable amount of time.

The overall sensitivities of a sensor are often affected by mass transport limitations, such as when the substance is transported to the receptor by diffusion mechanism. The limiting factor in saturating the receptor for a specific geometry of the sensor is the time required by the analytes to reach the sensor area. As the receptor area becomes smaller, the time grows higher. The decreased number of analytes obstructs the detection of extremely dilute solutions. Many practical solutions have been tested to concentrate the dilute solution straight onto the sensor surface. For example, Melli et al. formulated a unique solution with a series of micropillars with superhydrophobic surfaces to simplify the management of samples with a low concentration of analytes. A diluted solution can be placed on these micropillars, the target molecule will become concentrated by the evaporation of liquid, and the detection time is significantly decreased [72].

Halo et al. used the same theory to detect colorectal cells by mRNA quantification. They developed a NanoFlare platform comprising a single layer of single-stranded DNA (ssDNA) coated on spherical AuNP. A fluorescent reporter was added to a short DNA counterpart that was hybridized to the ssDNA recognition sequence. The reporter fluorophore was slaked while it was close to the AuNPs, but it broadened when the target mRNA displaced the DNA, allowing the fluorescence reading. The detection limit of the nanosensor was about 100 cancer cells/mL blood [72].

A particular type of RNA aptamer that causes small-molecule fluorophores to emit fluorescence is called a light-up aptamer. Several new light-up aptamers have been developed that can bind to different biocompatible fluorogenic ligands and make their way to the design of modern RNA-based molecular strategies for sensing applications. The Systemic Evolution of Ligands by EXponential enrichment (SELEX) is a combinatorial procedure for identifying such aptamers. RNA aptamer libraries were exposed to recurrent cycles of collection and amplification in this procedure, resulting in RNA aptamers having the strongest selectivity for the target ligand. Various RNA NPs can be coupled with light-up aptamers having programmable sensing capabilities to create dynamic reporting entities [73].

2.3. RNA Nano-Biosensors

RNA/DNA nano-biosensors can measure the responses produced by aptamer hybridization or nucleic acid conversion processes as promising diagnostic tools [74]. As aptamers, RNA or DNA are single-stranded nucleic acid oligomers whose structure is extremely organized and complex, forming a strong connection with the target molecules (Figure 1) [75]. The RNA with a functional group usually reacts with molecular labels with an orthogonal reactive group in a suitable reaction environment. Fluorescent compounds activated by NHS are attached directly to an NH_2 group of the RNA fragment, and similarly, thiol to maleimide and alkyne to azide, etc. RNA was directly manufactured, and classical pairing reactive groups, e.g., amino $-NH_2$-COOH, azide, maleimide, alkyne, and thiol, were incorporated by similar methods for the production of polyvalent RNA NPs [76]. Single-stranded RNA NPs that are used for many purposes, such as biodistribution and diagnostic studies, are the subsequent derivatives of fluorescent dyes (FITC, Cy5, Cy3, and AF-647) attached to RNA [77].

RNA polymer is conjugated with NPs (such as iron oxide NPs, quantum dots, and gold NPs), and a sequence of research studies used siRNA, pRNA, and phi29 for this purpose. The siRNA was coupled with many nano-based imaging agents to form multifunctional NPs exhibiting diagnostic and therapeutic moieties.

Bhatia et al. established tumor-highlighted peptides (F3) and siRNA coupled to the PEGylated QDs core as a framework. The F3 peptide was conjugated to amine group-modified QD via sulfo-LC-SPDP (sulfosuccinimidyl 6-(3′-(2-pyridyldithio)-propionamido) hexanoate as a heterofunctional cross-linker, and thiol modified siRNA used sulfo-SMCC (sulfosuccinimidyl 4-(N-maleimidomethyl) cyclohexane-1-carboxylate) for conjugation. The QD-siRNA/F3 conjugate NPs were proficiently transported to HeLa cells and unconfined from their endosomal setup, which provided the demolished EGFP signal. These siRNA-NPs conjugates exhibited both imaging and therapeutic properties. Furthermore, the siRNA was coupled to iron oxide NPs, which showed magnetic characteristics for biomedical applications. The linkage of siRNA to iron NPs exhibited a double response, such as the in-vivo delivery of siRNA and gathering of siRNA in the tumor by MRI and the near-infrared fluorescent (NIRF) in-vivo optical imaging. The amine groups of iron oxide NPs were treated with m-maleimidobenzoyl N-hydroxysuccinimde ester (MBS) for linkage between siRNA and magnetic NPs. After that, the reduced thiol group of RNA was treated, and magnetic NPs were coupled with and near-infrared Cy5.5 dye and membrane translocation peptides. The MRI and NIRF were used simultaneously to see the siRNA-magnetic NP uptake. The coupling of AuNPs with RNA is studied to enhance the accessibility of tethered RNA splicing enhancers. Guo et al., in 2007, produced a pRNA of the phi29 and DNA-packaging motor linkage to AuNPs to study the phage assembly. In this case, the SH-labeled DNA oligonucleotide was merged with 3′ terminal of pRNA to incorporate the thiol (-SH) group into pRNA. Then, the thiol-treated pRNA was conjugated to gold NPs. The pRNA/AuNPs conjugate was attached to procapsid by an in-vitro phage assembly. Guo's group proved that the RNA polymer was coupled with the AuNPs effortlessly, and this procedure can be used for imaging purposes. Currently, pRNA-3WJ was coupled straight to the quantum dot for resistive biomemory applications. They introduced a sephadex G-100 resin-recognizing RNA aptamer in the biotin-labeled pRNA/3WJ (SEP_{apt}/3WJ/Bio) theme for the conjugation of pRNA-3WJ to the quantum dot (QD). Firstly, the SEPapt/3WJ/Bio was attached to G-100 resin, and then the streptavidine-labeled quantum dot was linked to SEP_{apt}/3WJ/Bio by streptavidine-biotin coupling on Sephadex G-100 and STV/QD-SEP_{apt}/3WJ/Bio conjugates were subjected to dissociation in the elution buffer. Later, the STV/QD-Bio/$3WJ_b$, $3WJ_a$, and SEP_{apt}/$3WJ_c$ were split by elution buffer. After that, the STV/QD-Bio/$3WJ_b$ fragment was filtered and reconvened with pRNA $3WJ_a$ and thiol-labeled pRNA $3WJ_c$ for resistive biomemory application. Hence, the RNA polymer can be easily incorporated with other NPs and used for various diagnostic, therapeutic, and bio-electronic fields [77]. Table 1 summarizes the role of RNA-based nanostructures in diagnosing cancers.

Table 1. Summary of RNA-based nanostructures in diagnosis of cancers.

RNA-Based Nanoparticles	Key Feature
Immune-Magnetic Exosome RNA (iMER)	Exosomal analysis of glioblastoma multiforme (GBM).
Anti-RNA aptamer	Initial detection and analysis of residual GBM.
RNA tetrahedrons	Target triple-negative breast cancer cells.
Oligonucleotide-treated Au-NPs	Analyzing circulating tumor cells (CTCs) of the prostate.
miR-122 mimicked using cationic lipid NPs	Theranostic agent against hepatocellular carcinoma.
Superparamagnetic iron oxide NPs (PEG-g-PEI-SPION)	Initial detection and analysis of gastric cancer.

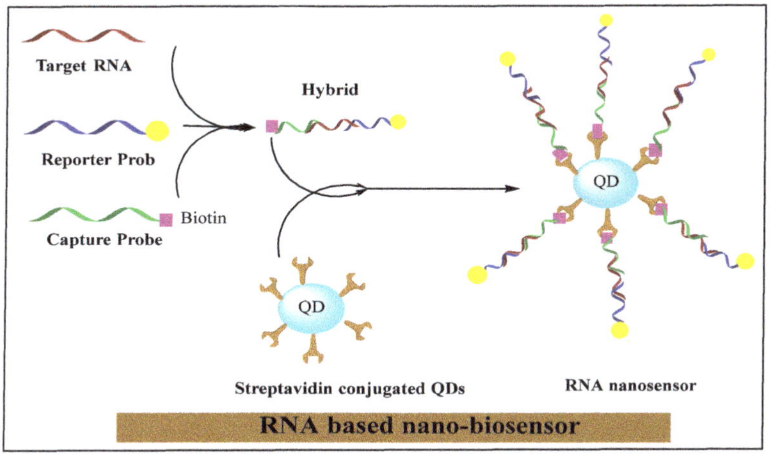

Figure 2. RNA-based nano-biosensor. Reprinted with permission from [75]. Copyright 2021 Elsevier.

Graphene consists of a single-atom dense two-dimensional honeycomb framework made of sp^2-bonded carbon atoms. Many graphite structures, such as nanotubes, graphite, and fullerenes are produced using graphene. British scientists Andre Geim and Konstantin Novoselov were awarded Nobel Prize in Physics in 2010 at the University of Manchester for their revolutionary research on graphene.

Presently, novel functional materials can be manufactured, and siRNA can be delivered to cancer cells by immobilizing RNA on graphenes. Hu et al. produced polydisperse and stable RNA-graphene oxide nanosheets by covalently immobilizing an RNA aptamer on grapheme oxide. Sharifi's group exfoliated graphene flakes from nano-crystalline graphite to yield conducting and transparent RNA-graphene-labeled thin films by using RNA as a surfactant. These thin films are used in a lot of electronic applications [77]. Proteins and peptides are used to form nucleic acid carriers because the nucleotides are shortened by electrostatic linkages with positively charged amino acids of proteins, which are used to transport nucleic acids and small non-coding RNA molecules. On the other hand, the amino acids impart bioreversible polyplex stabilization of the system, endosomal escape, and targeted delivery [78].

2.4. RNA Nanotechnology in Diagnosis of Different Cancers

2.4.1. GBM

Shao et al. developed a microfluidic platform entitled immune-Magnetic Exosome RNA (iMER) for the exosomal analysis of GBM. Three functional sections were combined by iMER, i.e., real-time RNA analysis, targeted up-gradation of extracellular vesicles, and on-chip RNA isolation. The up-gradation or enrichment process detached the cancer exosomes immunomagnetically from host-derived exosomes, and the later analysis was executed on enriched populations. Later, a glass-bead filter was used to pass the lysate collected after the lysation of exosomes. RNA was immobilized onto glass beads by forming electrostatic bonds during this method and then extracted and quantified by qPCR. Magnetic microbeads with anti-epidermal growth factor receptor (EGFR) antibodies were used to identify and enrich GBM-derived exosomes. These GBM-derived exosomes were incubated with beads, and the whole surface seemed thickly covered. A large quantity of mRNA was found in these vesicles, having mRNA of nuclear proteins as well [72].

Spherical nucleic acids SNA are a type of NPs that are made up of a NP core treated with oligonucleotide structures and comprised of RNA interference RNAi reporter molecules and therapeutics. The exclusive 3D SNA structures are resilient to nuclease degradation and co-opt into the cells despite the lack of transfection agents. SNAs act as diagnostic agents and have the potential to detect two unique mRNA sites at a time inside of a cell [77]. These SNAs can effectively cross blood-brain and blood–tumor barriers and are broadly experimented against the most violent and widespread type of malignant brain cancer, i.e., GBM models [79]. Currently, much attention has been given to EGFRvIII, which is an EGFR receptor variant and is associated with GBM progression. DNA and RNA aptamers have been fabricated and used for GBM detection by the involvement of EGFR.

Iqbal et al. sequestered an anti-RNA aptamer from purified human protein by creative selection. The authors confirmed the 'aptamer's ability to detect and seize murine and human GBM cells after its immobilization on a glass substrate. Aptamer binds with wild-type and mutant EGFR with excellent specificity and affinity (Kd = 2.4 nM). This technique is used for the initial detection and analysis of residual disease. Similarly, Iqbal's group fabricated a flow-through lab-on-chip tool that used surface-attached aptamer's affinity for GBM's overexpressed biomarker, i.e., EGFR, to prove that a microfluidic-based technique can be used to detect and capture GBM cells. Later, the same group took radical steps in diagnostics related to anti-EGFR aptamers and came up with two succeeding articles on tracing the differential dynamics of GBM cell structure on substrates grafted with aptamers. They analyzed the dynamic morphology of GBM in computational single-cell metrics to identify and capture tumor cells [80,81].

Choulier et al. linked cell-SELEX and protein to separate RNA aptamers. These aptamers have the potential to bind specifically to integrin $\alpha5\beta1$, which is an $\alpha\beta$ heterodimeric receptor related to cancer angiogenesis and GBM ferociousness. The authors used in histo-fluorescence analysis on patient-based xenografts and a fluorescence-related analysis on cell lines to verify the diagnostic capability of tagged aptamers [82]. Recently, some specific aptamers were considered as markers for metastasis and recurrence of GBM stem cells (GSCs) and brain tumor-initiating cells (TICs).

Rich et al. designed a pool of DNA aptamers distinguishing TICs with an extremely low dissociation constant (Kd between 0.12 and 3.75), and an aptamer-Cy3-related fluorescence proved the binding. In 2019, Affinito et al. used an RNA library on basic human GSCs to design a 20-F-RNA aptamer A40s, which was explicitly related to the GBM cells. The authors established the detection of A40s-based analysis in GBM cells and GSCSs of human tissue sections [83]. Here, the aptamer showed a great affinity (Kd = 41.92 nM) for target cells in a less nanomolar range. As stem cells play an important role in chemo-resistance and metastasis, these aptamers can be used in the clinical field to detect violent areas and analyze GBM treatments [81].

2.4.2. Breast Cancer

Currently, three-dimensional RNA NPs with tetrahedral structures have been designed. RNA tetrahedrons are used in various applications in nanomedicine and nanomaterials because of their structural permanence and mechanical rigidity. The EGFR aptamer was bonded with the RNA tetrahedron structure to target triple-negative breast cancer cells.

After IV administration, the drug-loaded NPs face a sequence of biological blockades. NPs face quick opsonization and succeeding sequestration by local macrophages under physiological conditions. As a result, healthy organs, such as the spleen and liver, accumulate very high levels of NPs [84]. To address these issues, RNA NPs were used to specifically target tumor cells and avoid renal clearance and organ accumulation. For example, Prna-3WJ NPs were fixed with an RNA aptamer specific for EGFR. These NPs specifically targeted triple-negative breast cancer (TNBC). Fluorescence confocal microscopy was used for the histological analysis of tumors to detect the precise directing and retention of RNA NPs in a cancer-frozen cross-section. Both treated and control groups were compared, and it became evident that pRNA-3WJ-EGFR showed extraordinary accumulation at tumor cells without disturbing the healthy organs [69].

In another study, QD-mi RNA let-7a-gold NPs (QD-RNA-Au NP) were conjugated with Chitosan-based nano-formulation, including negatively charged poly (g-glutamic acid) (PGA) for transfer to breast cancer cells. In the cells, the QDs were separated by dicer-mediated release and showed fluorescence for theranostic applications [85].

Mediley et al. described the use of an AuNP aggregation-related colorimetric sensor for straight cancer cell detection. However, the signal produced by aptamers after binding with cancerous cells was too low to detect CCRF-CM cells due to their weak binding affinity. To solve this problem, Lu et al. used S6 RNA aptamer-linked multifunctional oval-shaped AuNPs and the monoclonal anti-HER2/c-erb-2 antibody for multivalent attachment of AuNPs with target cells for extremely sensitive analysis of SK-BR-3 breast cancer cells. AuNP-linked colorimetric techniques are used for initial and sensitive in-vitro recognition of cancerous cells [86].

2.4.3. Prostate Cancer

Sioss et al. fabricated a nanowire-resonator array sensor for signal detection using oligonucleotide-treated AuNPs to analyze RNA in circulating tumor cells (CTCs). In this method, RNAs were attached to AuNPs previously immobilized on the sensor by hybridization. The resonance frequency of the sensor was changed by AuNPs after adding mass to it. The authors calculated this change in resonance frequency and detected PCA3 RNA, a nucleic acid prostate cancer marker. After measurement of RNA and volume used, they calculated that the level of development was 1 CTC/10 mL blood. The sensors used were extremely sensitive and specific, showing a single-nucleotide inconsistency calculation. AuNPs have a large surface area, and they are used as a support to increase signal formation by enhancing molecular binding events [87].

The prostate-specific antigen (PSA) was spotted in prostate cancer cells and the neovasculature of many types of malicious neoplasms, such as breast cancer [88], lung [89], and some other tumor cells [90]. One study reported ligand-receptor relation of RNA NPs with PSA aptamers. The 3WJ-RNA NPs were attached to PSA aptamer (PSA-RNA), indicating extraordinary tumor accumulation in the bio-distribution analysis [77].

Mohamadi et al. fabricated a microelectrode biosensor based on the PSA mRNA and magnetic NPs-based circulating tumor cells. Another study demonstrated aptamer-based in-vivo targeting by pRNA-3WJ NPs containing the anti-prostate-specific membrane antigen (PSA) RNA aptamer. On the other hand, folate can be coupled with pRNA-3WJ NPs as an innovative strategy for improving nanocarrier distribution.

The pRNA-3WJ-folate conjugates have been used in many cancer cells in which FR is overexpressed, such as colorectal, gastric, head and neck cancers, and glioblastoma. The in-vivo distribution of RNA NPs was studied using pRNA-X NPs containing fluorophore or folate. These NPs were injected into athymic mice with KB cells xenografts. Whole-body images were taken at different time intervals and indicated the high accumulation of RNA NPs in cancer cells within 4 h. Specific localization of pRNA-X NPs in the tumor was calculated at the 8th hour of organ imaging, without accumulating in healthy organs [69].

2.4.4. Liver Cancer

Cationic lipids (CLs) are used for liposome-oriented DNA/miRNA transport. These lipids are made up of a linker to which a hydrophobic part is linked to a cationic head part. The positively charged head group is attached to the negatively charged phosphate group of nucleic acids [91]. Liposomes have advantages of less risk of immunological reaction, less toxicity, and are easy to handle, making them a useful tool for non-viral drug delivery and diagnostics. Lipid NPs containing lactosylated gramicidin were exploited to transfer anti-mir-155 to hepatocellular carcinoma. Hsu et al. transferred miR-122 mimic using cationic lipid NPs to suppress miR-122 in hepatocellular carcinoma [92].

2.4.5. Gastric Cancer

Chen et al. developed an MRI-visible system based on superparamagnetic iron oxide NPs (PEG-g-PEI-SPION) and polyethylenimine conjugated to polyethylene glycol (PEG) for the delivery of siRNA to gastric cancer. Despite its use in cancer gene down-regulation, this PEG-g-PEI-SPION has verified itself as an extremely effective contrast agent for in-vivo MRI scanning as well. In the same way, Sun et al. designed Micro-RNA-16-loaded magnetic NPs to solve drug resistance challenges in the mouse gastric cancer model. Polyethylene glycol (PEG)-coated iron oxide (Fe_3O_4) NPs were used in this study. These magnetic NPs showed highly efficient in-vivo imaging along with increased (human gastric cancer cell line 7901) SGC7901 sensitivity to the drug Adriamycin [93].

Rychahou et al. developed the precise delivery of folate-linked pRNA NPs into colorectal cancer. After IV injection, the RNA NPs showed accumulation in metastatic cells of colorectal cancer, lung, liver, and lymph node cancer, but no accumulation was found inside healthy organs [69].

3. RNA-Nanomaterials for Targeted Therapy of Different Cancers

Extracellular matrix (ECM) is composed of glycoproteins, elastin, collagen, and hyaluronic for providing solid structural support for cellular processes, i.e., proliferation, cell migration, and growth [94]. Moreover, ECM also systematizes the intracellular communication of cytokines and growth factors and acts as a source of physical barrier against the tumor microenvironment [95]. However, in solid metastatic tumors, the ECM equilibrium in maintaining homeostasis gets affected, leading to disorganization of the physicochemical and biochemical features, as shown in Figure 3.

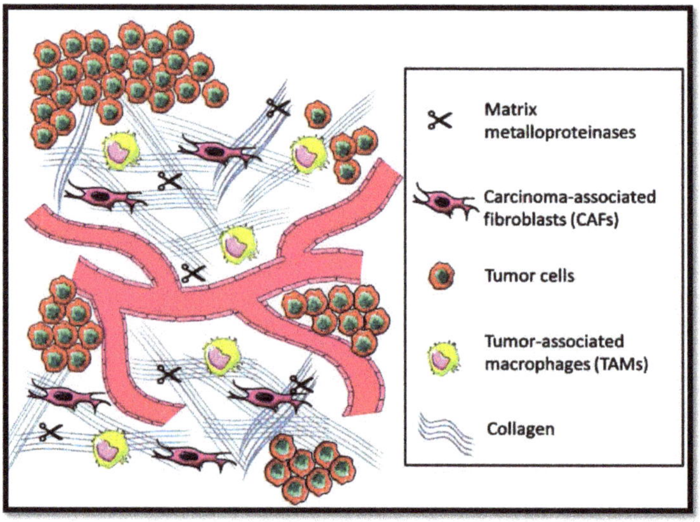

Figure 3. Tumor microenvironment prevalence in the extracellular matrix.

Conventional cancer treatment has failed to mitigate the impact of malignancies [96]. Therefore, multi-functionalized NPs were preferred to be synthesized for targeted action via attachment of specified ligands to target the tissues that are highly over-expressed in certain diseases. The development of targeted and personalized therapeutics based on RNA interference has also been made possible due to advances in molecular genetics [97]. The modification in RNA therapies utilized small RNA subunits in the form of small interfering RNAs (siRNA) for overexpressing the specific genes of the related cancers [98]. The RNAi technology can be exploited to change the oncogenic characteristics of breast cancer cells, making them highly conducive to apoptosis cell death. Combination therapies are useful approaches to generating apoptotic effects mediated via carcinogenic pathways and help to overcome drug resistance [99].

RNA-based drugs are often entrapped or attached to the surface of different nanovehicles to deliver the cargo to the cells. These nanovehicles can be modified with various moieties, including polyethylene glycol (PEG) and cholesterol, which enhance them by the membrane of target cells, where the nanovehicle can enter the target cells via endocytosis (Figure 4A). Meanwhile, some RNA-based therapeutics are conjugated to moieties directly, which facilitates their transmembrane transport (Figure 4B). In another therapeutic approach, the synthesized RNA-therapeutics are chemically modified to enhance their binding affinity, stability, and biocompatibility (Figure 4C).

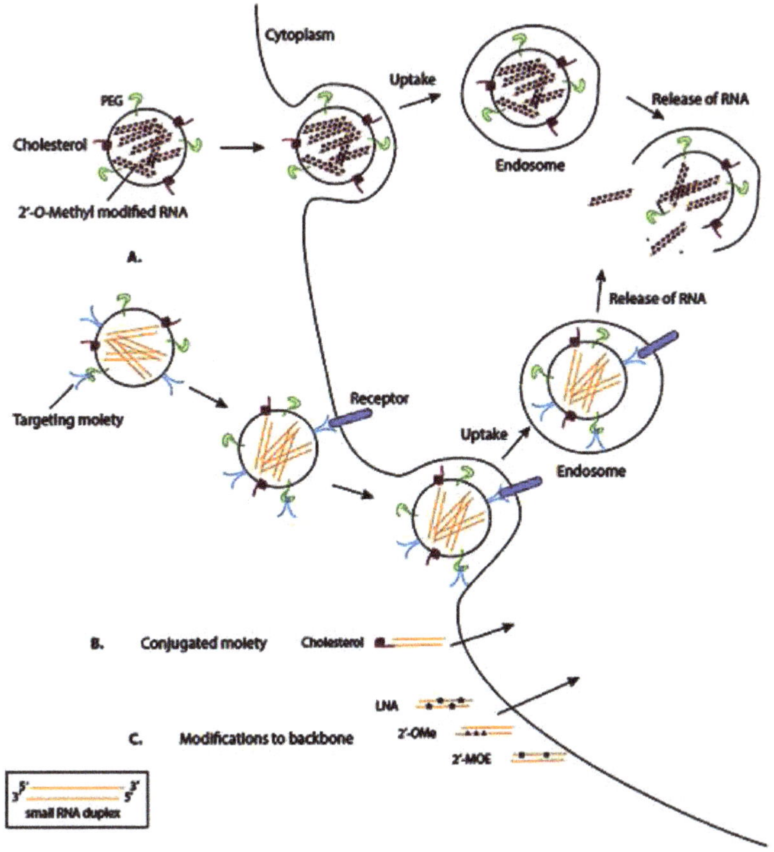

Figure 4. Schematic representation of common delivery methods for RNA-based therapies. LNA: locked nucleic acid (2′4′-methylene; 2′OMe: 2′-O-methyl; 2′MOE: 2′-O-methoxyethyl. Nanocarriers can enter the target cells via endocytosis (**A**), direct conjugation to moieties (**B**), and chemical modification (**C**). Reprinted from [100].

The miRNA or siRNA possesses the capability to bind with the enzyme-containing molecule RNA-induced silencing complex (RISC), resulting in the enzymatic cleavage of the target mRNA [101]. Targeted mRNA has the capability of silencing over-expressed genes as well as inducing programmed death-ligand 1 (PD-L1) efficacy towards programmed apoptosis against the deadliest cancer cells, as shown in Figure 5. The targeted role of mRNA as a ligand in various cancers is shown in Table 2.

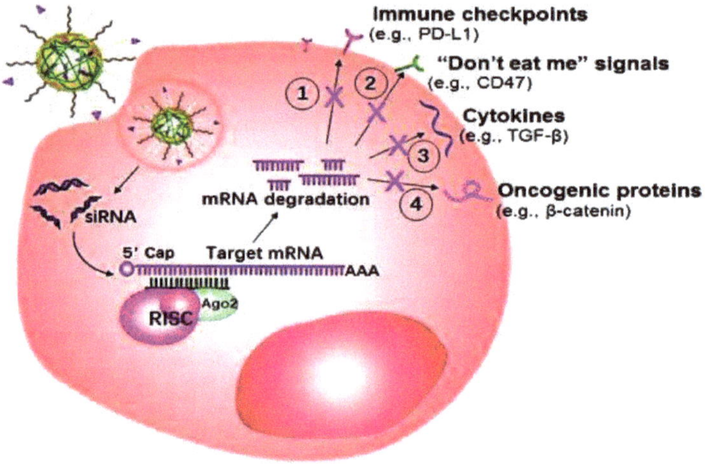

Figure 5. Small interfering RNAs (siRNA) mechanism of action for overexpressing the specific genes of the related cancers.

Table 2. Summary of RNA-based nanostructures in treatment of cancers.

Nanostructure	Key Feature	Ref
Ultra-thermostable RNA NPs	RNA ligand proved to dramatically inhibit the growth of breast cancer with non-detectable toxicity and immune responses in mice.	[102]
Selenium-siRNA NPs	Small interfering RNA (siRNA) showed great potential in advanced therapeutics because of its highly sequential ability for silencing HeLa genes for cervical cancer	[103]
MSN-anti-miR-155 NPs	miR-155 was highly over-expressed in colorectal tissues and cell lines as compared to the control groups and showed enhanced therapeutic efficacy.	[104]
Survivin-siRNA NPs	The novel nanocarrier system was able to initiate a specified and safe cellular uptake with increased transfection efficacy, promoting the downregulation of HCC cells.	[105]
Enveloped siRNA NPs	siRNA multi-functionalized nano-enveloped carriers can strongly silence target genes expressions as well as strongly pre-dominant genes, such as prohibitin 1 (PHB1), resulting in significantly culminating prostate tumor growth	[106]
FA-PEI-Fe_3O_4-siRNA NPs	Effective targeted PD-L1-knockdown therapy as well as a diagnosis in gastric cancers, thus favoring towards the best theranostic approach	[107]
PLL-siRNA-MSN NPs	MSNPs-PLL proved to be an accomplished candidate for non-invasive transdermal drug delivery in alleviating skin cancer cells division	[108]

Moreover, all the RNA-based nanoparticles tend to adopt various advanced strategies to provide the targeted and efficacious treatment against various metastatic cancer, as shown in Figure 6.

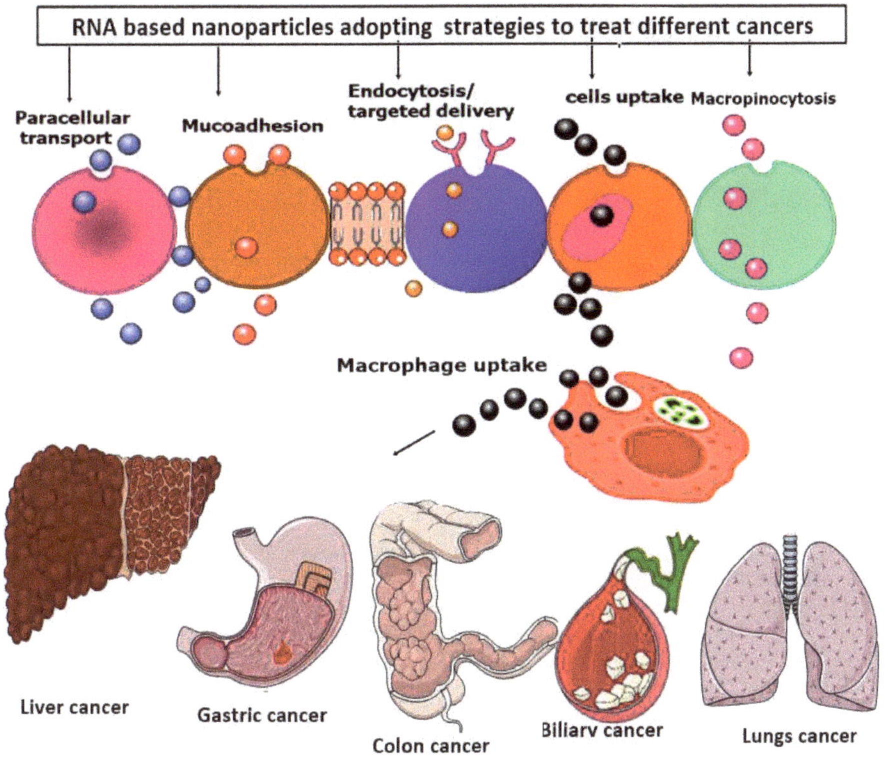

Figure 6. RNA-based nanoparticles various advanced strategies for providing the targeted and efficacious treatment against various metastatic cancers.

3.1. RNA NPs

Cancer of the breast is the deadliest disease in the world [109]. Breast cancer therapy includes radiation therapy, chemotherapy, adjunct therapy (radiation therapy and chemotherapy), endocrine, and ligand-mediated therapy [110]. Guo et al. (2020) [102] conjugated PTX with RNA via the synthesizing prodrug for PTX as PTX-N3. PTX-N3 prodrug was synthesized by admixing the PTX, and other substituents in a 20 mL DCM solvent. The reaction mixture was then stirred at room temperature for 36 h following filtration and rotary evaporation to attain the yield's crude product. The crude product was then purified by silica gel chromatography using n-Hexane: ethyl acetate as an eluent. RNA-6 alkynes oligomers were synthesized via a stranded solid-phase RNA synthesis and purified through desalting. RNA sequences of nine types were isolated and conjugated to PTX using copper (I)-catalyzed alkyne-azide cycloaddition by click addition. The reaction mixture was then diluted with ethanol followed by overnight incubation for RNA precipitation in DEPC-H_2O. The precipitated were re-dissolved and purified by the ion-pair reverse phase HPLC in a PTX-labeled RNA for assembling NPs. Finally, an RNA four-way junction NP (4WJ-X nanostructure) with ultra-thermodynamic stability to solubilize and load PTX for targeted cancer therapy was developed by a 3D computational model generated using Swiss PDB Viewer and PyMOL Molecular Graphics System. Assembly of NPs was confirmed using a specified buffer solution. EGFR NPs were also conjugated with one of the aptameric RNA oligomers and 4WJ-X-24 PTXs. It was observed that each RNA NPs was found to be successfully attached via covalent interaction to twenty-four molecules of PTX as a prodrug. The developed RNA-PTX complex was found to be structurally stable and rigid. It was concluded that RNA NPs proved to help increase the water solubility of the

BCS class II drug PTX by 32,000-fold, which possesses the issue of low water solubility. This treatment strategy of RNA functionalization in the form of intravenous injections resulted in the specified cancer targeting. RNA ligand proved to dramatically inhibit the growth of breast cancer with non-detectable toxicity and immune responses in mice. Moreover, no mortalities were observed at the LD_{50} dose of PTX [102].

3.2. Nanotechnology for Transfer of Therapeutic RNAs

By using nanotechnology for RNA, we are able to overcome many shortcomings of naked RNA molecules, such as poor chemical stability, easy degradation by nucleases, and extremely short half-lives for in-vivo applications. NPs act as a multipurpose and targeted system for the safe transfer of naked RNA molecules. The NP-based protect RNA from enzymatic cleavage and immune system threats. Because of their EPR properties, these nanostructures facilitate excellent RNA accumulation at the tumor site. Nowadays, the nanocarriers used for RNA delivery are lipid-based nanosystems [111], polymeric nanomaterials, bio-inspired nanovesicles, and inorganic NPs [112].

RNA NPs can easily include targeting ligands, counting RNA aptamers, and chemical ligands to impart specific targeting against proteins, fluorescent chemicals, and cell surface receptors. Chemical ligands, e.g., folate, are attached to the end of the RNA strand while the RNA is synthesized by modified RNA phosphoamidites. These ligands can also be attached to the RNA strand after RNA synthesis by the chemical conjugation method. The RNA aptamer strand, such as EGFR aptamer, can be manufactured by elongation of the scaffold strand.

The binding efficiency in-vitro and in-vivo of RNA NPs with ligands is tested by labeling them with radioisotopes and fluorescent dyes [113]. The labeled RNA NPs are incubated with cells for in-vitro binding and internalization studies. After incubation, these NPs are quantified by flow cytometry or fluorescence confocal microscopy [77]. Folic acid, FA, has attracted much attention for targeted siRNA delivery due to its small size, outstanding in-vivo stability, low immunogenicity, high binding affinity for folate receptors FRs, and great specificity to cancer cells. Folate receptors play an important role in diagnosing and therapeutics of inflammatory diseases and carcinoma. Folate receptors are overexpressed in cancer cells, and folic acid-decorated siRNA transporters bind precisely to FRs. The site-specific distribution of FAsiRNA conjugates and enhances siRNA concentration at the target site [112,114].

Folate receptors are overexpressed in epithelial cancer cells surfaces, permitting the folic acid conjugated NPs to target these cancer cells at a frequency higher than the normal cells.

RNA NPs can efficiently target cancer metastasis that is difficult to target because of the spread of cancerous cells to far-off cells and lymph nodes. Folic acid was used as a targeting agent by RNA NPs to simultaneously target colon cancer cells in the chief sites of metastasis, such as lungs, liver, and lymph nodes [49]. The 3WJ-based design procedure was used to form highly branched RNA dendrimers. RNA dendrimers used pRNA nanosquare as a symmetrical core for their formation. The formation of higher-ordered structures may face steric hindrance that is reduced by the square shape. Targeted delivery of NPs highly lessens the off-target toxicity and accumulation of NPs in healthy organs. RNA aptamers are an emergent field of therapeutics in which single-stranded RNA sequences form 3D structures by folding up and binding to extracellular domains of cell surface receptors with high selectivity and affinity. Nowadays, cell surface receptors are targeted by tens of RNA aptamers, such as prostate cancer (e.g., PSA), colon cancer (e.g., EpCAM), ovarian cancer (e.g., E-selection), glioblastoma (e.g., EGFRvIII), lymphoma (e.g., CD19) and breast cancer (e.g., EGFR, HER2, HER3) [49].

3.3. Small Interfering RNA-Selenium NPs

Small interfering RNA (siRNA) showed great potential in advanced therapeutics because of its highly sequential ability for silencing genes [115]. One of the most common causes of death in women is cervical cancer [116]. As a result, the right medication is necessary to minimize the severity of this cancer. For this purpose, Yu Xia synthesized biocompatible selenium NPs (SeNPs) and loaded them with the arginyl glycyl aspartic acid (RGD) Fc peptide for the sake of active targeting [103]. The RGDfC peptide bore a rich cationic charge and functionalized SeNPs for enhanced gene delivery. RGDfC-SeNPs have the capability of binding with HeLa cervical cancer lines.

Furthermore, Derlin 1-siRNA can be adjunct to the formulated conjugate of RGDfC-SeNPs via electrostatic interaction. The RGDfC-Se@siRNA successful conjugation and synthesis was confirmed via size determination by a zeta sizer, transmission electron microscopy (TEM), and Fourier transform infrared spectroscopy (FTIR). Elemental compositions of RGDfC-SeNPs were studied by energy dispersive spectroscopy (EDS). RGDfC-Se@siRNA characterization results proved that it followed Clathrin-mediated endocytosis for specifically reaching HeLa cancer cell lines and exhibited triggered siRNA release in a tumor microenvironment as compared to the biological microenvironment. However, in qPCR and Western blotting assays, both techniques showed eminent chances of gene silencing in HeLa cells. RGDfC-Se@siRNA was found to suppress the tumor invasion and division in HeLa cells via triggering the apoptosis pathway. Moreover, another crucial mechanistic approach of mitochondrial membrane disruption and reactive oxygen species generation (ROS) in HeLa cells was quite convincing in understanding that mitochondrial dysfunction mediated by ROS might play a significant role in RGDfC-Se@siRNA-induced apoptosis. Interestingly, advanced nanotherapeutics also presented substantial antitumor activity in a HeLa tumor-bearing mouse model [103].

3.4. *siRNA-Polymeric NPs*

Prostate cancer malignancy is one of the significant reasons for mortality worldwide. Prostate cancer is a non-skin malignancy causing the second biggest number of deaths in men when differentiated with different tumors. Conventional therapies are expected to cause erectile dysfunction, libido, obesity, and bone mass loss. Nanotechnology has modernized the field of medication to sidestep conventional therapies against metastatic cancers and different intracellular infections [117]. Xiao ding Xu et al. (2017) [106] synthesized siRNA multi-functionalized enveloped NPs for prostate cancer advanced therapy. SiRNA was self-assembled in the NPs via utilizing the library of oligo arginine and sharp pH-responsive polymers. Moreover, siRNA-functionalized and self-assembled NPs resulted in prolonged blood circulation, and the pH triggered a drug release through the activation of the endosomal membrane penetration. Furthermore, modification of the synthesized nanocarrier was done via attaching a specified molecular ligand that can recognize the PSA receptor. Synthesized nano-enveloped particles were characterized for size and zeta potential via DLS. The morphology of NPs was determined by TEM, fluorescence intensity, encapsulation efficiency, in-vitro siRNA release, luciferase silencing, endosomal escape, flow cytometry, in-vitro PHB1 silencing, digestion assay, Western blot, immunofluorescence staining, in-vitro inhibition of cell proliferation, a xenograft tumor model, and pharmacokinetic studies. siRNA multi-functionalized nano-enveloped carriers have the capability to strongly silence target genes expressions as well as strongly pre-dominant genes, such as prohibitin 1 (PHB1), resulting in significantly culminating tumor growth. Moreover, these advanced NPs also possess great potential for a robust siRNA delivery vehicle for prostate cancer-targeted therapy [106].

3.5. *siRNA-Superparamagnetic Iron Oxide NPs*

Globally, death from gastric cancer accounts for the majority of deaths. Conventional treatments have not been able to alleviate the consequences of this deadly disease. The development of targeted therapies for gastric cancer requires new technologies [118]. The most novel approach being considered is finding the primary targets, such as immune checkpoints such as PD-L1, in gastric cancer [119]. PD-L1 presents an over-expression on the activated T cells, resulting in the endosomal escape by cancer cells. The mechanism of inhibiting cancer cell regulation via PD-L1 cells lies in promoting and maintaining the T-cell responses in a controlled manner [120]. Xin Luo promoted siRNA delivery system for knocking down PD-L1 by developing folic acid (FA) and disulfide (SS)-polyethylene glycol (PEG)-conjugated polyethylenimine (PEI), complexed with superparamagnetic iron oxide Fe_3O_4 NPs (SPIONs) [107]. SPIONs were encapsulated with FA-PEG-SS-PEI via a ligand-exchange method, and then this conjugate was combined with synthesized siRNA-complexed cationic micelles. Furthermore, synthesized NPs were characterized based on binding capability, cytotoxicity, cellular internalization, and transfection efficacy. Cell viability assays demonstrated negligible toxicity and maximum cellular uptake as well. Cellular magnetic resonance imaging (MRI) presented that the NPs depicted maximum contrast for the T2 weight for cancer MRI. Furthermore, PD-L1 siRNAs displayed nominal knockdown of PD-L1 in the PD-L1-overexpressing gene. However, the co-culture model of activated T cells and the over-expressed gene cells represented an increased level of secreted cytokines. Therefore, these findings highlight the potential of this class of multi-functionalized polyplexed NPs for effective targeted PD-L1-knockdown therapy and diagnosis in gastric cancers, thus favoring the best theranostic approach [107].

Hepatocellular carcinoma (HCC) is the most common malignancy of the liver and the most common cause of morbidity and mortality [121,122]. A few molecular targeting drugs, such as sorafenib (SO), have been approved for advanced HCC, which also show a peripheral survival chance compared to conventional therapeutics. Unfortunately, its efficacy for HCC patients stayed substandard. Thus, the development of new methods for diagnosing and managing HCC is of the utmost importance. Gene delivery is the most preferred therapy involving RNA interference for the purpose of post-transcriptional gene silencing. Zhuo Wu et al. (2017) [105] developed a new class of amylose NPs, where the cationic amylose was used as the backbone functionalized with folate for targeting. Furthermore, SPIONs were utilized for the purpose of imaging and diagnosis for delivering specified surviving siRNA to hepatocellular carcinoma cells. The synthesized siRNA and multi-functionalized NPs were characterized based on zeta sizing, NMR, FTIR, cytotoxicity studies, cellular uptake, gene silencing, apoptosis signaling, and magnetic resonance imaging (MRI), and the results of these characterization techniques assured the successful conjugation of a new class of amylose NPs for the targeted delivery to HCC cells. Moreover, the novel nanocarrier system was able to initiate a specified and safe cellular uptake with increased transfection efficacy, promoting the downregulation of HCC cells. The resulting conjugate biocompatible complex based on cationic amylose could be used as a well-organized, prompt, and innocuous gene delivery vector. Furthermore, upon SPION addition, it holds great potential as a theranostic carrier for the gene therapy of HCC [105].

3.6. *RNA-Mesoporous Silica NPs*

Timely diagnosis and therapy are the prime responsibility of the healthcare system [123]. Several advancements in molecular genetics have allowed pathogenic mutations in diagnosing and classifying unique skin infections, such as psoriasis, atopic dermatitis, and skin cancer [124]. Advancements in molecular genetics also allowed the development of targeted and personalized therapeutics based on RNA interference [97]. Modification in RNA therapies utilized small RNA subunits in the form of small interfering RNAs (siRNA) for overexpressing the specific genes of the related disorders [98]. Therefore, Daniel Chin Shiuan Lio et al. developed mesoporous silica NPs nucleotide complexes of 200 nm with a pore size of 4 nm and mixed it with oligonucleotides of RNA, followed by overnight stirring. After stirring, the mesoporous silica NPs-oligonucleotide conjugate were coated with poly-L-lysine (PLL) in 1:1 for 10 min [108]. The excessive amount of PLL was removed via centrifugation at the speed of 10,000 rpm for 15 min. Excessive PLL-removed NPs were again re-suspended in the phosphate-buffered saline (PBS). Characterization of the PLL-coated nanocarriers was done via size analysis, labeling cells with MSNPs for the time-course study, and post-treatment in-vivo studies with a xenograft tumor model, primer sequence, real-time-polymerase chain reaction (RT-PCR), histological sectioning, Western blot, and flow cytometry. Results concluded that the loading of NPs-based oligonucleotides by poly-L-lysine resulted in improved transdermal drug delivery, increased zeta potential, and enhanced stability [125]. The MSNPs-PLL evaluation on skin squamous cell carcinoma (SCC) cells in-vitro showed a safety profile and increased penetration. Therefore, we are bound to believe that MSNPs-PLL proved to be accomplished candidates for non-invasive transdermal drug delivery in alleviating the skin cancer cells division [108].

Mesoporous silica NPs (MSNs) have been considered the most promising nanocarriers for attached targeted moieties owing to unique features of tunable pore structure, greater surface area, and pore volume [126–128]. These flexible features of the MSNs are accountable for successful conjugation, thermal stability, and biocompatibility. Colorectal cancer is a heterogeneous and lethal disease, proceeding towards the development of malignant tumors in the inner walls of the colon and rectum in the form of polyps. MicroRNA-155 (miR-155) is an oncogenic microRNA, and is over-expressed in many cancers, including colorectal cancer (CRC). Therefore, targeting the approach in treating cancers by using miR-155 is a nominal strategy for treating cancer. Therefore, Yang Li (2018) [104] developed anti-miR-155-loaded MSNs functionalized with polymeric dopamine (PDA) and AS1411 aptameric (MSNs-anti-miR-155@PDA-Apt) for the targeted therapy of CRC. The prepared NPs were characterized based on size, cell uptake studies, in-vitro cytotoxicity, xenograft tumor model, in-vivo imaging and biodistribution, in-vivo antitumor efficacy, and systemic toxicity. The results showed that miR-155 was highly over-expressed in CRC tissues, resulting in a significantly high targeted therapy and enhanced therapeutic efficacy [104].

4. Advantages and Limitations of RNA-Based Nano-Theranostic Systems

Thanks to their selectivity and enhanced sensitivity, RNA-based theragnostic approaches have tremendously improved current methods for diagnosis and provided versatile delivery systems for the treatment of a variety of cancers. This is important because in-vivo applications of the naked RNA molecule face many challenges that need to be tackled. For instance, the naked RNA molecules have poor chemical solubility, extremely short half-lives, and are easily degraded by nucleases [129–131]. Interestingly, developed nanovehicles protect the RNA molecule from immune system threats and degradation by nuclease enzymes and enhance the EPR effect, thus enabling the accumulation of the RNA molecule in the cancerous sites [130,132]. In terms of cancer therapy, however, there are benefits and drawbacks to applying such innovative modalities. RNA-nanovehicles functionalized with different moieties have enhanced cargo delivery without adverse side effects, and thus, can be considered efficient and targeted strategies for delivering chemotherapeutic agents to malignant cells.

A number of nanovehicles have been studied for their ability to deliver RNA molecules to target locations. For instance, it has been documented that lipid-based nanostructures (i.e., liposomes, solid lipid NPs, and lipid emulsions), polymer-based nanomaterials, inorganic NPs, and bio-inspired nanovehicles can be used for the targeted delivery of nucleic acids, such as RNA. The majority of these nanocarriers have the advantages of easy preparation, good biocompatibility and bioavailability, low cost of production, controlled release, easy modification, and easy uptake. However, they have shown poor solubility, rapid clearance, easy leakage of cargo, dose-dependent toxicity, and stability concerns [112].

The application of RNA technology has a major advantage of exploiting cellular machinery that permits the targeting of complementary transcripts, resulting in a dramatic decrease in the expression of a target gene. The main limitations of this approach are ineffective in-vivo delivery, off-target effects, and the induction of type I interferon responses. [129].

The RNA molecule itself is unable to cross cell membranes. In addition, prompt degradation of the therapeutic agent in the endocytic pathway, called the endosomal escape, could be a great challenge in ligand-mediated endocytosis for the specific delivery of siRNA [133]. A crucial limitation for in-vivo delivery of siRNAs is the size of the synthesized RNA NPs. Basically, the average size of an RNA nanostructure is about

20 to 40 nm, which enhances its biodistribution in the blood circulation system, while the average size of a normal single siRNA molecule is less than 10 nm. Moreover, non-formulated siRNAs can be easily excreted through the urinary system [133–135]. Chemical and thermodynamic instability, short in-vivo half-life, undesirable in-vivo toxicity of RNA NPs along with targeting problems, high production costs, and a low yield are the primary drawbacks of the application of RNA nanostructures for theranostic purposes [133]. Using targeting moieties, such as aptamers or receptor-targeting ligands (i.e., folate), can improve RNA delivery. Still, endosomal escape might be a fundamental limitation that needs to be overcome shortly. Designing novel targeting RNA NPs with the ability to escape endosomes will lead the way toward more promising RNA therapeutics. mRNA-based vaccines impart immunoglobulins and immune responses, leading towards phagocytosis, as shown in Figure 7. Moreover, a list and links to information about clinical trials regarding RNA-based nanomaterials against cancer is now given in Table 3.

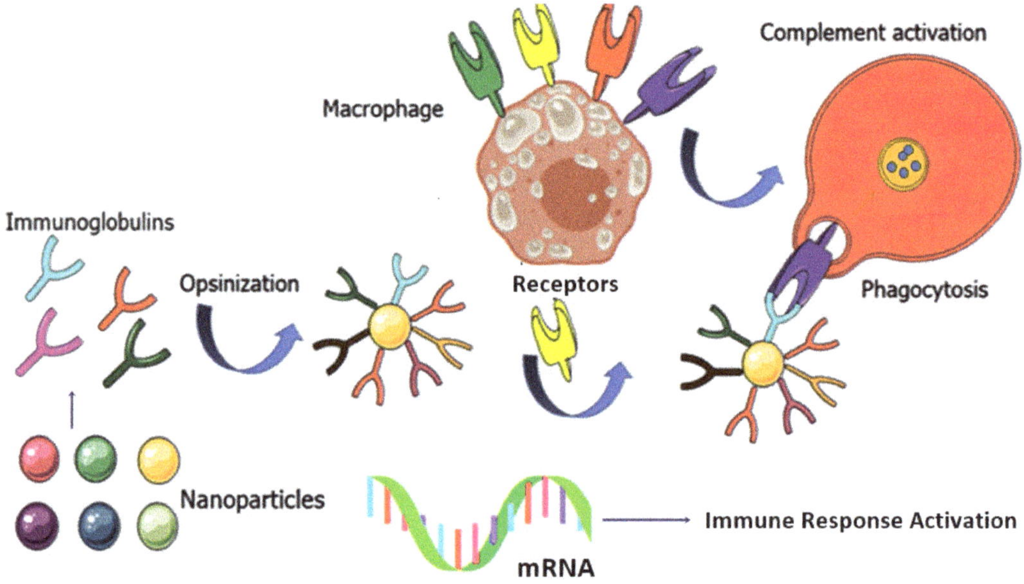

Figure 7. mRNA-based vaccines impart immunoglobulins and immune responses, leading towards phagocytosis.

Table 3. List and links to information about clinical trials regarding RNA-based nanomaterials against cancer.

RNA Based Nanomaterials	Clinical Trials	Ref.
The self-delivering RNA (sd RNA)	Combination of immunotherapy and chemotherapy for cancer treatment in pre-clinical trials.	[136]
Single mRNA-4157vaccine	Preclinical phase 2 against melanoma.	[137]
Adjuvant claudin mRNA cells	Pre-clinical stages against metastatic breast cancer.	[138]
mRNA 5671 based NPs	Pre-clinical stages against colorectal cancer, lungs cancer, and pancreatic cancer.	[139]
mRNA 2416 based NPs	Pre-clinical stages against solid tumors in ovarian cancer.	[140]

5. Conclusions and Future Prospective

In this review, we highlighted advancements in the use of RNA nanotechnology in cancer diagnosis and treatment. The development of novel RNA nanotechnology-based tools for cancer diagnosis and treatment has been studied extensively in recent years.

Compared to presently available cancer diagnosis and treatment in clinics, a number of RNA-based NPs demonstrated improved sensitivity and selectivity or provided whole new capabilities that could not be attained with conventional techniques. Compared with conventional diagnosis and therapies, an RNA functionalized nanocarrier-mediated anticancer drug delivery leads to high therapeutic efficacy, targeted binding with the ligand, more accurate diagnosis, lower toxicity, and site-specific delivery, resulting in cytotoxicity management and cost-effectiveness. The barriers to the diagnosis and treatment of cancers and the killing of healthy cells have been minimized using biocompatible polymers ligands. Promising advances in cancer treatment and detection, such as RNA-functionalization and biocompatible ligands, are rapidly paved the way in addressing the disadvantages of existing approaches by effectively enhancing the treatment and diagnosis of metastatic tumors. Recently, significant advances have been made in the area of RNA nanotechnology for cancer diagnosis and treatment, and our knowledge of this topic has significantly expanded. However, just a few RNA-based NPs have progressed to clinical trials, and RNA nanotechnology is expected to enter the clinic in the coming years with collaborative efforts among scientists, technologists, and therapists. RNA nanotechnology, with its excellent specificity, accuracy, and multiplexed measuring capabilities, has significant potential for improving cancer diagnosis and treatment, finally leading to a higher cancer patients' chance of survival.

Author Contributions: Conceptualization, S.S. and A.R.; writing-original draft preparation, R.A., I.F., S.S. and M.K.-J.; writing-review and editing, S.S., S.P., M.B. and R.A.; supervision, A.R. and A.M.D.-P. All authors have read and agreed to the published version of the manuscript.

Funding: Financial support from the Community of Madrid within the framework of the multi-year agreement with the University of Alcalá in the line of action "Stimulus to Excellence for Permanent University Professors", Ref. EPU-INV/2020/012, is gratefully acknowledged.

Data Availability Statement: Data sharing is not applicable to this article as no new data were created or analyzed in this study.

Conflicts of Interest: The authors declare no conflict of interest.

Abbreviations

NPs	Nanoparticles
siRNAs	Short interfering RNAs
PTX	Paclitaxel
SPR	Surface plasmon resonance
PRRs	Pattern recognition receptors
EPR	Enhanced permeability and retention
AuNPs	Gold nanoparticles
QDs	Quantum dots
ssDNA	Single-stranded DNA
SELEX	Systemic Evolution of Ligands by EXponential enrichment
NIRF	Near-infrared fluorescent
MBS	m-maleimidobenzoyl N-hydroxysuccinimde ester
GBM	Glioblastoma multiforme
iMER	Immune-Magnetic Exosome RNA
EGFR	Epidermal growth factor receptor
GSCs	GBM stem cells
TICc	Tumor-initiating cells
TNBC	Triple-negative breast cancer
CTCs	Circulating tumor cells
PSA	Prostate-specific antigen
CLs	Cationic lipids
PEG	Polyethylene glycol
ECM	Extracellular matrix

PD-L1	Programmed death-ligand 1
TEM	Transmission electron microscopy
FTIR	Fourier transform infrared spectroscopy
EDS	Energy dispersive spectroscopy
ROS	Reactive oxygen species
PHB1	Prohibitin 1
FA	Folic acid
MRI	Magnetic resonance imaging
HCC	Hepatocellular carcinoma
CRC	Colorectal cancer
PLL	Poly-L-lysine
PBS	Phosphate-buffered saline
RT-PCR	Real-time-polymerase chain reaction
MSNs	Mesoporous silica nanoparticles
PDA	Polymeric dopamine

References

1. Jemal, A.; Bray, F.; Center, M.M.; Ferlay, J.; Ward, E.; Forman, D. Global cancer statistics. *CA Cancer J. Clin.* **2011**, *61*, 69–90. [CrossRef] [PubMed]
2. Moumaris, M.; Bretagne, J.-M.; Abuaf, N. Nanomedical Devices and Cancer Theranostics. *Open Nanomed. Nanotechnol. J.* **2020**, *61*, 69–90. [CrossRef]
3. Barani, M.; Nematollahi, M.H.; Zaboli, M.; Mirzaei, M.; Torkzadeh-Mahani, M.; Pardakhty, A.; Karam, G.A. In silico and in vitro study of magnetic niosomes for gene delivery: The effect of ergosterol and cholesterol. *Mater. Sci. Eng. C* **2019**, *94*, 234–246. [CrossRef] [PubMed]
4. Mora-Huertas, C.E.; Fessi, H.; Elaissari, A. Polymer-based nanocapsules for drug delivery. *Int. J. Pharm.* **2010**, *385*, 113–142. [CrossRef] [PubMed]
5. Mashayekhi, S.; Rasoulpoor, S.; Shabani, S.; Esmaeilizadeh, N.; Serati-Nouri, H.; Sheervalilou, R.; Pilehvar-Soltanahmadi, Y. Curcumin-loaded mesoporous silica nanoparticles/nanofiber composites for supporting long-term proliferation and stemness preservation of adipose-derived stem cells. *Int. J. Pharm.* **2020**, *587*, 119656. [CrossRef] [PubMed]
6. Sheervalilou, R.; Shirvaliloo, M.; Sargazi, S.; Ghaznavi, H. Recent advances in iron oxide nanoparticles for brain cancer theranostics: From in vitro to clinical applications. *Expert Opin. Drug Deliv.* **2021**, *31*, 1–29. [CrossRef]
7. Kafshdooz, L.; Pourfathi, H.; Akbarzadeh, A.; Kafshdooz, T.; Razban, Z.; Sheervalilou, R.; Ebrahimi Sadr, N.; Khalilov, R.; Saghfi, S.; Kavetskyy, T. The role of microRNAs and nanoparticles in ovarian cancer: A review. *Artif. Cells Nanomed. Biotechnol.* **2018**, *46*, 241–247. [CrossRef]
8. Irajirad, R.; Ahmadi, A.; Najafabad, B.K.; Abed, Z.; Sheervalilou, R.; Khoei, S.; Shiran, M.B.; Ghaznavi, H.; Shakeri-Zadeh, A. Combined thermo-chemotherapy of cancer using 1 MHz ultrasound waves and a cisplatin-loaded sonosensitizing nanoplatform: An in vivo study. *Cancer Chemother. Pharmacol.* **2019**, *84*, 1315–1321. [CrossRef]
9. Raghubir, M.; Rahman, C.N.; Fang, J.; Matsui, H.; Mahajan, S.S. Osteosarcoma growth suppression by riluzole delivery via iron oxide nanocage in nude mice. *Oncol. Rep.* **2020**, *43*, 169–176. [CrossRef]
10. Mishra, S.; Sharma, S.; Javed, M.N.; Pottoo, F.H.; Barkat, M.A.; Alam, M.S.; Amir, M.; Sarafroz, M. Bioinspired nanocomposites: Applications in disease diagnosis and treatment. *Pharm. Nanotechnol.* **2019**, *7*, 206–219. [CrossRef]
11. Haghiralsadat, F.; Amoabediny, G.; Naderinezhad, S.; Nazmi, K.; De Boer, J.P.; Zandieh-Doulabi, B.; Forouzanfar, T.; Helder, M.N. EphA2 targeted doxorubicin-nanoliposomes for osteosarcoma treatment. *Pharm. Res.* **2017**, *34*, 2891–2900. [CrossRef] [PubMed]
12. Ma, H.; He, C.; Cheng, Y.; Li, D.; Gong, Y.; Liu, J.; Tian, H.; Chen, X. PLK1shRNA and doxorubicin co-loaded thermosensitive PLGA-PEG-PLGA hydrogels for osteosarcoma treatment. *Biomaterials* **2014**, *35*, 8723–8734. [CrossRef]
13. Shakeri-Zadeh, A.; Zareyi, H.; Sheervalilou, R.; Laurent, S.; Ghaznavi, H.; Samadian, H. Gold nanoparticle-mediated bubbles in cancer nanotechnology. *J. Control. Release* **2020**, *330*, 49–60. [CrossRef] [PubMed]
14. Rauf, A.; Tabish, T.A.; Ibrahim, I.M.; ul Hassan, M.R.; Tahseen, S.; Sandhu, M.A.; Shahnaz, G.; Rahdar, A.; Cucchiarini, M.; Pandey, S. Design of Mannose-Coated Rifampicin nanoparticles modulating the immune response and Rifampicin induced hepatotoxicity with improved oral drug delivery. *Arab. J. Chem.* **2021**, *14*, 103321. [CrossRef]
15. Niazi, M.; Zakeri-Milani, P.; Najafi Hajivar, S.; Soleymani Goloujeh, M.; Ghobakhlou, N.; Shahbazi Mojarrad, J.; Valizadeh, H. Nano-based strategies to overcome p-glycoprotein-mediated drug resistance. *Expert Opin. Drug Metab. Toxicol.* **2016**, *12*, 1021–1033. [CrossRef]
16. Rahdar, A.; Hasanein, P.; Bilal, M.; Beyzaei, H.; Kyzas, G.Z. Quercetin-loaded F127 nanomicelles: Antioxidant activity and protection against renal injury induced by gentamicin in rats. *Life Sci.* **2021**, *276*, 119420. [CrossRef]
17. Javad Farhangi, M.; Es-Haghi, A.; Taghavizadeh Yazdi, M.E.; Rahdar, A.; Baino, F. MOF-Mediated Synthesis of CuO/CeO_2 Composite Nanoparticles: Characterization and Estimation of the Cellular Toxicity against Breast Cancer Cell Line (MCF-7). *J. Funct. Biomater.* **2021**, *12*, 53. [CrossRef]

18. Arshad, R.; Tabish, T.A.; Kiani, M.H.; Ibrahim, I.M.; Shahnaz, G.; Rahdar, A.; Kang, M.; Pandey, S. A Hyaluronic Acid Functionalized Self-Nano-Emulsifying Drug Delivery System (SNEDDS) for Enhancement in Ciprofloxacin Targeted Delivery against Intracellular Infection. *Nanomaterials* **2021**, *11*, 1086. [CrossRef]
19. Bilal, M.; Qindeel, M.; Raza, A.; Mehmood, S.; Rahdar, A. Stimuli-responsive nanoliposomes as prospective nanocarriers for targeted drug delivery. *J. Drug Deliv. Sci. Technol.* **2021**, 102916. [CrossRef]
20. Rahdar, A.; Taboada, P.; Aliahmad, M.; Hajinezhad, M.R.; Sadeghfar, F. Iron oxide nanoparticles: Synthesis, physical characterization, and intraperitoneal biochemical studies in Rattus norvegicus. *J. Mol. Struct.* **2018**, *1173*, 240–245. [CrossRef]
21. Sivasankarapillai, V.S.; Somakumar, A.K.; Joseph, J.; Nikazar, S.; Rahdar, A.; Kyzas, G.Z. Cancer theranostic applications of MXene nanomaterials: Recent updates. *Nano-Struct. Nano-Objects* **2020**, *22*, 100457. [CrossRef]
22. Bertrand, N.; Wu, J.; Xu, X.; Kamaly, N.; Farokhzad, O.C. Cancer nanotechnology: The impact of passive and active targeting in the era of modern cancer biology. *Adv. Drug Deliv. Rev.* **2014**, *66*, 2–25. [CrossRef] [PubMed]
23. Paris, J.L.; Vallet-Regí, M. Mesoporous Silica Nanoparticles for Co-Delivery of Drugs and Nucleic Acids in Oncology: A Review. *Pharmaceutics* **2020**, *12*, 526. [CrossRef]
24. Chaturvedi, V.K.; Singh, A.; Singh, V.K.; Singh, M.P. Cancer nanotechnology: A new revolution for cancer diagnosis and therapy. *Curr. Drug Metab.* **2019**, *20*, 416–429. [CrossRef] [PubMed]
25. Eivazzadeh-Keihan, R.; Pashazadeh-Panahi, P.; Baradaran, B.; Maleki, A.; Hejazi, M.; Mokhtarzadeh, A.; de la Guardia, M. Recent advances on nanomaterial based electrochemical and optical aptasensors for detection of cancer biomarkers. *TrAC Trends Anal. Chem.* **2018**, *100*, 103–115. [CrossRef]
26. Datta, L.P.; Manchineella, S.; Govindaraju, T. Biomolecules-derived biomaterials. *Biomaterials* **2020**, *230*, 119633. [CrossRef]
27. Wang, J.; Li, Y.; Nie, G. Multifunctional biomolecule nanostructures for cancer therapy. *Nat. Rev. Mater.* **2021**, *19*, 1–18. [CrossRef]
28. Guo, P. The emerging field of RNA nanotechnology. *Nat. Nanotechnol.* **2010**, *5*, 833–842. [CrossRef]
29. Mansouri, S.; Cuie, Y.; Winnik, F.; Shi, Q.; Lavigne, P.; Benderdour, M.; Beaumont, E.; Fernandes, J.C. Characterization of folate-chitosan-DNA nanoparticles for gene therapy. *Biomaterials* **2006**, *27*, 2060–2065. [CrossRef]
30. Mao, H.-Q.; Roy, K.; Troung-Le, V.L.; Janes, K.A.; Lin, K.Y.; Wang, Y.; August, J.T.; Leong, K.W. Chitosan-DNA nanoparticles as gene carriers: Synthesis, characterization and transfection efficiency. *J. Control. Release* **2001**, *70*, 399–421. [CrossRef]
31. Liu, G.; Li, D.; Pasumarthy, M.K.; Kowalczyk, T.H.; Gedeon, C.R.; Hyatt, S.L.; Payne, J.M.; Miller, T.J.; Brunovskis, P.; Fink, T.L. Nanoparticles of compacted DNA transfect postmitotic cells. *J. Biol. Chem.* **2003**, *278*, 32578–32586. [CrossRef]
32. Mastorakos, P.; Da Silva, A.L.; Chisholm, J.; Song, E.; Choi, W.K.; Boyle, M.P.; Morales, M.M.; Hanes, J.; Suk, J.S. Highly compacted biodegradable DNA nanoparticles capable of overcoming the mucus barrier for inhaled lung gene therapy. *Proc. Natl. Acad. Sci. USA* **2015**, *112*, 8720–8725. [CrossRef] [PubMed]
33. Da Silva, A.L.; Oliveira, G.P.; Kitoko, J.Z.; Blanco, N.G.; Suk, J.S.; Hanes, J.; Olsen, P.C.; Morales, M.M.; Rocco, P.R. Thymulin Gene Therapy Delivered By A New Biodegradable Dna Nanoparticle In Experimental Chronic Allergic Asthma. In *D21. Asthma Treatment: Glucocorticoids, Biologicals and Beyond*; American Thoracic Society: New York, NY, USA, 2016; p. A6490.
34. Haque, F.; Pi, F.; Zhao, Z.; Gu, S.; Hu, H.; Yu, H.; Guo, P. RNA versatility, flexibility, and thermostability for practice in RNA nanotechnology and biomedical applications. *Wiley Interdiscip. Rev. RNA* **2018**, *9*, e1452. [CrossRef]
35. Guo, P. The Emerging Field of RNA Nanotechnology. In *Nano-Enabled Medical Applications*; Jenny Stanford Publishing: Dubai, United Arab Emirates, 2020; pp. 131–157.
36. Arshad, R.; Barani, M.; Rahdar, A.; Sargazi, S.; Cucchiarini, M.; Pandey, S.; Kang, M. Multi-Functionalized Nanomaterials and Nanoparticles for Diagnosis and Treatment of Retinoblastoma. *Biosensors* **2021**, *11*, 97. [CrossRef] [PubMed]
37. Barani, M.; Sargazi, S.; Mohammadzadeh, V.; Rahdar, A.; Pandey, S.; Jha, N.K.; Gupta, P.K.; Thakur, V.K. Theranostic Advances of Bionanomaterials against Gestational Diabetes Mellitus: A Preliminary Review. *J. Funct. Biomater.* **2021**, *12*, 54. [CrossRef] [PubMed]
38. Barani, M.; Zeeshan, M.; Kalantar-Neyestanaki, D.; Farooq, M.A.; Rahdar, A.; Jha, N.K.; Sargazi, S.; Gupta, P.K.; Thakur, V.K. Nanomaterials in the Management of Gram-Negative Bacterial Infections. *Nanomaterials* **2021**, *11*, 2535. [CrossRef] [PubMed]
39. Fiume, E.; Magnaterra, G.; Rahdar, A.; Verné, E.; Baino, F. Hydroxyapatite for Biomedical Applications: A Short Overview. *Ceramics* **2021**, *4*, 542–563. [CrossRef]
40. Hashemi, B.; Firouzi-amandi, A.; Amirazad, H.; Dadashpour, M.; Shirvaliloo, M.; Nasrabadi, D.; Ahmadi, M.; Sheervalilou, R.; Reza, M.A.S.; Ghazi, F. Emerging importance of nanotechnology-based approaches to control the COVID-19 pandemic; focus on nanomedicine iterance in diagnosis and treatment of COVID-19 patients. *J. Drug Deliv. Sci. Technol.* **2021**, 102967. [CrossRef] [PubMed]
41. Hassanisaadi, M.; Bonjar, G.H.S.; Rahdar, A.; Pandey, S.; Hosseinipour, A.; Abdolshahi, R. Environmentally Safe Biosynthesis of Gold Nanoparticles Using Plant Water Extracts. *Nanomaterials* **2021**, *11*, 2033. [CrossRef]
42. Mukhtar, M.; Sargazi, S.; Barani, M.; Madry, H.; Rahdar, A.; Cucchiarini, M. Application of Nanotechnology for Sensitive Detection of Low-Abundance Single-Nucleotide Variations in Genomic DNA: A Review. *Nanomaterials* **2021**, *11*, 1384. [CrossRef] [PubMed]
43. Norouzi, M.; Yasamineh, S.; Montazeri, M.; Dadashpour, M.; Sheervalilou, R.; Abasi, M.; Pilehvar-Soltanahmadi, Y. Recent advances on nanomaterials-based fluorimetric approaches for microRNAs detection. *Mater. Sci. Eng. C* **2019**, *104*, 110007. [CrossRef]

44. Rahdar, A.; Hajinezhad, M.R.; Bilal, M.; Askari, F.; Kyzas, G.Z. Behavioral effects of zinc oxide nanoparticles on the brain of rats. *Inorg. Chem. Commun.* **2020**, *119*, 108131. [CrossRef]
45. Sargazi, S.; Mukhtar, M.; Rahdar, A.; Barani, M.; Pandey, S.; Díez-Pascual, A.M. Active targeted nanoparticles for delivery of poly (ADP-ribose) polymerase (PARP) inhibitors: A preliminary review. *Int. J. Mol. Sci.* **2021**, *22*, 10319. [CrossRef] [PubMed]
46. Shirvalilou, S.; Khoei, S.; Esfahani, A.J.; Kamali, M.; Shirvaliloo, M.; Sheervalilou, R.; Mirzaghavami, P. Magnetic Hyperthermia as an adjuvant cancer therapy in combination with radiotherapy versus radiotherapy alone for recurrent/progressive glioblastoma: A systematic review. *J. Neuro-Oncol.* **2021**, *13*, 1–10. [CrossRef] [PubMed]
47. Zhang, F.; Nangreave, J.; Liu, Y.; Yan, H. Structural DNA nanotechnology: State of the art and future perspective. *J. Am. Chem. Soc.* **2014**, *136*, 11198–11211. [CrossRef]
48. Ohno, H.; Akamine, S.; Saito, H. RNA nanostructures and scaffolds for biotechnology applications. *Curr. Opin. Biotechnol.* **2019**, *58*, 53–61. [CrossRef]
49. Jasinski, D.; Haque, F.; Binzel, D.W.; Guo, P. Advancement of the emerging field of RNA nanotechnology. *ACS Nano* **2017**, *11*, 1142–1164. [CrossRef]
50. Shu, Y.; Pi, F.; Sharma, A.; Rajabi, M.; Haque, F.; Shu, D.; Leggas, M.; Evers, B.M.; Guo, P. Stable RNA nanoparticles as potential new generation drugs for cancer therapy. *Adv. Drug Deliv. Rev.* **2014**, *66*, 74–89. [CrossRef]
51. Guo, S.; Xu, C.; Yin, H.; Hill, J.; Pi, F.; Guo, P. Tuning the size, shape and structure of RNA nanoparticles for favorable cancer targeting and immunostimulation. *Wiley Interdiscip. Rev. Nanomed. Nanobiotechnol.* **2020**, *12*, e1582. [CrossRef] [PubMed]
52. Li, M.; Zheng, M.; Wu, S.; Tian, C.; Liu, D.; Weizmann, Y.; Jiang, W.; Wang, G.; Mao, C. In vivo production of RNA nanostructures via programmed folding of single-stranded RNAs. *Nat. Commun.* **2018**, *9*, 1–9. [CrossRef]
53. Farokhzad, O.C.; Jon, S.; Khademhosseini, A.; Tran, T.-N.T.; LaVan, D.A.; Langer, R. Nanoparticle-aptamer bioconjugates: A new approach for targeting prostate cancer cells. *Cancer Res.* **2004**, *64*, 7668–7672. [CrossRef]
54. Germer, K.; Leonard, M.; Zhang, X. RNA aptamers and their therapeutic and diagnostic applications. *Int. J. Biochem. Mol. Biol.* **2013**, *4*, 27.
55. Shu, Y.; Shu, D.; Diao, Z.; Shen, G.; Guo, P. Fabrication of polyvalent therapeutic RNA nanoparticles for specific delivery of siRNA, ribozyme and drugs to targeted cells for cancer therapy. In Proceedings of the 2009 IEEE/NIH Life Science Systems and Applications Workshop, Bethesda, MD, USA, 8–10 April 2009; pp. 9–12.
56. Yin, H.; Xiong, G.; Guo, S.; Xu, C.; Xu, R.; Guo, P.; Shu, D. Delivery of anti-miRNA for triple-negative breast cancer therapy using RNA nanoparticles targeting stem cell marker CD133. *Mol. Ther.* **2019**, *27*, 1252–1261. [CrossRef] [PubMed]
57. Ghimire, C.; Wang, H.; Li, H.; Vieweger, M.; Xu, C.; Guo, P. RNA nanoparticles as rubber for compelling vessel extravasation to enhance tumor targeting and for fast renal excretion to reduce toxicity. *ACS Nano* **2020**, *14*, 13180–13191. [CrossRef]
58. Wang, H.; Guo, P. Radiolabeled RNA Nanoparticles for Highly Specific Targeting and Efficient Tumor Accumulation with Favorable In Vivo Biodistribution. *Mol. Pharm.* **2021**, *18*, 2924–2934. [CrossRef] [PubMed]
59. Kim, T.; Hyun, H.N.; Heo, R.; Nam, K.; Yang, K.; Kim, Y.M.; Lee, Y.S.; An, J.Y.; Park, J.H.; Choi, K.Y. Dual-targeting RNA nanoparticles for efficient delivery of polymeric siRNA to cancer cells. *Chem. Commun.* **2020**, *56*, 6624–6627. [CrossRef]
60. Xu, Y.; Pang, L.; Wang, H.; Xu, C.; Shah, H.; Guo, P.; Shu, D.; Qian, S.Y. Specific delivery of delta-5-desaturase siRNA via RNA nanoparticles supplemented with dihomo-γ-linolenic acid for colon cancer suppression. *Redox Biol.* **2019**, *21*, 101085. [CrossRef]
61. Haque, F.; Shu, D.; Shu, Y.; Shlyakhtenko, L.S.; Rychahou, P.G.; Evers, B.M.; Guo, P. Ultrastable synergistic tetravalent RNA nanoparticles for targeting to cancers. *Nano Today* **2012**, *7*, 245–257. [CrossRef] [PubMed]
62. Metkar, S.K.; Girigoswami, K. Diagnostic biosensors in medicine—A review. *Biocatal. Agric. Biotechnol.* **2019**, *17*, 271–283. [CrossRef]
63. El Harrad, L.; Bourais, I.; Mohammadi, H.; Amine, A. Recent advances in electrochemical biosensors based on enzyme inhibition for clinical and pharmaceutical applications. *Sensors* **2018**, *18*, 164. [CrossRef]
64. Bagherzade, G.; Tavakoli, M.M.; Namaei, M.H. Green synthesis of silver nanoparticles using aqueous extract of saffron (*Crocus sativus* L.) wastages and its antibacterial activity against six bacteria. *Asian Pac. J. Trop. Biomed.* **2017**, *7*, 227–233. [CrossRef]
65. Usman, A.I.; Aziz, A.A.; Noqta, O.A. Application of green synthesis of gold nanoparticles: A review. *J. Teknol.* **2019**, *81*. [CrossRef]
66. Shanaa, O.A.; Rumyantsev, A.; Sambuk, E.; Padkina, M. In Vivo Production of RNA Aptamers and Nanoparticles: Problems and Prospects. *Molecules* **2021**, *26*, 1422. [CrossRef] [PubMed]
67. Yin, H.; Wang, H.; Li, Z.; Shu, D.; Guo, P. RNA micelles for the systemic delivery of anti-miRNA for cancer targeting and inhibition without ligand. *ACS Nano* **2018**, *13*, 706–717. [CrossRef]
68. Zhang, Y.; Leonard, M.; Shu, Y.; Yang, Y.; Shu, D.; Guo, P.; Zhang, X. Overcoming tamoxifen resistance of human breast cancer by targeted gene silencing using multifunctional pRNA nanoparticles. *ACS Nano* **2017**, *11*, 335–346. [CrossRef]
69. Xu, C.; Haque, F.; Jasinski, D.L.; Binzel, D.W.; Shu, D.; Guo, P. Favorable biodistribution, specific targeting and conditional endosomal escape of RNA nanoparticles in cancer therapy. *Cancer Lett.* **2018**, *414*, 57–70. [CrossRef]
70. Shi, J.; Kantoff, P.W.; Wooster, R.; Farokhzad, O.C. Cancer nanomedicine: Progress, challenges and opportunities. *Nat. Rev. Cancer* **2017**, *17*, 20–37. [CrossRef] [PubMed]
71. Guo, S.; Li, H.; Ma, M.; Fu, J.; Dong, Y.; Guo, P. Size, shape, and sequence-dependent immunogenicity of RNA nanoparticles. *Mol. Ther. Nucleic Acids* **2017**, *9*, 399–408. [CrossRef]
72. Salvati, E.; Stellacci, F.; Krol, S. Nanosensors for early cancer detection and for therapeutic drug monitoring. *Nanomedicine* **2015**, *10*, 3495–3512. [CrossRef]

73. Rossetti, M.; Del Grosso, E.; Ranallo, S.; Mariottini, D.; Idili, A.; Bertucci, A.; Porchetta, A. Programmable RNA-based systems for sensing and diagnostic applications. *Anal. Bioanal. Chem.* **2019**, *411*, 4293–4302. [CrossRef] [PubMed]
74. Chandrasekaran, A.R. DNA nanobiosensors: An outlook on signal readout strategies. *J. Nanomater.* **2017**, *2017*, 2820619. [CrossRef]
75. Shandilya, R.; Bhargava, A.; Bunkar, N.; Tiwari, R.; Goryacheva, I.Y.; Mishra, P.K. Nanobiosensors: Point-of-care approaches for cancer diagnostics. *Biosens. Bioelectron.* **2019**, *130*, 147–165. [CrossRef] [PubMed]
76. Bartosik, K.; Debiec, K.; Czarnecka, A.; Sochacka, E.; Leszczynska, G. Synthesis of Nucleobase-Modified RNA Oligonucleotides by Post-Synthetic Approach. *Molecules* **2020**, *25*, 3344. [CrossRef] [PubMed]
77. Binzel, D.W.; Li, X.; Burns, N.; Khan, E.; Lee, W.-J.; Chen, L.-C.; Ellipilli, S.; Miles, W.; Ho, Y.S.; Guo, P. Thermostability, Tunability, and Tenacity of RNA as Rubbery Anionic Polymeric Materials in Nanotechnology and Nanomedicine—Specific Cancer Targeting with Undetectable Toxicity. *Chem. Rev.* **2021**, *121*, 1322–1335. [CrossRef] [PubMed]
78. Fernandez-Piñeiro, I.; Badiola, I.; Sanchez, A. Nanocarriers for microRNA delivery in cancer medicine. *Biotechnol. Adv.* **2017**, *35*, 350–360. [CrossRef] [PubMed]
79. Tommasini-Ghelfi, S.; Lee, A.; Mirkin, C.A.; Stegh, A.H. Synthesis, physicochemical, and biological evaluation of spherical nucleic acids for RNAi-based therapy in glioblastoma. In *RNA Interference and Cancer Therapy*; Springer: Berlin/Heidelberg, Germany, 2019; pp. 371–391.
80. Hasan, M.R.; Hassan, N.; Khan, R.; Kim, Y.-T.; Iqbal, S.M. Classification of cancer cells using computational analysis of dynamic morphology. *Comput. Methods Programs Biomed.* **2018**, *156*, 105–112. [CrossRef]
81. Nuzzo, S.; Brancato, V.; Affinito, A.; Salvatore, M.; Cavaliere, C.; Condorelli, G. The role of RNA and dna aptamers in glioblastoma diagnosis and therapy: A systematic review of the literature. *Cancers* **2020**, *12*, 2173. [CrossRef]
82. Fechter, P.; Da Silva, E.C.; Mercier, M.-C.; Noulet, F.; Etienne-Seloum, N.; Guenot, D.; Lehmann, M.; Vauchelles, R.; Martin, S.; Lelong-Rebel, I. RNA Aptamers Targeting Integrin $\alpha5\beta1$ as Probes for Cyto-and Histofluorescence in Glioblastoma. *Mol. Ther. Nucleic Acids* **2019**, *17*, 63–77. [CrossRef]
83. Affinito, A.; Quintavalle, C.; Esposito, C.L.; Roscigno, G.; Vilardo, C.; Nuzzo, S.; Ricci-Vitiani, L.; De Luca, G.; Pallini, R.; Kichkailo, A.S. The discovery of RNA aptamers that selectively bind glioblastoma stem cells. *Mol. Ther. Nucleic Acids* **2019**, *18*, 99–109. [CrossRef]
84. Hadjidemetriou, M.; Kostarelos, K. Evolution of the nanoparticle corona. *Nat. Nanotechnol.* **2017**, *12*, 288–290. [CrossRef]
85. Ganju, A.; Khan, S.; Hafeez, B.B.; Behrman, S.W.; Yallapu, M.M.; Chauhan, S.C.; Jaggi, M. miRNA nanotherapeutics for cancer. *Drug Discov. Today* **2017**, *22*, 424–432. [CrossRef] [PubMed]
86. Huang, X.; Liu, Y.; Yung, B.; Xiong, Y.; Chen, X. Nanotechnology-enhanced no-wash biosensors for in vitro diagnostics of cancer. *ACS Nano* **2017**, *11*, 5238–5292. [CrossRef]
87. Huang, X.; O'Connor, R.; Kwizera, E.A. Gold nanoparticle based platforms for circulating cancer marker detection. *Nanotheranostics* **2017**, *1*, 80. [CrossRef]
88. Matsuda, M.; Ishikawa, E.; Yamamoto, T.; Hatano, K.; Joraku, A.; Iizumi, Y.; Masuda, Y.; Nishiyama, H.; Matsumura, A. Potential use of prostate specific membrane antigen (PSMA) for detecting the tumor neovasculature of brain tumors by PET imaging with 89 Zr-Df-IAB2M anti-PSMA minibody. *J. Neuro-Oncol.* **2018**, *138*, 581–589. [CrossRef]
89. Tanjore Ramanathan, J.; Lehtipuro, S.; Sihto, H.; Tóvári, J.; Reiniger, L.; Téglási, V.; Moldvay, J.; Nykter, M.; Haapasalo, H.; Le Joncour, V. Prostate-specific membrane antigen expression in the vasculature of primary lung carcinomas associates with faster metastatic dissemination to the brain. *J. Cell. Mol. Med.* **2020**, *24*, 6916–6927. [CrossRef]
90. Engur, C.O.; Turoglu, H.T.; Ozguven, S.; Tanidir, Y.; Erdil, T.Y. 68Ga-Prostate-Specific Membrane Antigen PET-Positive Paget Bone Disease With Metastatic Prostatic Carcinoma. *Clin. Nucl. Med.* **2020**, *45*, e425–e426. [CrossRef]
91. Anwer, K.; Meaney, C.; Kao, G.; Hussain, N.; Shelvin, R.; Earls, R.M.; Leonard, P.; Quezada, A.; Rolland, A.P.; Sullivan, S.M. Cationic lipid-based delivery system for systemic cancer gene therapy. *Cancer Gene Ther.* **2000**, *7*, 1156–1164. [CrossRef]
92. Chaudhary, V.; Jangra, S.; Yadav, N.R. Nanotechnology based approaches for detection and delivery of microRNA in healthcare and crop protection. *J. Nanobiotechnol.* **2018**, *16*, 1–18. [CrossRef]
93. Li, R.; Liu, B.; Gao, J. The application of nanoparticles in diagnosis and theranostics of gastric cancer. *Cancer Lett.* **2017**, *386*, 123–130. [CrossRef] [PubMed]
94. Amorim, S.; Reis, C.A.; Reis, R.L.; Pires, R.A. Extracellular matrix mimics using hyaluronan-based biomaterials. *Trends Biotechnol.* **2021**, *39*, 90–104. [CrossRef] [PubMed]
95. Jaakonmäki, R.; Simons, A.; Müller, O.; vom Brocke, J. ECM implementations in practice: Objectives, processes, and technologies. *J. Enterp. Inf. Manag.* **2018**, *5*, 704–723. [CrossRef]
96. Marchandet, L.; Lallier, M.; Charrier, C.; Baud'huin, M.; Ory, B.; Lamoureux, F. Mechanisms of Resistance to Conventional Therapies for Osteosarcoma. *Cancers* **2021**, *13*, 683. [CrossRef] [PubMed]
97. Arnett, A.B.; Wang, T.; Eichler, E.E.; Bernier, R.A. Reflections on the genetics-first approach to advancements in molecular genetic and neurobiological research on neurodevelopmental disorders. *J. Neurodev. Disord.* **2021**, *13*, 1–10. [CrossRef]
98. Falese, J.P.; Donlic, A.; Hargrove, A.E. Targeting RNA with small molecules: From fundamental principles towards the clinic. *Chem. Soc. Rev.* **2021**, *50*, 2224–2243. [CrossRef]
99. Saraswathy, M.; Gong, S. Recent developments in the co-delivery of siRNA and small molecule anticancer drugs for cancer treatment. *Mater. Today* **2014**, *17*, 298–306. [CrossRef]

100. Bajan, S.; Hutvagner, G. RNA-based therapeutics: From antisense oligonucleotides to miRNAs. *Cells* **2020**, *9*, 137. [CrossRef]
101. Elcheva, I.A.; Spiegelman, V.S. Targeting RNA-binding proteins in acute and chronic leukemia. *Leukemia* **2021**, *35*, 360–376. [CrossRef]
102. Guo, S.; Vieweger, M.; Zhang, K.; Yin, H.; Wang, H.; Li, X.; Li, S.; Hu, S.; Sparreboom, A.; Evers, B.M.; et al. Ultra-thermostable RNA nanoparticles for solubilizing and high-yield loading of paclitaxel for breast cancer therapy. *Nat. Commun.* **2020**, *11*, 972. [CrossRef]
103. Xia, Y.; Tang, G.; Wang, C.; Zhong, J.; Chen, Y.; Hua, L.; Li, Y.; Liu, H.; Zhu, B. Functionalized selenium nanoparticles for targeted siRNA delivery silence Derlin1 and promote antitumor efficacy against cervical cancer. *Drug Deliv.* **2020**, *27*, 15–25. [CrossRef]
104. Li, Y.; Duo, Y.; Bi, J.; Zeng, X.; Mei, L.; Bao, S.; He, L.; Shan, A.; Zhang, Y.; Yu, X. Targeted delivery of anti-miR-155 by functionalized mesoporous silica nanoparticles for colorectal cancer therapy. *Int. J. Nanomed.* **2018**, *13*, 1241. [CrossRef]
105. Wu, Z.; Xu, X.-L.; Zhang, J.-Z.; Mao, X.-H.; Xie, M.-W.; Cheng, Z.-L.; Lu, L.-J.; Duan, X.-H.; Zhang, L.-M.; Shen, J. Magnetic cationic amylose nanoparticles used to deliver survivin-small interfering RNA for gene therapy of hepatocellular carcinoma in vitro. *Nanomaterials* **2017**, *7*, 110. [CrossRef] [PubMed]
106. Xu, X.; Wu, J.; Liu, Y.; Saw, P.E.; Tao, W.; Yu, M.; Zope, H.; Si, M.; Victorious, A.; Rasmussen, J. Multifunctional envelope-type siRNA delivery nanoparticle platform for prostate cancer therapy. *ACS Nano* **2017**, *11*, 2618–2627. [CrossRef]
107. Luo, X.; Peng, X.; Hou, J.; Wu, S.; Shen, J.; Wang, L. Folic acid-functionalized polyethylenimine superparamagnetic iron oxide nanoparticles as theranostic agents for magnetic resonance imaging and PD-L1 siRNA delivery for gastric cancer. *Int. J. Nanomed.* **2017**, *12*, 5331. [CrossRef] [PubMed]
108. Lio, D.C.S.; Liu, C.; Oo, M.M.S.; Wiraja, C.; Teo, M.H.Y.; Zheng, M.; Chew, S.W.T.; Wang, X.; Xu, C. Transdermal delivery of small interfering RNAs with topically applied mesoporous silica nanoparticles for facile skin cancer treatment. *Nanoscale* **2019**, *11*, 17041–17051. [CrossRef]
109. Batool, A.; Arshad, R.; Razzaq, S.; Nousheen, K.; Kiani, M.H.; Shahnaz, G. Formulation and evaluation of hyaluronic acid-based mucoadhesive self nanoemulsifying drug delivery system (SNEDDS) of tamoxifen for targeting breast cancer. *Int. J. Biol. Macromol.* **2020**, *152*, 503–515. [CrossRef] [PubMed]
110. Marta, G.; Ramiah, D.; Kaidar-Person, O.; Kirby, A.; Coles, C.; Jagsi, R.; Hijal, T.; Sancho, G.; Zissiadis, Y.; Pignol, J.-P. The financial impact on reimbursement of moderately hypofractionated postoperative radiation therapy for breast cancer: An international consortium report. *Clin. Oncol.* **2021**, *33*, 322–330. [CrossRef]
111. Sato, Y.; Hashiba, K.; Sasaki, K.; Maeki, M.; Tokeshi, M.; Harashima, H. Understanding structure-activity relationships of pH-sensitive cationic lipids facilitates the rational identification of promising lipid nanoparticles for delivering siRNAs in vivo. *J. Control. Release* **2019**, *295*, 140–152. [CrossRef]
112. Lin, Y.-X.; Wang, Y.; Blake, S.; Yu, M.; Mei, L.; Wang, H.; Shi, J. RNA nanotechnology-mediated cancer immunotherapy. *Theranostics* **2020**, *10*, 281. [CrossRef]
113. Jasinski, D.L.; Yin, H.; Li, Z.; Guo, P. Hydrophobic effect from conjugated chemicals or drugs on in vivo biodistribution of RNA nanoparticles. *Hum. Gene Ther.* **2018**, *29*, 77–86. [CrossRef]
114. Gangopadhyay, S.; Nikam, R.R.; Gore, K.R. Folate Receptor-Mediated Small Interfering RNA Delivery: Recent Developments and Future Directions for RNA Interference Therapeutics. *Nucleic Acid Ther.* **2021**, *31*, 245–270. [CrossRef]
115. Mendonça, M.C.; Kont, A.; Aburto, M.R.; Cryan, J.F.; O'Driscoll, C.M. Advances in the Design of (Nano) Formulations for Delivery of Antisense Oligonucleotides and Small Interfering RNA: Focus on the Central Nervous System. *Mol. Pharm.* **2021**, *18*, 1491–1506. [CrossRef] [PubMed]
116. Arbyn, M.; Gultekin, M.; Morice, P.; Nieminen, P.; Cruickshank, M.; Poortmans, P.; Kelly, D.; Poljak, M.; Bergeron, C.; Ritchie, D. The European response to the WHO call to eliminate cervical cancer as a public health problem. *Int. J. Cancer* **2021**, *148*, 277–284. [CrossRef]
117. Barani, M.; Sabir, F.; Rahdar, A.; Arshad, R.; Kyzas, G.Z. Nanotreatment and Nanodiagnosis of Prostate Cancer: Recent Updates. *Nanomaterials* **2020**, *10*, 1696. [CrossRef]
118. Chiang, T.-H.; Chang, W.-J.; Chen, S.L.-S.; Yen, A.M.-F.; Fann, J.C.-Y.; Chiu, S.Y.-H.; Chen, Y.-R.; Chuang, S.-L.; Shieh, C.-F.; Liu, C.-Y. Mass eradication of Helicobacter pylori to reduce gastric cancer incidence and mortality: A long-term cohort study on Matsu Islands. *Gut* **2021**, *70*, 243–250.
119. Lei, M.; Siemers, N.O.; Pandya, D.; Chang, H.; Sanchez, T.; Harbison, C.; Szabo, P.M.; Janjigian, Y.; Ott, P.A.; Sharma, P. Analyses of PD-L1 and Inflammatory Gene Expression Association with Efficacy of Nivolumab±Ipilimumab in Gastric Cancer/Gastroesophageal Junction Cancer. *Clin. Cancer Res.* **2021**, *27*. [CrossRef]
120. Wu, L.; Cai, S.; Deng, Y.; Zhang, Z.; Zhou, X.; Su, Y.; Xu, D. PD-1/PD-L1 enhanced cisplatin resistance in gastric cancer through PI3K/AKT mediated P-gp expression. *Int. Immunopharmacol.* **2021**, *94*, 107443. [CrossRef]
121. Llovet, J.M.; Kelley, R.K.; Villanueva, A.; Singal, A.G.; Pikarsky, E.; Roayaie, S.; Lencioni, R.; Koike, K.; Zucman-Rossi, J.; Finn, R.S. Hepatocellular carcinoma. *Nat. Rev. Dis. Primers* **2021**, *7*, 6. [CrossRef] [PubMed]
122. Chew, X.H.; Sultana, R.; Mathew, E.N.; Ng, D.C.E.; Lo, R.H.; Toh, H.C.; Tai, D.; Choo, S.P.; Goh, B.K.P.; Yan, S.X. Real-World Data on Clinical Outcomes of Patients with Liver Cancer: A Prospective Validation of the National Cancer Centre Singapore Consensus Guidelines for the Management of Hepatocellular Carcinoma. *Liver Cancer* **2021**, *7*, 1–16. [CrossRef]
123. uz Zaman, S.; Arshad, R.; Tabish, T.A.; Naseem, A.A.; Shahnaz, G. Mapping the potential of thiolated pluronic based nanomicelles for the safe and targeted delivery of vancomycin against staphylococcal blepharitis. *J. Drug Deliv. Sci. Technol.* **2021**, *61*, 102220.

124. Nguyen, T.V.; Damiani, G.; Orenstein, L.A.; Hamzavi, I.; Jemec, G. Hidradenitis suppurativa: An update on epidemiology, phenotypes, diagnosis, pathogenesis, comorbidities and quality of life. *J. Eur. Acad. Dermatol. Venereol.* **2021**, *35*, 50–61. [CrossRef]
125. Arshad, R.; Tabish, T.A.; Naseem, A.A.; Hassan, M.R.u.; Hussain, I.; Hussain, S.S.; Shahnaz, G. Development of poly-L-lysine multi-functionalized muco-penetrating self- emulsifying drug delivery system (SEDDS) for improved solubilization and targeted delivery of ciprofloxacin against intracellular Salmonella typhi. *J. Mol. Liq.* **2021**, *333*, 115972. [CrossRef]
126. Olivieri, F.; Castaldo, R.; Cocca, M.; Gentile, G.; Lavorgna, M. Mesoporous silica nanoparticles as carriers of active agents for smart anticorrosion organic coatings. A critical review. *Nanoscale* **2021**, *13*, 9091–9111. [CrossRef] [PubMed]
127. López, V.; Villegas, M.R.; Rodríguez, V.; Villaverde, G.; Lozano, D.; Baeza, A.; Vallet-Regí, M. Janus mesoporous silica nanoparticles for dual targeting of tumor cells and mitochondria. *ACS Appl. Mater. Interfaces* **2017**, *9*, 26697–26706. [CrossRef]
128. Arshad, R.; Pal, K.; Sabir, F.; Rahdar, A.; Bilal, M.; Shahnaz, G.; Kyzas, G.Z. A review of the nanomaterials use for the diagnosis and therapy of salmonella typhi. *J. Mol. Struct.* **2021**, *1230*, 129928. [CrossRef]
129. Aagaard, L.; Rossi, J.J. RNAi therapeutics: Principles, prospects and challenges. *Adv. Drug Deliv. Rev.* **2007**, *59*, 75–86. [CrossRef]
130. Pecot, C.V.; Calin, G.A.; Coleman, R.L.; Lopez-Berestein, G.; Sood, A.K. RNA interference in the clinic: Challenges and future directions. *Nat. Rev. Cancer* **2011**, *11*, 59–67. [CrossRef]
131. Mura, S.; Nicolas, J.; Couvreur, P. Stimuli-responsive nanocarriers for drug delivery. *Nat. Mater.* **2013**, *12*, 991–1003. [CrossRef]
132. Xiong, Q.; Lee, G.Y.; Ding, J.; Li, W.; Shi, J. Biomedical applications of mRNA nanomedicine. *Nano Res.* **2018**, *11*, 5281–5309. [CrossRef]
133. Guo, P.; Haque, F.; Hallahan, B.; Reif, R.; Li, H. Uniqueness, advantages, challenges, solutions, and perspectives in therapeutics applying RNA nanotechnology. *Nucleic Acid Ther.* **2012**, *22*, 226–245. [CrossRef]
134. Abdelmawla, S.; Guo, S.; Zhang, L.; Pulukuri, S.M.; Patankar, P.; Conley, P.; Trebley, J.; Guo, P.; Li, Q.-X. Pharmacological characterization of chemically synthesized monomeric phi29 pRNA nanoparticles for systemic delivery. *Mol. Ther.* **2011**, *19*, 1312–1322. [CrossRef] [PubMed]
135. de Fougerolles, A.; Vornlocher, H.-P.; Maraganore, J.; Lieberman, J. Interfering with disease: A progress report on siRNA-based therapeutics. *Nat. Rev. Drug Discov.* **2007**, *6*, 443–453. [CrossRef]
136. Kim, J.; Eygeris, Y.; Gupta, M.; Sahay, G. Self-assembled mRNA vaccines. *Adv. Drug Deliv. Rev.* **2021**, *170*, 83–112. [CrossRef] [PubMed]
137. Zhong, S.; Breton, B.; Zheng, W.; McFadyen, I.; Hopson, K.; Frederick, J.; Meehan, R.S.; Zaks, T. Bioinformatics algorithm of mRNA-4157 identifies neoantigens with pre-existing TIL reactivities in colorectal tumors. In Proceedings of the AACR Annual Meeting, Philadelphia, PA, USA, 22–24 June 2020.
138. Paquet-Fifield, S.; Koh, S.L.; Cheng, L.; Beyit, L.M.; Shembrey, C.; Mølck, C.; Behrenbruch, C.; Papin, M.; Gironella, M.; Guelfi, S. Tight junction protein claudin-2 promotes self-renewal of human colorectal cancer stem-like cells. *Cancer Res.* **2018**, *78*, 2925–2938. [CrossRef] [PubMed]
139. Kowalski, P.S.; Rudra, A.; Miao, L.; Anderson, D.G. Delivering the messenger: Advances in technologies for therapeutic mRNA delivery. *Mol. Ther.* **2019**, *27*, 710–728. [CrossRef]
140. Lim, S.A.; Cox, A.; Tung, M.; Chung, E.J. Clinical progress of nanomedicine-based RNA therapies. *Bioact. Mater.* **2021**, in press. [CrossRef]

nanomaterials

MDPI

Review

Advances in Nanomaterials Used in Co-Delivery of siRNA and Small Molecule Drugs for Cancer Treatment

Shei Li Chung [1,2], Maxine Swee-Li Yee [1,*], Ling-Wei Hii [3,4], Wei-Meng Lim [3,4], Mui Yen Ho [5,6], Poi Sim Khiew [1] and Chee-Onn Leong [3,4,*]

1 Nanotechnology Research Group, Faculty of Science and Engineering, University of Nottingham Malaysia Campus, Jalan Broga, Semenyih 43500, Selangor, Malaysia; edxsc1@nottingham.edu.my (S.L.C.); Poisim.Khiew@nottingham.edu.my (P.S.K.)

2 Department of Mechanical, Materials & Manufacturing Engineering, Faculty of Engineering, University of Nottingham Malaysia Campus, Jalan Broga, Semenyih 43500, Selangor, Malaysia

3 Center for Cancer and Stem Cell Research, Institute for Research, Development and Innovation (IRDI), International Medical University, Kuala Lumpur 57000, Malaysia; lingweihii@imu.edu.my (L.-W.H.); weimeng_lim@imu.edu.my (W.-M.L.)

4 School of Pharmacy, International Medical University, Kuala Lumpur 57000, Malaysia

5 Department of Materials Engineering, Faculty of Engineering and Technology, Tunku Abdul Rahman University College, Jalan Genting Kelang, Kuala Lumpur 53300, Malaysia; homy@tarc.edu.my

6 Centre of Advanced Materials, Faculty of Engineering and Technology, Tunku Abdul Rahman University College, Jalan Genting Kelang, Kuala Lumpur 53300, Malaysia

* Correspondence: Maxine.Yee@nottingham.edu.my (M.S.-L.Y.); cheeonn_leong@imu.edu.my (C.-O.L.)

Abstract: Recent advancements in nanotechnology have improved our understanding of cancer treatment and allowed the opportunity to develop novel delivery systems for cancer therapy. The biological complexities of cancer and tumour micro-environments have been shown to be highly challenging when treated with a single therapeutic approach. Current co-delivery systems which involve delivering small molecule drugs and short-interfering RNA (siRNA) have demonstrated the potential of effective suppression of tumour growth. It is worth noting that a considerable number of studies have demonstrated the synergistic effect of co-delivery systems combining siRNA and small molecule drugs, with promising results when compared to single-drug approaches. This review focuses on the recent advances in co-delivery of siRNA and small molecule drugs. The co-delivery systems are categorized based on the material classes of drug carriers. We discuss the critical properties of materials that enable co-delivery of two distinct anti-tumour agents with different properties. Key examples of co-delivery of drug/siRNA from the recent literature are highlighted and discussed. We summarize the current and emerging issues in this rapidly changing field of research in biomaterials for cancer treatments.

Keywords: cancer treatment; drug-siRNA co-delivery systems; multifunctional nanocarrier; siRNA delivery

Citation: Chung, S.L.; Yee, M.S.-L.; Hii, L.-W.; Lim, W.-M.; Ho, M.Y.; Khiew, P.S.; Leong, C.-O. Advances in Nanomaterials Used in Co-Delivery of siRNA and Small Molecule Drugs for Cancer Treatment. *Nanomaterials* **2021**, *11*, 2467. https://doi.org/10.3390/nano11102467

Academic Editors: Fiore P. Nicoletta and Francesca Iemma

Received: 11 August 2021
Accepted: 3 September 2021
Published: 22 September 2021

1. Introduction

Cancer is a large group of dreaded diseases in which abnormal cells proliferate at an uncontrolled rate. Metastasis is the invasion of malignant tumour cells to adjacent parts or other organs [1]. Globally, cancer is the major leading cause of premature deaths. In 2020 alone, 19.3 million new cancer cases were reported worldwide, with over 10 million deaths. A significant rise in cancer incidence of 47% is expected in 2040 [2]. Critically, provision of cancer care is one-half of a two-pronged global effort to reduce the overall burden of cancer [1,2].

Current cancer therapies include radiation, surgery, chemotherapy, targeted therapy or personalized medicine, as well as immunotherapy. Cancer is caused by mutations of genes that lead to activation of proto-oncogenes or inactivation of tumour suppressor genes [3]. Biomarker testing is a profiling of the tumour genetics, which enables treatments which target tumours with genetic changes in particular genes [4].

Chemotherapy drugs are commonly delivered directly through the oral, intravenous, or topical routes as well as through direct injections to the cancer site [4]. However, current chemotherapy targets only a limited subset of tumour-related molecules and pathways such as kinases, and DNA damage. Targeted therapy involves deploying small molecule drugs and monoclonal antibodies which targets the molecules which regulate tumour growth.

Recent advances in drug delivery systems have improved the efficacy of existing chemotherapy drugs by increasing their bioavailability to tumour sites. However, the fact remains that these drugs only target a few tumour-related factors but exclude tumour transcription factors. The majority of the signalling molecules in cancer is regulated by transcription factors (e.g., kRAS, p53, cJUN). Although drug molecules are unable to target these transcription factors, small- or short-interfering ribonucleic acid (siRNA) is able to interfere with their function.

siRNA has proven to be a useful tool to inhibit specific targets within tumour cells [5–8]. However, there are two significant issues with the use of siRNA alone for cancer treatment. Firstly, due to the specificity of siRNA, there is a high chance of tumour cells developing acquired resistance through further mutation. Secondly, as tumours are heterogenous in nature, even with the successful delivery of siRNA to a tumour, this alone does not guarantee the reduction of the tumour [8–10].

A viable solution for cancer treatment, therefore, consists of a combination of chemotherapy drugs and siRNA. A chemotherapy drug targets the bulk of a tumour (based on tumour-related phenotypes such as cell proliferation, DNA repair, etc.), whereas siRNA targets a specific mutation in a tumour. This justifies further development into co-delivery systems integrating small molecule chemotherapy drugs (molecular weight < 500 according to the National Cancer Institute) and siRNA.

Currently, various co-delivery systems are in various developmental stages. They consist of various classes of materials, including liposomes, dendrimers, as well as nanoparticles of polymeric, inorganic and metallic origins [11,12]. This review paper will summarize the desirable properties of these co-delivery systems, followed by a detailed report of different classes of co-delivery systems which host a combination of small molecule drugs and siRNA. A discussion of the advantages and disadvantages of each class of co-delivery system is presented as well. We also discuss current challenges faced in the systemic co-delivery of small molecule drugs-siRNA, and the strategies employed to overcome them. Finally, concluding remarks are provided to comment on the state of the art in this highly evolving and multidisciplinary field.

2. Desirable Properties of Co-Delivery Systems

A co-delivery system is a system which allows the concurrent delivery of more than one therapeutic substance. The therapeutic substances include chemotherapy drugs, deoxyribonucleic acids (DNAs), ribonucleic acids (RNAs), antibodies, etc.

Drug delivery systems enable the manipulation of drug properties, for instance, pharmaco-kinetics, pharmaco-dynamics, therapeutic index, biodistribution and cellular uptake of therapeutic agents. Various chemotherapeutics with poor water solubility, including Paclitaxel, require suitable delivery vehicles with high loading efficiency in order to successfully reach tumour cells [13].

In order to construct an effective and successful drug delivery system, several fundamental prerequisites are required: (i) the vehicles should be biocompatible, biodegradable, non-immunogenic, (ii) high loading capacity of and preservation of guest molecules, (iii) zero premature release before reaching and optimal uptake at the target site, (iv) effective endosomal escape, (v) controllable release rate, and (vi) active targeting of cells and tissues [5,14].

However, the requirements of a co-delivery system extend beyond the criteria listed above, as two agents with different physicochemical properties are incorporated in one drug carrier.

The vehicle for gene-delivery systems has additional criteria: (i) it must be capable of evading reticuloendothelial uptake [6,8], (ii) it must not engage in interaction with vascular endothelial cells and blood components, i.e., possess good stability or persistence in blood [5,6], and (iii) the system must be compact and stable enough to penetrate through the cell membrane and not degrade before reaching the nucleus [6,8,15]. The release mechanism from the vehicle is contingent on the types of oligonucleotides being transported. For instance, antisense oligodeoxyribonucleotides (ODNs) and siRNA should be targeted to be released in the cytosol to inhibit mRNA expression, while delivery of plasmid DNA should reach the nucleus as ectopic DNA [16]. In this review, the focus will be on the co-delivery of small molecule drugs and siRNA.

As cancer belongs to the category of genetic diseases, siRNA-employed cancer treatment was reported to have significant results; however, the inherent characteristics of siRNA have elevated the difficulty of this type of therapy. Nucleic acid has an anionic hydrophilic structure with ~38 to 50 phosphate groups, which renders them impermeable through biological membranes [5,6,8]. Systemic delivered, unmodified naked siRNA is vulnerable to degradation through enzymatic digestion [17]. Besides that, siRNA has a molecular weight of ~13 kDa which is small enough for it to be easily removed through the kidney. However, its molecular weight is comparatively larger than small molecule drugs; this difference eventually affects the co-delivery system in terms of biodistribution and pharmacokinetics [7,8].

3. Classes of Co-Delivery Systems

Various combinations of small molecule drug-siRNA pairs have been reported in the literature to combat the highly heterogeneous and extremely complicated microtumour environment. There are strategies to formulate small molecule drug-siRNA combinations to improve the therapeutic efficacy of anticancer drugs: (i) to synergize the efficacy of anticancer drugs through synthetic lethality, (ii) to overcome multidrug resistance (MDR) silencing drug efflux pump-related gene, (iii) to enhance efficiency of cancer treatment by promoting apoptosis in cancer cells, and (iv) to promote anti-tumour therapy by targeting metastasis and angiogenesis [7].

The emerging technology in nanomaterials development has introduced various options of drug carriers to be employed in co-delivery systems. Each class of materials has their own unique properties, conferring advantages as well as disadvantages for co-delivery. Figure 1 shows the structure and size range of various classes of materials commonly used in co-delivery applications. Each vehicle possesses distinctive structural features and morphology, which promote various advantages for co-delivery applications. Hollow or porous structures have better capability to encapsulate drugs, while surface modification of structures enables the surface attachment of drugs or siRNA. However, a common characteristic of all these structures is high surface-area-to-volume ratio.

The diversity and heterogeneity of cancer cells has made a tailored combination of chemotherapy drug-siRNA a highly promising treatment option. However, an appropriate vehicle is vital to a successful co-delivery system. Recent material choices for small molecule drug-siRNA co-delivery are reviewed in the following sections. A summary of the advantages and disadvantages of each co-delivery system is presented in Table 1.

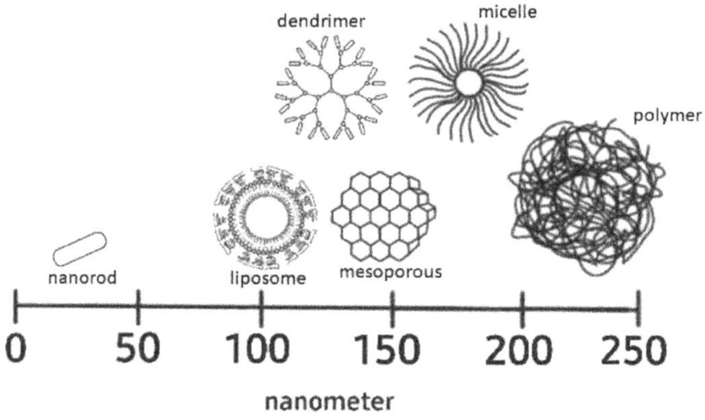

Figure 1. The structures and size range of vehicles used to co-deliver drug molecules and siRNA (not drawn to scale).

Table 1. Summary of advantages and disadvantages of various systems used for co-delivery of small molecule drugs and siRNA.

Delivery System	Advantages	Disadvantages
Mesoporous Silica Nanoparticles	• Naturally abundant • Modifiable, nano-sized mesoporous structure • Large pore volume • High specific surface area • Dual-functional exterior and interior surfaces • Good biocompatibility and biodegradability • Can be incorporated in magnetic applications • Protects guest molecules before reaching target site	• Extra modifications or coatings on surface for specific functions (targeted delivery, co-delivery of drugs and siRNA, sustained release of drug, etc.) • Unsuitable for encapsulating drug molecules larger than the size of its pores • Extra coating needed to enhance stability of drug in the carrier
Polymeric Materials	• Cationic polymer encourages attachment of siRNA • Hydrophobic core eases encapsulation • Can be modified to have controlled release behaviours • Can be designed to be sensitive to pH, temperature, chemical substances and enzymes • Preferential accumulation in tumour tissues • Various co-polymer combinations can be done to explore various potentials • Protects guest molecules before reaching target site	• Toxicity dependent on the chemical structure of polymers • Appropriate design of polymer drug carrier is vital • Modifications are needed for better performance • Structure modification is required in certain conditions to reduce the cytotoxicity
Liposomes	• Enables encapsulation of active agents • Shielding effect reduces toxicity of chemotherapy drugs • Improves cellular uptake • Protects guest molecules before reaching target site • Confers good drug stability • Good controlled release profile • Can be modified into cationic liposome for siRNA conjugation • Manipulatable size • Smart liposomes (sensitive to pH, temperature, enzymes and magnetic field) • Imaging agents can be attached • Good biocompatibility and biodegradability • Low immunogenicity • Able to penetrate the stratum corneum	• Restricted by rapid clearance from the bloodstream in the reticuloendothelial system (RES) • Modifications are needed to prolong circulation time in blood • Ligand attachment is needed for site-specific delivery • Appropriate selection is needed depending on the application

Table 1. *Cont.*

Delivery System	Advantages	Disadvantages
Dendrimers	• Unique and precise molecular structure • Conjugation of drugs through various types of bonding • Uniform globular structure and molecular weight • Wide range of generation number can be chosen for specific applications • Manipulatable size and lipophilicity • Can be engineered with ligands • Low systemic toxicity • Protects guest molecules before reaching target site	• Properties are highly dependent on functional group of outer shell • Performance of carrier is dependent on the appropriate functional group chosen • Appropriate selection and method of encapsulation is crucial
Gold nanoparticles	• Various morphologies available for different applications • Variable optical properties • Suitable for bioimaging applications • Passive and preferential accumulation at tumour site • High specific surface area • Antimicrobial properties • Good biocompatibility and biodegradability • Least reactive element • Ligands can be conjugated for various applications	• High cost • Rare • Ligand attachment is needed for site-specific delivery

3.1. Mesoporous Silica Nanoparticles

3.1.1. General Properties

Mesoporous silica nanoparticles (MSNPs) are solid materials, in which hundreds of empty nanoscale channels are arranged in a two-dimensional network where the mesoporous structure is described as honeycomb-like [18,19]. By definition, the pores of mesoporous materials have a narrow pore size distribution ranging between 2–50 nm [20]. For instance, the MCM-41 mesoporous silica has a diameter of approximately 20 nm [21]. MSNPs have a manipulatable mesoporous structure, large pore volume, high specific surface area, and dual-functionality on the exterior and interior surfaces [19,22–26]. Their mesoporous inner structure provides for high drug loading capacity and enable control over drug release behaviour [19,25–27]. The unique properties of MSNPs enable the encapsulation of a wide range of therapeutic agents for targeted delivery, while preventing their premature release and degradation.

Silica is an abundant natural mineral having good biological compatibility. It is declared as "Generally Recognized As Safe" (GRAS) by the United States Food and Drug Authority (FDA). Silica materials are conventionally employed in industries such as cosmetics and food additives owing to their good biocompatibility [28]. Surface modification of the silica structure assists bioavailability and cellular uptake of the platform, avoids unwanted biological interactions, and bypasses monitoring by the immune system [22,27]. The theory of immune surveillance describes a patrol system to recognize and destroy invading pathogens as well as potential cancerous host cells [22,29,30].

Generally, the high surface-area-to-volume ratio and well-ordered mesoporous structure of MSNPs has been shown to facilitate improved drug loading capacity, biocompatibility and biodegradability [31–33]. The easily-functionalized surface properties of MSNPs allow their surfaces to be functionalized with siRNA to impart excellent colloidal stability [31,32]. These properties make MSNPs a promising vehicle for small molecule drug and siRNA co-delivery.

3.1.2. Applications in Cancer Treatment

MSNPs offer the possibility of carrying drugs to the target site without degradation. However, it has been found that the drug released from the mesopores of unmodified MSNPs are often distributed off-target, an undesirable outcome for targeted drug delivery systems. Hence, the potential of MSNPs as co-delivery platforms have been expanded by the introduction of various surface modifications. The functionality of MSNPS has been enhanced from static drug vehicles into multifunctional delivery systems [18,19,25,26]. The ease of functionalizing the surfaces of MSNPs has resulted in MSNP-based systems which demonstrate properties such as sensitivity to pH, heat and enzymes, capability to undergo redox reactions, and responsiveness to magnetic fields. For instance, Yu et al. reported integrating superparamagnetic iron oxide nanocrystals in the core of MSNPs for magnetic hyperthermia cancer therapy applications [34].

Song et al. [31] developed an MSNP vehicle loaded with myricetin (Myr) and MRP1-siRNA (Figure 2). The MSNP which was conjugated with folic acid (FA) showed increased specificity of Myr towards non-small cell lung cancer (NSCLC) compared with a non-targeted siRNA control. Sustained release of therapeutics was observed in the FA-conjugated vehicle compared with free Myr and the non-FA conjugated co-delivery system. Furthermore, the FA-conjugated co-delivery system caused increased apoptosis of lung cancer cells.

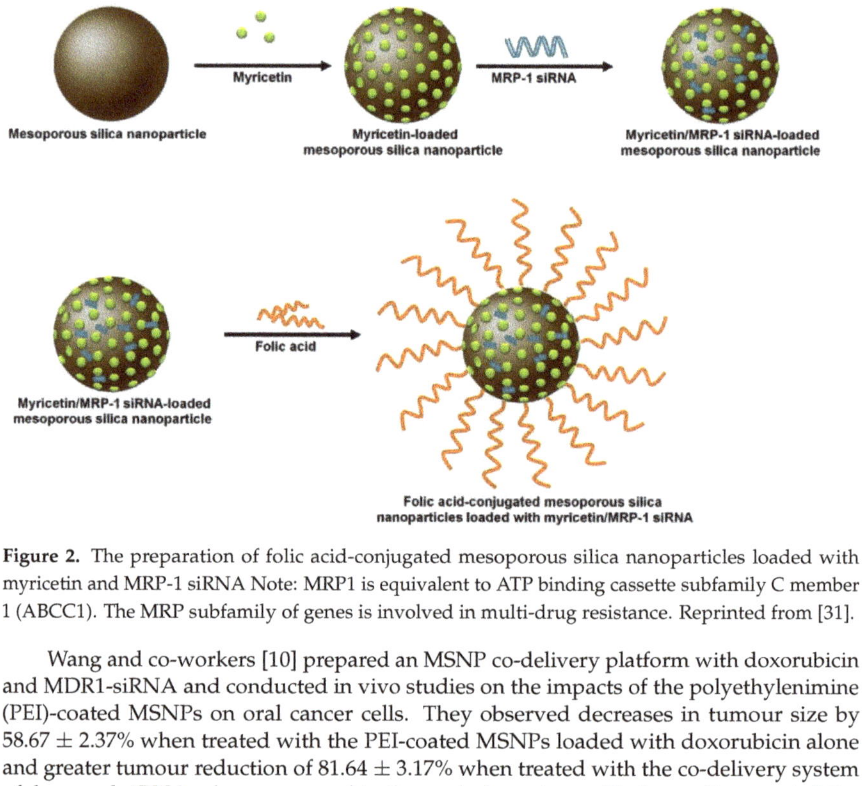

Figure 2. The preparation of folic acid-conjugated mesoporous silica nanoparticles loaded with myricetin and MRP-1 siRNA Note: MRP1 is equivalent to ATP binding cassette subfamily C member 1 (ABCC1). The MRP subfamily of genes is involved in multi-drug resistance. Reprinted from [31].

Wang and co-workers [10] prepared an MSNP co-delivery platform with doxorubicin and MDR1-siRNA and conducted in vivo studies on the impacts of the polyethylenimine (PEI)-coated MSNPs on oral cancer cells. They observed decreases in tumour size by $58.67 \pm 2.37\%$ when treated with the PEI-coated MSNPs loaded with doxorubicin alone and greater tumour reduction of $81.64 \pm 3.17\%$ when treated with the co-delivery system of drug and siRNA when compared to the control specimen. Their result suggested the effective inhibition of tumour growth in vivo when a co-delivery approach was applied.

Zhao et al. [32] found that an MSNP co-delivery system was able to improve tumour targeting efficiency of breast cancers. Their MSNP vehicle was modified with a disulfide group and loaded with doxorubicin and siRNA targeting BCL2 protein. It was reported that the doxorubicin content in tumour tissue after the co-delivery treatment was approximately 2.6-times greater than that of the free doxorubicin treatment. This indicated the potential of co-delivery systems in improving the enhanced permeability and retention (EPR) effect in breast tumour tissues. EPR describes a preferred accumulation of therapeutic agents in tumour cells [35,36].

However, Park et al. [36] stated that the heterogenous nature of tumours remain a challenge in targeted cancer therapy, even for nanoscale co-delivery platforms. Crucial factors that can enhance the EPR effect of cancer therapies include the application of molecular markers and enzymes that are specific to the tumour microenvironment (TME), as well as physical transformation of the TME.

Zhou et al. [37] reported a greater apoptotic rate (36.88%) when breast cancer cells were treated with an MSNP co-delivery system of doxorubicin and BCL2 siRNA, when compared with approximately 14% of apoptotic cells observed for the group treated with only doxorubicin loaded in the delivery system. The MSNP delivery system was modified with a copolymer of polyethylenimine-polylysine (PEI-PLL). The disulfide bonds on the copolymer were conjugated with a folate-linked poly(ethylene glycol) (FA-PEG). The resulting positively charged vehicle was able to bind with the negatively charged siRNA while encapsulating doxorubicin within the MSNP mesopores. Their results show that the co-delivery of doxorubicin and BCL2 siRNA offered synergistic cell apoptotic impacts on MDA-MB-231cells [37]. Table 2 shows various other examples of using MSNPs for co-delivery of small molecule drug and siRNA for the treatment of different types of cancer.

Table 2. Summary of co-delivery of small molecule drug and siRNA by mesoporous silica nanoparticles for cancer therapy.

Delivery System	Small Molecule Drug	siRNA Target	Type of Cancer	Cell Line	Testing Stage	Ref.
Mesoporous silica nanoparticles modified with polyethylenimine	Doxorubicin	ABCB1 (or MDR1)	Oral squamous carcinoma	KBV	*In vitro* *In vivo*	[10]
Folic acid (FA)-conjugated mesoporous silica nanoparticles	Myricetin	ABCC1 (or MRP1)	Non-small cell lung cancer	A594 NCI-H1299	*In vitro* *In vivo*	[31]
Mesoporous silica nanoparticles (MSNPs)	Doxorubicin	BCL2	Breast Cancer	MCF-7 HEK 293	*In vitro* *In vivo*	[32]
Acid-sensitive calcium phosphate/silica dioxide (CAP/SiO_2) composite	Doxorubicin	ABCB1 (or MDR1)	Leukaemia	K562/ADR	*In vitro*	[33]
Mesoporous silica nanoparticles modified with polyethylenimine–polylysine and folate-linked poly(ethylene glycol)	Doxorubicin	BCL2	Breast Cancer	MDA-MB-231 RAW 264.7	*In vitro*	[37]

Note: ABCB1, ATP binding cassette subfamily B member 1; ABCC1, ATP binding cassette subfamily C member 1; BCL2, BCL2 apoptosis regulator; MDR1, multidrug resistance protein 1; MRP1, multidrug resistance associated protein 1.

3.2. *Polymeric Materials*

3.2.1. General Properties

Polymers are a large class of natural as well as synthetic materials, with a wide range of properties, which are directly related to their molecular weight, molecular structure and elemental composition. These macromolecules are made up of multiple repeating sub-units (mers). Polymer-based delivery systems are one of the most well-established platforms used to deliver therapeutic agents [38]. Their success is due to their versatility and tuneable characteristics in order to achieve the desired pharmaceutical and biomedical requirements [39].

Among the natural polymers which have been reported for use in drug delivery include derivatives of arginine, chitosan, cyclodextrin, polyglycolic acid (PGA), polylactic acid (PLA), and polysaccharides. These are generally favourable options due to their nontoxicity, biocompatibility and biodegradability [40,41]. PGA, PLA, polycaprolactone and polydioxanone are used as polymeric implant materials due to their biodegradable and bioabsorbable qualities [42].

On the other hand, man-made polymers such as poly(2-hydroxyethyl methacrylate (PHEMA) hydrogel, poly(n-isoproply acrylamide, polyethyleimine (PEI), and poly(n-(2-

hydropropyl) methacrylamide) are preferred over the natural polymers as drug delivery systems due to their immunogenicity [42].

Amphiphilic block copolymers are a versatile class of self-assembling structures which can be formulated to produce polymeric nanoparticles such as. micelles and polymersomes [43]. Nanopolymeric delivery systems in micellar form alleviate the adverse effects of drugs, in which their hydrophobic core enables effective encapsulation of multiple anticancer agents including hydrophobic drugs [44,45]. The use of micelles was reported to lengthen drug retention time in the blood circulation and selectively accumulated in tumour tissue through the EPR effect [44,46].

Polymers bearing a positive charge or containing cationic functional groups in their structure are termed polycationic polymers. They have been extensively studied as a form of injectable, nonviral delivery agent for nucleic acid and drugs [40,47]. They also have wide application in biomedical areas, for instance as antimicrobial agents due to their affinity for the negatively-charged membrane of microorganisms, hence causing lysis of cell walls [48]. The most well-investigated synthetic linear cationic polymers include PEI, polyvinylpyrrolidone (PVP), and poly-L-lysine (PLL).

As a transfection agent, polycationic polymers are able to compress nucleic acid to form polyplexes for gene therapy. In general, to encapsulate and deliver a higher concentration of siRNA, polycations are engineered with greater charge densities, which are often associated with carrier-induced toxicity [49]. The toxicity of polymeric cation-mediated gene therapy is strongly linked to their chemical structure [50]. For instance, PEI with a low molecular weight (11.9 kDa) and a moderately branched structure showed about 100 times higher delivery efficiency, coupled with low toxicity, in comparison to a higher molecular weight PEI vector [51]. After delivery of the genetic payload, PEI is liberated of their cargo and has the potential to negatively interact with cellular components. Hence, an appropriate and suitable design of the polymeric drug carrier is vital. One solution to address delayed polymer toxicity issues is by incorporating PEG [50].

Copolymers of PLL-PEG are another example of thoughtful polymeric nanocarrier design. PLL are biodegradable, linear polypeptides in which the repeating unit is the amino acid lysine. PLL-nucleic acid polyplexes require the addition of chloroquine in order to improve their gene delivery efficiency. However, this results in increased cytotoxicity. With the use of higher molecular weight PLL, there is a trade-off between greater transfection efficiency and the accompanying higher cytotoxicity [52]. It has been shown that grafting of PEG to PLL can reduce cytotoxicity without reduction of transfection efficiency [53].

3.2.2. Applications in Cancer Treatment

An emerging set of polymer-based drug delivery systems are being designed with moieties that are sensitive to changes in pH, temperature, the presence of glutathione, reactive oxygen species, and enzymes [54]. In the absence of disease, the bodily pH level is homeostatically maintained within a range of 7.35–7.45 [55]. Polymeric delivery systems can be designed to control the release of their therapeutic cargo by sensing the slightly acidic-pH levels linked to tumour microenvironments (pH 6.5–6.8) [44,46,56]. This particular property has gained popularity in the design of a controlled and targeted anticancer therapy system.

For instance, Suo et al. [56] applied a reversible addition-fragmentation chain transfer (RAFT) polymerization to obtain a triblock copolymer nanocarrier to mediate the delivery of doxorubicin and Bcl-2 siRNA to target breast cancer cells (MCF-7). Apart from offering a carrier for the delivery of the hydrophobic drug doxorubicin and siRNA, this synthetic method imparts excellent control of the nanocarrier physical properties, including uniformity in size, molecular weight, structure and reproducibility. The nanopolymeric carriers were able to achieve dual-release of its cargo in a reducing and acidic environment [56]. Their results demonstrated that a higher cellular uptake (51.6%) was noted in the cells treated with folate-modified co-delivery system, compared with the unmodified co-delivery system with only a 9.2% uptake. It was deduced that the targeted co-delivery promoted cellular uptake while anticancer effects were synergistically achieved by co-delivering two types of therapeutic agents.

Li et al. [46] also reported a triblock copolymer delivery system in which the micelle size is controlled by changes in pH values; at acidic levels, the therapeutic agents are released in the tumour target. A schematic of the process of the self-assembly and therapeutics release of the co-delivery system designed by Li et al. [46] is shown in Figure 3. The triblock copolymer consisted of a PEG shell, a cationic PLL intermediate layer, and a pH-sensitive core of poly(aspartyl) (Benzylamine-co-(Diisopropylamino)ethylamine.

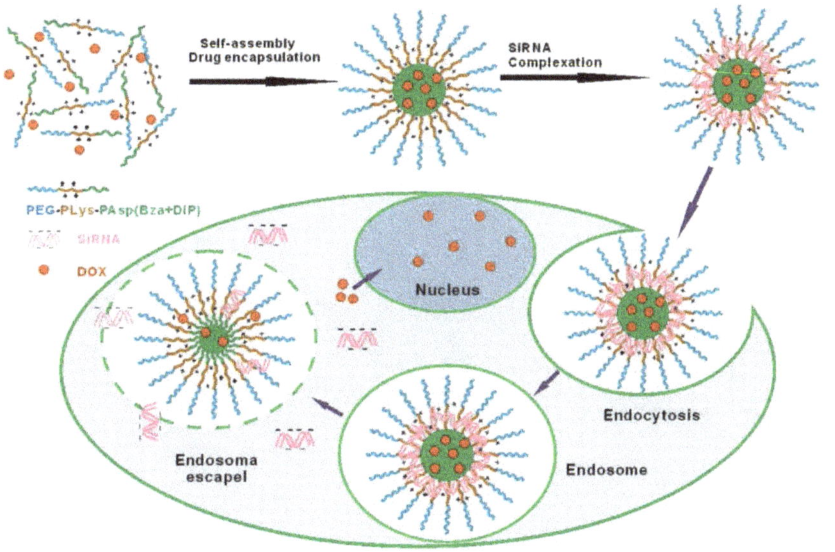

Figure 3. Illustration of self-assembly at pH 7.4 and intracellular releasing behaviour of amphiphilic block copolymer of monomethoxylpoly(ethylene glycol), poly(l-lysine), and poly(aspartyl(Benzylamine-co-(Diisopropylamino)ethylamine))mPEG-PLLys-PLAsp(BzA-co-DIP), abbreviated as PELABD micelles, for combinatorial delivery of hydrophobic anticancer drugs (Doxorubicin) and siRNA. Reprinted with permission from [46]. Copyright 2020 Springer Nature.

Additionally, Wu et al. [57] had constructed a type of co-delivering drug carrier with the use of triblock copolymer nanomicelles. The nanomicelles consisted of polyethylene glycol (PEG), polycaprolactone (PCL), and polyethylenimine (PEI). These core-shell nanomicelles were functionalized with folic acid (FA), then loaded with doxorubicin and P-glycoprotein (P-gp) siRNA to induce cell apoptosis in breast cancer cells. In vitro studies were carried out on MCF-7/ADR cell line. The flow cytometry results showed that the co-delivery system had increased the apoptosis level by 69.6% compared to the cells treated with free doxorubicin (85.3% vs. 15.7%) while 44% of increment in apoptosis level when compared to cells treated without the presence of P-gp siRNA (85.3% vs. 41.3%). These results confirm that this co-delivery system is able to deliver small molecule drug synchronously with siRNA. Furthermore, the release of siRNA was effective and able to interfere with the targeted expression [57].

In experiments conducted by Norouzi et al. [58], the best apoptosis induction was observed in dual functional polymeric micelles containing NF-κB (RELA) siRNA and gemcitabine (GemC18). Apoptosis was induced in 78% of MCF-7 cell lines and 95.4% in AsPC-1 cancer cells, illustrating co-delivery as the optimum delivery system for synergistic anticancer effects [58]. In another study, Yan et al. [44] developed a pH-responsive, chitosan-based polymer co-delivery system that delivered doxorubicin and Bcl-2 siRNA to actively target liver hepatocellular carcinoma cells, HepG2.

Table 3 shows a summary of the current combination of small molecule drug and siRNA co-delivery by using polymer-based nanomaterials for different types of cancer treatment.

Table 3. Summary of co-delivery of small molecule drug and siRNA mediated by polymer-based nanomaterials for cancer therapy.

Delivery System	Small Molecule Drug	siRNA Target	Type of Cancer	Cell Line	Testing Stage	Ref.
Chitosan based pH-responsive polymeric prodrug vector (GA-CS-PEI-HBA-DOX) *where GA-CS-PEI-HBA-DOX is prodrug chitosan-polyethylenimine-4-hydrazino-benzoic acid doxorubicin*	Doxorubicin	BCL2	Liver cancer	HUVEC, HepG2	*In vitro* *In vivo*	[44]
Amphiphilic block copolymer of monomethoxylpoly(ethylene glycol), poly(l-lysine), and poly(aspartyl(Benzylamine-co-(Diisopropylamino)ethylamine)) mPEG-PLLys-PLAsp(BzA-co-DIP), abbreviated as PELABD micelles	Doxorubicin	siRNA	Ovarian cancer	SKOV3	*In vitro*	[46]
Triblock copolymer nanocarrier of PAH-b-PDMAPMA-b-PAH *where PAH-b-PDMAPMA-b-PAH is poly(acrylhydrazine)-block-poly(3-dimethylaminopropyl methacrylamide)-block-poly(acrylhydrazine) (PAH-b-PDMAPMA-b-PAH)*	Doxorubicin	BCL2	Breast cancer	MCF-7	*In vitro*	[56]
PEG-PCL-PEI triblock copolymer nanomicelle functionalized with folic acid	Doxorubicin	(P-gp) siRNA	Breast cancer	MCF-7/ADR	*In vitro*	[57]
Poly(ε-caprolactone), polyethylenimine and polyethylene glycol (PCL-PEI-PEG) copolymers	4-(N)-stearoyl gemcitabine	RELA	Pancreatic cancer and breast cancer	AsPC-1, MCF-7	*In vitro*	[58]
Lipid-polymer hybrid nanoparticles *where cationic ε-polylysine co-polymer nanoparticles (ENPs) are coated with PEGylated lipid bilayer resulted formation of LENPs, with reversed surface charge*	Gemcitabine	HIF1A	Pancreatic cancer	Panc-1	*In vitro* *In vivo*	[59]
pH/redox dual-sensitive polymeric materials (cPCPL) *where cPCPL is poly(ethylene glycol))x-(chitosan-polymine)y-(lipoic acid)z grafted with cRGDyC-PEG-NHS, cRGDyC is a kind of peptide, PEG is poly(ethylene glycol) and NHS is hydroxysuccinimide.*	Etoposide	EZH2	Orthotopic non-small-cell lung tumour	luc-A549	*In vitro* *In vivo*	[60]
Self-assembled polyjuglanin nanoparticles (PJAD-PEG) *where PJAD-PEG is poly(juglanin (Jug) dithiodipropionic acid (DA))-b-poly(ethylene glycol) (PEG)*	Doxorubicin	KRAS	Lung cancer	A549, H69	*In vitro* *In vivo*	[61]
Cationic polyethylenimine-block-polylactic acid (PEI-PLA)	Paclitaxel	BIRC5	Lung Adenocarcinoma	4T1, A549	*In vitro* *In vivo*	[62]
Lactic-co-glycolic acid (PLGA) nanoparticles	Paclitaxel	SPDYE7P	Cervical cancer	HeLa	*In vitro* *In vivo*	[63]
FeCo-polyethylenimine (FeCo-PEI) nanoparticles and polylactic acid-polyethylene glycol (PLA-PEG)	Paclitaxel	FAM group	Breast cancer	MCF-7, BT-474	*In vitro*	[64]

Table 3. *Cont.*

Delivery System	Small Molecule Drug	siRNA Target	Type of Cancer	Cell Line	Testing Stage	Ref.
Hypoxia-sensitive PEG-azobenzene-PEI-DOPE (PAPD) nanoparticles	Doxorubicin	ABCB1	Ovarian cancer and breast cancer	A2780 ADR, MCF7 ADR	*In vitro*	[65]
Chondroitin sulfate (CS)-coated β-cyclodextrin polyethylenemine polymer	Paclitaxel	MCAM	Breast Cancer	MDA-MB-231, MCF-7	*In vitro*	[66]
Targeted multifunctional polymeric micelle (TMPM) *where TMPM is made up of triblock copolymer poly(ε-caprolactone)-polyethyleneglycol-poly(L-histidine) (PCL-PEGPHIS)*	Paclitaxel	VEGF group	Breast Cancer	HUVECs, MCF-7	*In vitro*	[67]

Note: BCL2, BCL2 apoptosis regulator; HIF1A, hypoxia inducible factor 1 subunit alpha; EZH2, enhancer of zeste 2 polycomb repressive complex 2 subunit; KRAS, KRAS proto-oncogene, GTPase; BIRC5, baculoviral IAP repeat containing 5; SPDYE7P, speedy/RINGO cell cycle regulator family member E7 pseudogene; FAM group, long non-coding family with sequence group; ABCB1, ATP binding cassette subfamily B member 1; MCAM, melanoma cell adhesion molecule; VEGF group, vascular endothelial growth factor group; RELA, RELA proto-oncogene, NF-kB subunit.

3.3. Liposomes

3.3.1. General Properties

Liposomes are soft spherical vesicles that are built mainly from phospholipid layers. They can be formed by one or more phospholipid layers where the layers are also called lamellae [68–70]. Phospholipids are under the class of lipids and can be found naturally in egg yolks (cholesterol) [70], vegetable oils, and can also be created synthetically [68]. Variations in phospholipids variations are mostly based on the modifications of head group or bondings formed; these are generally categorized according to the backbone groups–glycerol (glycerophospholipids) and sphingosine (sphingomyelins) [68]. These unique lipids are able to encapsulate a wide range of agents owing to their basic molecular structure which is composed of a hydrophilic head (polar head) and two hydrophilic tails (hydrocarbon chains) [68,70,71]. Liposomes are used as carriers for active agents with differing properties in a wide range of applications, such as cosmetics, pharmaceuticals, food, and farming. The wide-ranging applications of liposomes are attributed to their biocompatibility, encapsulation capability and modifiable formulation for the desired purpose [68–72].

The liposomal delivery system has attracted considerable attention in cancer therapy due to its unique properties to incorporate dissimilar therapeutic agents [59,73–76]. There are four main types of liposomes: (i) conventional liposomes, which includes cationic, anionic, phospholipids (neutral) and cholesterol, (ii) steric-stable liposomes, (iii) ligand targeting liposomes, and (iv) a combination of the first three types of liposomes [69]. The conventional liposomes are basic shields that are capable of protecting the payload (chemotherapy drugs), and also act as barriers to reduce the toxicity of the compounds in vivo [69]. The encapsulation property of liposomes is expected to mitigate the side effects of chemotherapy drugs. Besides that, its structure which resembles the plasma membrane of cells, plays a vital role in improving the cellular uptake of the compounds [68,72]. Furthermore, it was reported that with the use of liposome delivery systems, retention time in circulation is prolonged, alongside higher drug stability. Other benefits include a better controlled release profile when compared to the effect of using free chemical drugs alone [59,75,76].

However, the therapeutic efficacy of conventional liposomes is restricted by rapid clearance from the bloodstream in the reticuloendothelial system (RES), specifically through the liver and spleen [69]. The formulation can be modified with the use of hydrophilic polymers, such as polyethylene glycol (PEG), to obtain steric-stable liposomes, which are able to reduce in vivo opsonization and prolong circulation time in the bloodstream [69].

These liposomes are also reported to preferentially accumulate in the pathological sites and tumour regions by the EPR effect [69,74]

Liposomal delivery systems are not limited to drug delivery; they have been adopted for the incorporation of nucleic acid delivery, as well as in co-delivery of drugs and siRNA hybrid combinations. In using cationic liposomes (CLs), siRNA is protected from degradation by nuclease accumulation in the tumour, one of the key concerns in siRNA delivery. Anionic siRNA can be electrostatically bound to CLs to form a stable complex and improve the cellular uptake of target cells [69,77].

For advanced applications, site-specific or targeted-delivery can be achieved by attachment of ligands onto liposomes [69,70]. The ligand selections are based on the specific over-expression of ligands or specifically expressed ligands at the target cells, organs or tumours. Antibodies, peptides, proteins and carbohydrates are the general types of ligands that are commonly involved in the formulation of ligand-targeted liposomes [69]. Based on this fundamental concept, various types of liposomes have been developed. For instance, imaging agents can be attached onto liposomes to incorporate imaging properties. This conceptual theory has triggered the development of the "smart liposome", in which the structure changes according to in vivo stimuli such as pH, temperature, enzyme, and magnetic field [77].

In general, all liposomes possess the fundamental properties required of delivery systems, such as, good biocompatibility, biodegradability, and low immunogenicity [59,76]. Liposome-based delivery systems also exhibit efficient loading of both hydrophobic drugs and genes, making them a promising candidate for co-delivery systems [76].

Liposomes were also reported as efficient in reducing cardiotoxicity of chemotherapy drugs such as doxorubicin, which damages heart muscle and function [75]. Besides that, liposomes are capable of penetrating the stratum corneum (outermost layer of skin) at different depths, compared to other nanocarriers [73,78]. The upper stratum corneum layer is the most penetrated layer while dioleoyl-phosphatidylethanolamine (DOPE)-based liposomes are able to achieve a deeper penetration [78]. Liposomes have been modified to improve their skin penetration ability for dermal drug delivery [79]. Deep penetration in the skin membrane is achievable via the use of transferosomes (liposomes with edge activators) or ethosomes (liposomes comprised of ethanol) [73]. These properties have made liposomes a great candidate for derma-related cancer treatment.

3.3.2. Applications in Cancer Treatment

Jose et al. used a cationic liposome, 1,2-dioleoyl-3-trimethylammonium propane (DOTAP), for the co-delivery of curcumin and STAT3 siRNA to treat skin cancer [73]. The results showed the highest cell growth inhibition ($76.3 \pm 4.0\%$) in mouse melanoma cells (B16F10) compared to either curcumin-loaded liposomes or free STAT3 siRNA. This result suggested that this form of co-delivery was effective in curbing melanoma tumour progression.

Oh et al. studied galactosylated liposomes as a co-delivery vehicle to target hepatocellular carcinoma by loading it with doxorubicin and siRNA (Gal-DOX/siRNA-L). The Gal-DOX/siRNA-L system showed better doxorubicin uptake than free doxorubicin [75]. Besides that, the accumulated Gal-DOX/siRNA-L in tumour tissues was 4.8 times higher than that of free doxorubicin.

Lin et al. developed a co-delivery system from GE-11 peptide conjugated liposomes loaded with gemcitabine and siRNA as a formulation to target pancreatic cancer treatment [59]. They found that the co-delivery system displayed a 4-fold reduction in tumour weight compared to the control. When compared to GE-11 peptide conjugated liposome loaded with gemcitabine only (GE-GML), the tumour weight reduction was 2-fold.

Collectively, these results demonstrate the effectiveness of liposomal co-delivery system as an anticancer therapy. Table 4 shows the current combination of small molecule drug and siRNA co-delivery by using liposome nanomaterials for different types of cancer treatments.

Table 4. Summary of co-delivery of small molecule drug and siRNA by liposomes for cancer therapy.

Delivery System	Small Molecule Drug	siRNA Target	Types of Cancer	Cell Line	Testing Stage	Ref.
GE-11 peptide conjugated liposome	Gemcitabine	HIF1A	Pancreatic cancer	Panc-1	*In vitro* *In vivo*	[59]
1,2-Dioleoyl-3-trimethylammonium propane (DOTAP) -based cationic liposomes	Curcumin	STAT3	Skin cancer	B16F10	*In vitro* *In vivo*	[73]
PEGylated liposomes	Docetaxel	BCL2	Lung cancer	A549, H226	*In vitro* *In vivo*	[74]
Galactosylated Liposomes	Doxorubicin	VIM	Hepatocellular Carcinoma	Huh7, A549	*In vitro* *In vivo*	[75]
Carbamate-linked cationic lipid (Cationic Liposome)	Paclitaxel	BIRC5	Lung Cancer	NCI-H460	*In vitro*	[76]

Note: STAT3, signal transducer and activator of transcription 3; BCL2, BCL2 apoptosis regulator; VIM, vimentin; HIF1A, hypoxia inducible factor 1 subunit alpha; BIRC5, baculoviral IAP repeat containing 5.

3.4. *Dendrimers*

3.4.1. General Properties

Dendrimers are a form of symmetrical, hyperbranched, low-molecular-weight polymers on the nanoscale, in which the architecture consists of a core, an inner shell, and an outer shell [80]. Its properties are dependent on functional groups which act as a capping agent on the outer shell [81]. The inner shell consists of several layers of repeating units (known as generations) built by a repetitive series of chemical reactions, while the outermost periphery contains multiple functional groups [82]. Polyamidoamine (PAMAM) dendrimers are one of the most well-studied dendrimers for delivery applications. Other types of dendrimers include peptide dendrimers (PPI), poly(l-lysine) dendrimers, and PAMAM-organosilicon dendrimers (PAMAMOS) [81].

Dendrimers are a versatile option as a vehicle for synergistic drug and gene combination therapy. Due to their unique and precise molecular structure, dendrimers can be used to deliver cancer drugs in any of the following ways: (1) cancer drugs can be covalently conjugated to the dendrimer outer shell to construct dendrimer prodrugs through direct coupling or cleavage linking, (2) encapsulating the drug within the central core to form a dendrimer-drug complex, and (3) exploiting the electrostatic interactions between functional groups on the dendrimer capping agent and the drug molecule [80,83].

Electrostatic interactions between the phosphate groups of siRNA and the cationic species on the dendrimer surface are crucial in the formation of complexes for effective delivery of gene therapy [84]. In comparison with conventional linear and branched polymers, dendrimers possess the following properties: (i) well-defined uniform spherical structure and manipulatable size in the nano-range, (ii) availability of numerous generations for different specific purposes [81,83], (iii) lipophilicity and suitable size enabling their diffusion through cell membranes [81], (iv) exceptional flexibility, and (v) low systemic toxicity [85,86]. Dendriplexes (dendrimers bound to nucleic acids) have an enhanced ability to evade endosomal entrapment due to the flexibility of dendrimers [81]. Endosomal escape is the process of siRNA exiting the endosome and entering the cytosol. This process is augmented by the "proton sponge" effect [81,85,87,88]. According to the "proton sponge" hypothesis, an influx of protons into the endosome results in increased buildup of osmotic pressure, causing destabilization of the endosomal membrane, leading to its rupture [88–90].

3.4.2. Applications in Cancer Treatment

Table 5 shows the current combination of small molecule drug and siRNA co-delivery by using dendrimer-based nanomaterials for different types of cancer treatment.

Table 5. Summary of co-delivery of small molecule drug and siRNA by dendrimer-based nanomaterials for cancer therapy.

Delivery System	Small Molecule Drug	siRNA Target	Type of Cancer	Cell Line	Testing Stage	Ref.
Amphiphilic dendrimer engineered nanocarrier system (ADENS) modified by tumour microenvironment-sensitive polypeptides (TMSP) (TMSP-ADENS)	Paclitaxel	FAM and VEGF group	Melanoma, prostate cancer	A375, PC-3, HT-1080	*In vitro* *In vivo*	[13]
PTP (plectin-1 targeted peptide, NH2-KTLLPTP-COOH), biomarker for pancreatic cancer, integrated with the PSPG vector to form peptide-conjugated PSPG (PSPGP) *where PSPG is branched poly(ethylene glycol) with G2 dendrimers through disulfide linkages*	Paclitaxel	NR4A1 (or TR3)	Pancreatic Cancer	Panc-1	*In vitro* *In vivo*	[83]
Hyaluronic acid (HA) modified MDMs *where MDMs is the PAMAM-PEG2k-DOPE co-polymer, together with mPEG2k-DOPE, was formulated into mixed dendrimer micelles, and PAMAM is the generation 4 polyamidoamine*	Doxorubicin	ABCB1 (or MDR1)	Ovarian cancer, colorectal carcinoma and breast cancer	A2780 ADR, HCT 116, MDA-MB-231	*In vitro*	[86]
PAMAM-OH derivative (PAMSPF)	Murine double minute 2 protein (MDM2) inhibitor RG7388	TP53	Breast cancer	P53-wild type MCF-7 cells (MCF-7/WT), MDA-MB-435	*In vitro* *In vivo*	[87]
Polyamidoamine (PAMAM) dendrimer	Curcumin	BCL2	Cervical cancer	HeLa	*In vitro*	[91]
Folate-polyethylene glycol appended dendrimer conjugate with glucuronylglucosyl-β cyclodextrin (Fol-PEG-GUG-β-CDE) (generation 3)	Doxorubicin	PLK1	Cervical cancer	KB	*In vitro* *In vivo*	[92]

Note: BCL2, BCL2 apoptosis regulator; NR4A1, nuclear receptor subfamily 4 group A member 1; TR3, thioredoxin reductase 3; ABCB1, ATP binding cassette subfamily B member 1; MDR1, multidrug resistance protein 1; FAM, long non-coding family with sequence group; VEGF, vascular endothelial growth factor group; TP53, tumour protein p53.

Ghaffari et al. [91] studied the apoptotic effects of curcumin (Cur) and BCL2 siRNA co-delivered using polyamidoamine (PAMAM) dendrimers on HeLa cells. BCL2 siRNA was grafted to the amine groups on the surface layer of PAMAM while Cur was enveloped within the core to form the co-delivery dendriplex. Cells treated with PAMAM-Cur/siRNA showed an improvement of 58.77% as compared to cells treated with free curcumin alone. Compared with various formulations, the PAMAM-Cur/siRNA co-delivery system had the highest percentage of apoptotic cells among all formulations [91].

Li et al. [13] developed a hollow core/shell amphiphilic dendrimer engineered nanocarrier system (also known as ADENS) to co-deliver the hydrophobic drug paclitaxel and the hydrophilic siRNA. siRNA was loaded in the polar hydrophilic cavity while the paclitaxel was grafted in the polylactic acid (PLA) interlayer. The outer layer was constructed of polyethylene glycol (PEG) to enhance the in vivo circulation time of the co-delivery system and to avoid uptake by the reticuloendothelial system. ADENS was further modified with polypeptides which respond to tumour signals from the surrounding microenvironment. Results revealed that the system inhibited up to 73% of vascular endothelial growth factor

(VEGF) at the mRNA level in A375 cell xenograft mice models when compared to controls; thus, demonstrating the synergistic effects of the co-delivery system.

3.5. Gold Nanoparticles

3.5.1. General Properties

Nanoparticles of noble metal elements such as gold, silver and palladium have widespread use in biomedical applications [93]. Gold nanoparticles (AuNPs) in particular, have been extensively studied due to their versatile properties. Gold is one of the least reactive chemical elements and this has contributed to the properties of biocompatibility and biodegradability [94–98]. AuNPs possess tuneable optical properties, such as light scattering, localized surface plasmon resonance (SPR) and photothermal effects, leading to the incorporation of AuNPs in numerous medical diagnostics including ultrasensitive bioimaging and photothermal therapy [95]. The antimicrobial properties of AuNPs combined with their non-immunogenic and biocompatible properties have led to potential applications in cancer diagnosis and therapy as well as treatment for HIV infection [93]. Nanoparticles have at least one dimension measuring <100 nm and are defined by their high specific surface area. Gold has been synthesized in a variety of morphologies including nanorods, nanospheres, nanocages, nanoshells nanostars, nanorattles, nanopopcorns and nanoaggregates [95,96,99].

3.5.2. Applications in Cancer Treatment

The surface of AuNPs can be modified with active targeting ligands or moieties for tumour-specific targeting [95,96,98]. Through this strategy, passive accumulation and preferential retention of AuNPs at the tumour sites were observed, due to the EPR effect [94,98]. Recently, the distinctive properties of AuNPs have been explored in drug-gene co-delivery systems in cancer treatment. Kotcherlakota and co-workers reported when human ovarian (SK-OV-3) cells were treated with an engineered bi-functional recombinant fusion protein TRAF(C) (TR) gold nanoparticles (AuNPs), loaded with doxorubicin and erbB2-siRNA, a 6-fold difference in the tumour volume was observed as compared to the untreated control [100]. The AuNPs were conjugated with the reactive moiety at the carboxyl-terminus of the tumour necrosis factor (TNF) receptor associated factors (TRAF) protein. The authors described the synergistic effects of co-delivered doxorubicin and siRNA in inhibition of cell proliferation and tumour suppression. Figure 4 shows the fabrication of their co-delivery system based on AuNPs. Table 6 shows the current combination of small molecule drug and siRNA co-delivery by using gold nanoparticles nanomaterials for different types of cancer treatment.

Table 6. Summary of co-delivery of small molecule drug and siRNA by gold nanomaterials for cancer therapy.

Delivery System	Small Molecule Drug	siRNA Target	Type of Cancer	Cell Line	Testing Stage	Ref.
Polyelectrolyte polymers coated gold nanorods (AuNRs)	Doxorubicin	KRAS	Pancreatic Cancer	Panc-1	*In vitro* *In vivo*	[95]
Gold nanoparticles (AuNPs) combined with an engineered bi-functional recombinant fusion protein TRAF(C) (TR)	Doxorubicin	ERBB2	Ovarian cancer	SK-OV-3, MDA-MB-231, A549, PANC-1, B16F10	*In vitro* *In vivo*	[100]
Layer-by-layer assembled gold nanoparticles (LbL-AuNP)	Imatinib mesylate	STAT3	Melanoma cancer	B16F10	*In vivo*	[101]

Note: KRAS, KRAS proto-oncogene, GTPase; ERBB2, erb-b2 receptor tyrosine kinase 2; STAT3, signal transducer and activator of transcription 3.

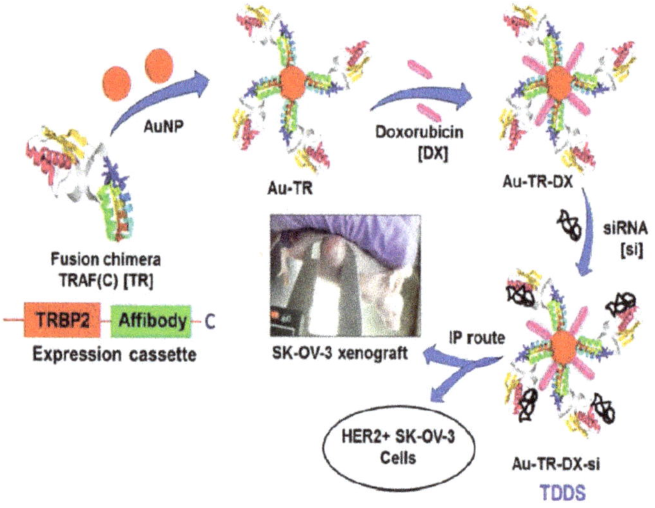

Figure 4. Fabrication of the AuNP-based targeted drug delivery system (TDDS) with an engineered bi-functional recombinant fusion protein TRAF(C) (TR), loaded with doxorubicin and ERBB2-siRNA (Au-TR-DX-si). It was used for further studies in vitro and *in vivo*. Reprinted with permission from [100]. Copyright Royal Society of Chemistry 2012.

4. Conclusions

There is an emerging trend in the use of combination therapy of drug-siRNA for cancer treatments. These experimental therapies rely increasingly on the use of nanostructured delivery vehicles to encapsulate and protect the therapeutics until it reaches the target. The most well-studied nanocarriers are mesoporous silica nanoparticles, dendrimers, polymers, liposomes and gold nanoparticles.

These structures fulfil the fundamental requirement of co-delivery systems of having low- or non-toxicity and biocompatibility. More complex challenges include tailoring their surface properties, designing suitable structures to entrap and protect the payload, prolonging their stability in the bloodstream, maximising targeted delivery and understanding their behaviour in the tumour environment. These are among the critical challenges and issues which need to be resolved before their successful implementation at the clinical level.

Given the advantage of co-delivering siRNA and drug simultaneously to achieve multi-target tumour therapy, there is a strong impetus for developing more materials that could improve the efficacy of these delivery systems. To conclude, the co-delivery of siRNA and small molecule drugs represent an emerging technology that warrants further investigation.

Funding: This research was funded by Ministry of Higher Education Malaysia (MOHE), grant number FRGS/1/2020/TK0/UNIM/03/3.

Acknowledgments: Chung, S.L. gratefully acknowledges the SEPRS scholarship from University of Nottingham Malaysia.

Conflicts of Interest: The authors declare no conflict of interest.

References

1. World Health Organization. *WHO Report on Cancer: Setting Priorities, Investing Wisely and Providing Care for All*; World Health Organization: Geneva, Switzerland, 2020.
2. Sung, H.; Ferlay, J.; Siegel, R.L.; Laversanne, M.; Soerjomataram, I.; Jemal, A.; Bray, F. Global Cancer Statistics 2020: GLOBOCAN Estimates of Incidence and Mortality Worldwide for 36 Cancers in 185 Countries. *CA Cancer J. Clin.* **2021**, *71*, 209–249. [CrossRef]

3. Alberts, B.; Johnson, A.; L., J. *The Molecular Basis of Cancer-Cell Behavior*, 4th ed.; Garland Science: New York, NY, USA, 2002.
4. Institute, N.C. Types of Cancer Treatment. Available online: https://www.cancer.gov/about-cancer/treatment/types (accessed on 28 August 2021).
5. Jin, J.O.; Kim, G.; Hwang, J.; Han, K.H.; Kwak, M.; Lee, P.C.W. Nucleic acid nanotechnology for cancer treatment. *Biochim. Biophys. Acta-Rev. Cancer* **2020**, *1874*, 188377. [CrossRef]
6. Lee, S.J.; Kim, M.J.; Kwon, I.C.; Roberts, T.M. Delivery strategies and potential targets for siRNA in major cancer types. *Adv. Drug Deliv. Rev.* **2016**, *104*, 2–15. [CrossRef] [PubMed]
7. Li, J.; Wang, Y.; Zhu, Y.; Oupický, D. Recent advances in delivery of drug-nucleic acid combinations for cancer treatment. *J. Control. Release* **2013**, *172*, 589–600. [CrossRef]
8. Dong, Y.; Siegwart, D.J.; Anderson, D.G. Strategies, design, and chemistry in siRNA delivery systems. *Adv. Drug Deliv. Rev.* **2019**, *144*, 133–147. [CrossRef] [PubMed]
9. Oh, Y.K.; Park, T.G. siRNA delivery systems for cancer treatment. *Adv. Drug Deliv. Rev.* **2009**, *61*, 850–862. [CrossRef] [PubMed]
10. Wang, D.; Xu, X.; Zhang, K.; Sun, B.; Wang, L.; Meng, L.; Liu, Q.; Zheng, C.; Yang, B.; Sun, H. Codelivery of doxorubicin and MDR1-siRNA by mesoporous silica nanoparticles-polymerpolyethylenimine to improve oral squamous carcinoma treatment. *Int. J. Nanomed.* **2018**, *13*, 187–198. [CrossRef] [PubMed]
11. Wang, M.; Wang, J.; Li, B.; Meng, L.; Tian, Z. Recent advances in mechanism-based chemotherapy drug-siRNA pairs in co-delivery systems for cancer: A review. *Colloids Surf. B Biointerfaces* **2017**, *157*, 297–308. [CrossRef] [PubMed]
12. Carvalho, B.G.; Vit, F.F.; Carvalho, H.F.; Han, S.W.; de la Torre, L.G. Recent advances in co-delivery nanosystems for synergistic action in cancer treatment. *J. Mater. Chem. B* **2021**, *9*, 1208–1237. [CrossRef]
13. Li, X.; Sun, A.N.; Liu, Y.J.; Zhang, W.J.; Pang, N.; Cheng, S.X.; Qi, X.R. Amphiphilic dendrimer engineered nanocarrier systems for co-delivery of siRNA and paclitaxel to matrix metalloproteinase-rich tumors for synergistic therapy. *NPG Asia Mater.* **2018**, *10*, 238–254. [CrossRef]
14. Malhotra, M.; Tomaro-Duchesneau, C.; Saha, S.; Prakash, S. Intranasal Delivery of Chitosan—siRNA Nanoparticle Formulation to the Brain. In *Drug Delivery System*; Jain, K.K., Ed.; Springer: New York, NY, USA, 2014; pp. 233–247. ISBN 978-1-4939-0363-4.
15. Ibraheem, D.; Elaissari, A.; Fessi, H. Gene therapy and DNA delivery systems. *Int. J. Pharm.* **2014**, *459*, 70–83. [CrossRef]
16. Scholz, C.; Wagner, E. Therapeutic plasmid DNA versus siRNA delivery: Common and different tasks for synthetic carriers. *J. Control. Release* **2012**, *161*, 554–565. [CrossRef]
17. Sajid, M.I.; Moazzam, M.; Tiwari, R.K.; Kato, S.; Cho, K.Y. Overcoming barriers for siRNA therapeutics: From bench to bedside. *Pharmaceuticals* **2020**, *13*, 294. [CrossRef]
18. Hsiao, I.L.; Fritsch-Decker, S.; Leidner, A.; Al-Rawi, M.; Hug, V.; Diabaté, S.; Grage, S.L.; Meffert, M.; Stoeger, T.; Gerthsen, D.; et al. Biocompatibility of Amine-Functionalized Silica Nanoparticles: The Role of Surface Coverage. *Small* **2019**, *15*, 1805400. [CrossRef] [PubMed]
19. Moreira, A.F.; Dias, D.R.; Correia, I.J. Stimuli-responsive mesoporous silica nanoparticles for cancer therapy: A review. *Microporous Mesoporous Mater.* **2016**, *236*, 141–157. [CrossRef]
20. Holban, A.M.; Grumezescu, A.M.; Andronescu, E. *Inorganic Nanoarchitectonics Designed for Drug Delivery and Anti-Infective Surfaces*; Elsevier Inc.: Amsterdam, The Netherlands, 2016; ISBN 9780323428613.
21. Holban, A.M.; Grumezescu, A. Nanocomposite Drug Carriers. In *Nanoarchitectonics for Smart Delivery and Drug Targeting*; William Andrew: Norwich, NY, USA, 2016; pp. 261–284, ISBN 9780323477222.
22. Mamaeva, V.; Sahlgren, C.; Lindén, M. Mesoporous silica nanoparticles in medicine-Recent advances. *Adv. Drug Deliv. Rev.* **2013**, *65*, 689–702. [CrossRef] [PubMed]
23. He, Y.; Liang, S.; Long, M.; Xu, H. Mesoporous silica nanoparticles as potential carriers for enhanced drug solubility of paclitaxel. *Mater. Sci. Eng. C* **2017**, *78*, 12–17. [CrossRef]
24. Santoso, A.V.; Susanto, A.; Irawaty, W.; Hadisoewignyo, L.; Hartono, S.B. Chitosan modified mesoporous silica nanoparticles as a versatile drug carrier with pH dependent properties. *AIP Conf. Proc.* **2019**, *2114*, 020011. [CrossRef]
25. Kumar, P.; Tambe, P.; Paknikar, K.M.; Gajbhiye, V. Mesoporous silica nanoparticles as cutting-edge theranostics: Advancement from merely a carrier to tailor-made smart delivery platform. *J. Control. Release* **2018**, *287*, 35–57. [CrossRef]
26. Elbialy, N.S.; Aboushoushah, S.F.; Sofi, B.F.; Noorwali, A. Multifunctional curcumin-loaded mesoporous silica nanoparticles for cancer chemoprevention and therapy. *Microporous Mesoporous Mater.* **2020**, *291*, 109540. [CrossRef]
27. Qi, S.S.; Sun, J.H.; Yu, H.H.; Yu, S.Q. Co-delivery nanoparticles of anti-cancer drugs for improving chemotherapy efficacy. *Drug Deliv.* **2017**, *24*, 1909–1926. [CrossRef]
28. Sun, B.; Wang, X.; Liao, Y.P.; Ji, Z.; Chang, C.H.; Pokhrel, S.; Ku, J.; Liu, X.; Wang, M.; Dunphy, D.R.; et al. Repetitive Dosing of Fumed Silica Leads to Profibrogenic Effects through Unique Structure-Activity Relationships and Biopersistence in the Lung. *ACS Nano* **2016**, *10*, 8054–8066. [CrossRef]
29. Hegde, S.; Krisnawan, V.E.; Herzog, B.H.; Zuo, C.; Breden, M.A.; Knolhoff, B.L.; Hogg, G.D.; Tang, J.P.; Baer, J.M.; Mpoy, C.; et al. Dendritic Cell Paucity Leads to Dysfunctional Immune Surveillance in Pancreatic Cancer. *Cancer Cell* **2020**, *37*, 289–307.e9. [CrossRef] [PubMed]
30. Seelige, R.; Searles, S.; Bui, J.D. Mechanisms regulating immune surveillance of cellular stress in cancer. *Cell. Mol. Life Sci.* **2018**, *75*, 225–240. [CrossRef] [PubMed]

31. Song, Y.; Zhou, B.; Du, X.; Wang, Y.; Zhang, J.; Ai, Y.; Xia, Z.; Zhao, G. Folic acid (FA)-conjugated mesoporous silica nanoparticles combined with MRP-1 siRNA improves the suppressive effects of myricetin on non-small cell lung cancer (NSCLC). *Biomed. Pharmacother.* **2020**, *125*, 109561. [CrossRef] [PubMed]
32. Zhao, S.; Xu, M.; Cao, C.; Yu, Q.; Zhou, Y.; Liu, J. A redox-responsive strategy using mesoporous silica nanoparticles for co-delivery of siRNA and doxorubicin. *J. Mater. Chem. B* **2017**, *5*, 6908–6919. [CrossRef] [PubMed]
33. Cai, Z.; Chen, Y.; Zhang, Y.; He, Z.; Wu, X.; Jiang, L.P. PH-sensitive CAP/SiO2 composite for efficient co-delivery of doxorubicin and siRNA to overcome multiple drug resistance. *RSC Adv.* **2020**, *10*, 4251–4257. [CrossRef]
34. Yu, X.; Zhu, Y. Preparation of magnetic mesoporous silica nanoparticles as a multifunctional platform for potential drug delivery and hyperthermia. *Sci. Technol. Adv. Mater.* **2016**, *17*, 229–238. [CrossRef]
35. Maeda, H. Polymer therapeutics and the EPR effect. *J. Drug Target.* **2017**, *25*, 781–785. [CrossRef]
36. Park, J.; Choi, Y.; Chang, H.; Um, W.; Ryu, J.H.; Kwon, I.C. Alliance with EPR effect: Combined strategies to improve the EPR effect in the tumor microenvironment. *Theranostics* **2019**, *9*, 8073–8090. [CrossRef] [PubMed]
37. Zhou, X.; Chen, L.; Nie, W.; Wang, W.; Qin, M.; Mo, X.; Wang, H.; He, C. Dual-responsive mesoporous silica nanoparticles mediated codelivery of doxorubicin and Bcl-2 SiRNA for targeted treatment of breast cancer. *J. Phys. Chem. C* **2016**, *120*, 22375–22387. [CrossRef]
38. Pandey, S.P.; Shukla, T.; Dhote, V.K.; Mishra, D.K.; Maheshwari, R.; Tekade, R.K. Use of Polymers in Controlled Release of Active Agents. In *Basic Fundamentals of Drug Delivery*; Academic Press: Cambridge, MA, USA, 2019; pp. 113–172. [CrossRef]
39. Deb, P.K.; Kokaz, S.F.; Abed, S.N.; Paradkar, A.; Tekade, R.K. Pharmaceutical and Biomedical Applications of Polymers. In *Basic Fundamentals of Drug Delivery*; Academic Press: Cambridge, MA, USA, 2019; pp. 203–267. [CrossRef]
40. Wechsler, M.E.; Clegg, J.R.; Peppas, N.A. *The Interface of Drug Delivery and Regenerative Medicine*; Elsevier Inc.: Amsterdam, The Netherlands, 2019; Volume 1–3, ISBN 9780128136997.
41. dos Santos Rodrigues, B.; Lakkadwala, S.; Sharma, D.; Singh, J. *Chitosan for Gene, DNA Vaccines, and Drug Delivery*; Elsevier Inc.: Amsterdam, The Netherlands, 2019; ISBN 9780128184332.
42. Sung, Y.K.; Kim, S.W. Recent advances in the development of gene delivery systems. *Biomater. Res.* **2019**, *23*, 8. [CrossRef] [PubMed]
43. Guo, X.; Wang, L.; Wei, X.; Zhou, S. Polymer-based drug delivery systems for cancer treatment. *J. Polym. Sci. Part A Polym. Chem.* **2016**, *54*, 3525–3550. [CrossRef]
44. Yan, T.; Zhu, S.; Hui, W.; He, J.; Liu, Z.; Cheng, J. Chitosan based pH-responsive polymeric prodrug vector for enhanced tumor targeted co-delivery of doxorubicin and siRNA. *Carbohydr. Polym.* **2020**, *250*, 116781. [CrossRef] [PubMed]
45. Hu, C.; Chen, Z.; Wu, S.; Han, Y.; Wang, H.; Sun, H.; Kong, D.; Leng, X.; Wang, C.; Zhang, L.; et al. Micelle or polymersome formation by PCL-PEG-PCL copolymers as drug delivery systems. *Chin. Chem. Lett.* **2017**, *28*, 1905–1909. [CrossRef]
46. Li, Z.; Feng, H.; Jin, L.; Zhang, Y.; Tian, X.; Li, J. Polymeric micelle with pH-induced variable size and doxorubicin and siRNA co-delivery for synergistic cancer therapy. *Appl. Nanosci.* **2020**, *10*, 1903–1913. [CrossRef]
47. Arote, R.B.; Jere, D.; Jiang, H.-L.; Kim, Y.-K.; Choi, Y.-J.; Cho, M.-H.; Cho, C.-S. Injectable polymeric carriers for gene delivery systems. In *Injectable Biomaterials*; Woodhead Publishing: Cambridge, UK, 2011; pp. 235–259. [CrossRef]
48. Gardini, D.; Lüscher, C.J.; Struve, C.; Krogfelt, K.A. *Tailored Nanomaterials for Antimicrobial Applications*; Elsevier Inc.: Amsterdam, The Netherlands, 2018; ISBN 9780323512558.
49. Zhao, X.; Li, F.; Li, Y.; Wang, H.; Ren, H.; Chen, J.; Nie, G.; Hao, J. Co-delivery of HIF1α siRNA and gemcitabine via biocompatible lipid-polymer hybrid nanoparticles for effective treatment of pancreatic cancer. *Biomaterials* **2015**, *46*, 13–25. [CrossRef]
50. Lv, H.; Zhang, S.; Wang, B.; Cui, S.; Yan, J. Toxicity of cationic lipids and cationic polymers in gene delivery. *J. Control. Release* **2006**, *114*, 100–109. [CrossRef]
51. Fischer, D.; Bieber, T.; Li, Y.; Elsässer, H.P.; Kissel, T. A novel non-viral vector for DNA delivery based on low molecular weight, branched polyethylenimine: Effect of molecular weight on transfection efficiency and cytotoxicity. *Pharm. Res.* **1999**, *16*, 1273–1279. [CrossRef]
52. Alinejad-Mofrad, E.; Malaekeh-Nikouei, B.; Gholami, L.; Mousavi, S.H.; Sadeghnia, H.R.; Mohajeri, M.; Darroudi, M.; Oskuee, R.K. Evaluation and comparison of cytotoxicity, genotoxicity, and apoptotic effects of poly-l-lysine/plasmid DNA micro- and nanoparticles. *Hum. Exp. Toxicol.* **2019**, *38*, 983–991. [CrossRef]
53. Rimann, M.; Lühmann, T.; Textor, M.; Guerino, B.; Ogier, J.; Hall, H. Characterization of PLL-g-PEG-DNA nanoparticles for the delivery of therapeutic DNA. *Bioconjug. Chem.* **2008**, *19*, 548–557. [CrossRef]
54. Hershberger, K.K.; Gauger, A.J.; Bronstein, L.M. Utilizing Stimuli Responsive Linkages to Engineer and Enhance Polymer Nanoparticle-Based Drug Delivery Platforms. *ACS Appl. Bio Mater.* **2021**, *4*, 4720–4736. [CrossRef]
55. Hopkins, E.; Sanvictores, T.; Sharma, S. Physiology, Acid Base Balance. In *Urolithiasis*; StatPearls Publishing: Treasure Island, FL, USA, 2020; pp. 19–22.
56. Suo, A.; Qian, J.; Xu, M.; Xu, W.; Zhang, Y.; Yao, Y. Folate-decorated PEGylated triblock copolymer as a pH/reduction dual-responsive nanovehicle for targeted intracellular co-delivery of doxorubicin and Bcl-2 siRNA. *Mater. Sci. Eng. C* **2017**, *76*, 659–672. [CrossRef]
57. Wu, Y.; Zhang, Y.; Zhang, W.; Sun, C.; Wu, J.; Tang, J. Reversing of multidrug resistance breast cancer by co-delivery of P-gp siRNA and doxorubicin via folic acid-modified core-shell nanomicelles. *Colloids Surf. B Biointerfaces* **2016**, *138*, 60–69. [CrossRef] [PubMed]

58. Norouzi, P.; Amini, M.; Dinarvand, R.; Arefian, E.; Seyedjafari, E.; Atyabi, F. Co-delivery of gemcitabine prodrug along with anti NF-κB siRNA by tri-layer micelles can increase cytotoxicity, uptake and accumulation of the system in the cancers. *Mater. Sci. Eng. C* **2020**, *116*, 111161. [CrossRef]
59. Lin, C.; Hu, Z.; Yuan, G.; Su, H.; Zeng, Y.; Guo, Z.; Zhong, F.; Jiang, K.; He, S. HIF1α-siRNA and gemcitabine combination-based GE-11 peptide antibody-targeted nanomedicine for enhanced therapeutic efficacy in pancreatic cancers. *J. Drug Target.* **2019**, *27*, 797–805. [CrossRef]
60. Yuan, Z.Q.; Chen, W.L.; You, B.G.; Liu, Y.; Li, J.Z.; Zhu, W.J.; Zhou, X.F.; Liu, C.; Zhang, X.N. Multifunctional nanoparticles co-delivering EZH2 siRNA and etoposide for synergistic therapy of orthotopic non-small-cell lung tumor. *J. Control. Release* **2017**, *268*, 198–211. [CrossRef]
61. Wen, Z.M.; Jie, J.; Zhang, Y.; Liu, H.; Peng, L.P. A self-assembled polyjuglanin nanoparticle loaded with doxorubicin and anti-Kras siRNA for attenuating multidrug resistance in human lung cancer. *Biochem. Biophys. Res. Commun.* **2017**, *493*, 1430–1437. [CrossRef] [PubMed]
62. Jin, M.; Jin, G.; Kang, L.; Chen, L.; Gao, Z.; Huang, W. Smart polymeric nanoparticles with pH-responsive and PEG-detachable properties for co-delivering paclitaxel and survivin siRNA to enhance antitumor outcomes. *Int. J. Nanomed.* **2018**, *13*, 2405–2426. [CrossRef] [PubMed]
63. Xu, C.; Liu, W.; Hu, Y.; Li, W.; Di, W. Bioinspired tumor-homing nanoplatform for co-delivery of paclitaxel and siRNA-E7 to HPV-related cervical malignancies for synergistic therapy. *Theranostics* **2020**, *10*, 3325–3339. [CrossRef] [PubMed]
64. Nasab, S.H.; Amani, A.; Ebrahimi, H.A.; Hamidi, A.A. Design and preparation of a new multi-targeted drug delivery system using multifunctional nanoparticles for co-delivery of siRNA and paclitaxel. *J. Pharm. Anal.* **2021**, *11*, 163–173. [CrossRef]
65. Joshi, U.; Filipczak, N.; Khan, M.M.; Attia, S.A.; Torchilin, V. Hypoxia-sensitive micellar nanoparticles for co-delivery of siRNA and chemotherapeutics to overcome multi-drug resistance in tumor cells. *Int. J. Pharm.* **2020**, *590*, 119915. [CrossRef] [PubMed]
66. Chen, Y.; Li, B.; Chen, X.; Wu, M.; Ji, Y.; Tang, G.; Ping, Y. A supramolecular co-delivery strategy for combined breast cancer treatment and metastasis prevention. *Chin. Chem. Lett.* **2020**, *31*, 1153–1158. [CrossRef]
67. Yang, Y.; Meng, Y.; Ye, J.; Xia, X.; Wang, H.; Li, L.; Dong, W.; Jin, D.; Liu, Y. Sequential delivery of VEGF siRNA and paclitaxel for PVN destruction, anti-angiogenesis, and tumor cell apoptosis procedurally via a multi-functional polymer micelle. *J. Control. Release* **2018**, *287*, 103–120. [CrossRef] [PubMed]
68. Ahmed, K.S.; Hussein, S.A.; Ali, A.H.; Korma, S.A.; Lipeng, Q.; Jinghua, C. Liposome: Composition, characterisation, preparation, and recent innovation in clinical applications. *J. Drug Target* **2019**, *27*, 742–761. [CrossRef]
69. Sercombe, L.; Veerati, T.; Moheimani, F.; Wu, S.Y.; Sood, A.K.; Hua, S. Advances and challenges of liposome assisted drug delivery. *Front. Pharmacol.* **2015**, *6*, 286. [CrossRef]
70. Akbarzadeh, A.; Rezaei-sadabady, R.; Davaran, S.; Joo, S.W.; Zarghami, N. Liposome: Classification, prepNew aspects of liposomesaration, and applications. *Nanoscale Res. Lett.* **2013**, *8*, 102. [CrossRef]
71. Kiaie, S.H.; Mojarad-Jabali, S.; Khaleseh, F.; Allahyari, S.; Taheri, E.; Zakeri-Milani, P.; Valizadeh, H. Axial pharmaceutical properties of liposome in cancer therapy: Recent advances and perspectives. *Int. J. Pharm.* **2020**, *581*, 119269. [CrossRef]
72. Filipczak, N.; Pan, J.; Yalamarty, S.S.K.; Torchilin, V.P. Recent advancements in liposome technology. *Adv. Drug Deliv. Rev.* **2020**, *156*, 4–22. [CrossRef]
73. Jose, A.; Labala, S.; Ninave, K.M.; Gade, S.K.; Venuganti, V.V.K. Effective Skin Cancer Treatment by Topical Co-delivery of Curcumin and STAT3 siRNA Using Cationic Liposomes. *AAPS PharmSciTech* **2018**, *19*, 166–175. [CrossRef]
74. Qu, M.H.; Zeng, R.F.; Fang, S.; Dai, Q.S.; Li, H.P.; Long, J.T. Liposome-based co-delivery of siRNA and docetaxel for the synergistic treatment of lung cancer. *Int. J. Pharm.* **2014**, *474*, 112–122. [CrossRef]
75. Oh, H.R.; Jo, H.Y.; Park, J.S.; Kim, D.E.; Cho, J.Y.; Kim, P.H.; Kim, K.S. Galactosylated liposomes for targeted co-delivery of doxorubicin/vimentin sirna to hepatocellular carcinoma. *Nanomaterials* **2016**, *6*, 141. [CrossRef]
76. Zhang, C.; Zhang, S.; Zhi, D.; Zhao, Y.; Cui, S.; Cui, J. Co-delivery of paclitaxel and survivin siRNA with cationic liposome for lung cancer therapy. *Colloids Surf. A Physicochem. Eng. Asp.* **2020**, *585*, 124054. [CrossRef]
77. Yao, Y.; Su, Z.; Liang, Y.; Zhang, N. pH-Sensitive carboxymethyl chitosan-modified cationic liposomes for sorafenib and siRNa co-delivery. *Int. J. Nanomed.* **2015**, *10*, 6185–6198. [CrossRef]
78. Lymberopoulos, A.; Demopoulou, C.; Kyriazi, M.; Katsarou, M.; Demertzis, N.; Hatziandoniou, S.; Maswadeh, H.; Papaioanou, G.; Demetzos, C.; Maibach, H.; et al. Liposome percutaneous penetration in vivo. *Toxicol. Res. Appl.* **2017**, *1*, 239784731772319. [CrossRef]
79. Subongkot, T.; Ngawhirunpat, T.; Opanasopit, P. Development of ultradeformable liposomes with fatty acids for enhanced dermal rosmarinic acid delivery. *Pharmaceutics* **2021**, *13*, 404. [CrossRef]
80. Jiang, T.; Jin, K.; Liu, X.; Pang, Z. *Nanoparticles for Tumor Targeting*; Elsevier Ltd.: Amsterdam, The Netherlands, 2017; ISBN 9780081019153.
81. Abedi-Gaballu, F.; Dehghan, G.; Ghaffari, M.; Yekta, R.; Abbaspour-Ravasjani, S.; Baradaran, B.; Ezzati Nazhad Dolatabadi, J.; Hamblin, M.R. PAMAM dendrimers as efficient drug and gene delivery nanosystems for cancer therapy. *Appl. Mater. Today* **2018**, *12*, 177–190. [CrossRef]
82. Thakore, S.I.; Solanki, A.; Das, M. *Exploring Potential of Polymers in Cancer Management*; Elsevier Inc.: Amsterdam, The Netherlands, 2019; ISBN 9780128184332.

83. Li, Y.; Wang, H.; Wang, K.; Hu, Q.; Yao, Q.; Shen, Y.; Yu, G.; Tang, G. Targeted Co-delivery of PTX and TR3 siRNA by PTP Peptide Modified Dendrimer for the Treatment of Pancreatic Cancer. *Small* **2017**, *13*, 1602697. [CrossRef] [PubMed]
84. Jain, K. *Dendrimers: Smart Nanoengineered Polymers for Bioinspired Applications in Drug Delivery*; Elsevier Ltd.: Amsterdam, The Netherlands, 2017; ISBN 9780081019153.
85. Li, J.; Liang, H.; Liu, J.; Wang, Z. Poly (amidoamine) (PAMAM) dendrimer mediated delivery of drug and pDNA/siRNA for cancer therapy. *Int. J. Pharm.* **2018**, *546*, 215–225. [CrossRef] [PubMed]
86. Zhang, X.; Pan, J.; Yao, M.; Palmerston Mendes, L.; Sarisozen, C.; Mao, S.; Torchilin, V.P. Charge reversible hyaluronic acid-modified dendrimer-based nanoparticles for siMDR-1 and doxorubicin co-delivery. *Eur. J. Pharm. Biopharm.* **2020**, *154*, 43–49. [CrossRef] [PubMed]
87. Chen, K.; Xin, X.; Qiu, L.; Li, W.; Guan, G.; Li, G.; Qiao, M.; Zhao, X.; Hu, H.; Chen, D. Co-delivery of p53 and MDM2 inhibitor RG7388 using a hydroxyl terminal PAMAM dendrimer derivative for synergistic cancer therapy. *Acta Biomater.* **2019**, *100*, 118–131. [CrossRef] [PubMed]
88. Du Rietz, H.; Hedlund, H.; Wilhelmson, S.; Nordenfelt, P.; Wittrup, A. Imaging small molecule-induced endosomal escape of siRNA. *Nat. Commun.* **2020**, *11*, 1–17. [CrossRef] [PubMed]
89. Bus, T.; Traeger, A.; Schubert, U.S. The great escape: How cationic polyplexes overcome the endosomal barrier. *J. Mater. Chem. B* **2018**, *6*, 6904–6918. [CrossRef]
90. Vermeulen, L.M.P.; Brans, T.; Samal, S.K.; Dubruel, P.; Demeester, J.; De Smedt, S.C.; Remaut, K.; Braeckmans, K. Endosomal Size and Membrane Leakiness Influence Proton Sponge-Based Rupture of Endosomal Vesicles. *ACS Nano* **2018**, *12*, 2332–2345. [CrossRef]
91. Ghaffari, M.; Dehghan, G.; Baradaran, B.; Zarebkohan, A.; Mansoori, B.; Soleymani, J.; Ezzati Nazhad Dolatabadi, J.; Hamblin, M.R. Co-delivery of curcumin and Bcl-2 siRNA by PAMAM dendrimers for enhancement of the therapeutic efficacy in HeLa cancer cells. *Colloids Surf. B Biointerfaces* **2020**, *188*, 110762. [CrossRef] [PubMed]
92. Mohammed, A.F.A.; Higashi, T.; Motoyama, K.; Ohyama, A.; Onodera, R.; Khaled, K.A.; Sarhan, H.A.; Hussein, A.K.; Arima, H. In Vitro and In Vivo Co-delivery of siRNA and Doxorubicin by Folate-PEG-Appended Dendrimer/Glucuronylglucosyl-β-Cyclodextrin Conjugate. *AAPS J.* **2019**, *21*, 54. [CrossRef] [PubMed]
93. Yaqoob, S.B.; Adnan, R.; Rameez Khan, R.M.; Rashid, M. Gold, Silver, and Palladium Nanoparticles: A Chemical Tool for Biomedical Applications. *Front. Chem.* **2020**, *8*, 376. [CrossRef]
94. Sztandera, K.; Gorzkiewicz, M.; Klajnert-Maculewicz, B. Gold Nanoparticles in Cancer Treatment. *Mol. Pharm.* **2019**, *16*, 1–23. [CrossRef]
95. Yin, F.; Yang, C.; Wang, Q.; Zeng, S.; Hu, R.; Lin, G.; Tian, J.; Hu, S.; Lan, R.F.; Yoon, H.S.; et al. A light-driven therapy of pancreatic adenocarcinoma using gold nanorods-based nanocarriers for co-delivery of doxorubicin and siRNA. *Theranostics* **2015**, *5*, 818–833. [CrossRef]
96. Beik, J.; Khateri, M.; Khosravi, Z.; Kamrava, S.K.; Kooranifar, S.; Ghaznavi, H.; Shakeri-Zadeh, A. Gold nanoparticles in combinatorial cancer therapy strategies. *Coord. Chem. Rev.* **2019**, *387*, 299–324. [CrossRef]
97. Dykman, L.A.; Khlebtsov, N.G. Immunological properties of gold nanoparticles. *Chem. Sci.* **2017**, *8*, 1719–1735. [CrossRef] [PubMed]
98. Yafout, M.; Ousaid, A.; Khayati, Y.; El Otmani, I.S. Gold nanoparticles as a drug delivery system for standard chemotherapeutics: A new lead for targeted pharmacological cancer treatments. *Sci. Afr.* **2021**, *11*, e00685. [CrossRef]
99. Singh, S.; Maurya, P.K. *Nanotechnology in Modern Animal Biotechnology: Recent Trends and Future Perspectives*; Springer: New York, NY, USA, 2019; pp. 29–65. ISBN 9789811360046.
100. Kotcherlakota, R.; Srinivasan, D.J.; Mukherjee, S.; Haroon, M.M.; Dar, G.H.; Venkatraman, U.; Patra, C.R.; Gopal, V. Engineered fusion protein-loaded gold nanocarriers for targeted co-delivery of doxorubicin and erbB2-siRNA in human epidermal growth factor receptor-2+ ovarian cancer. *J. Mater. Chem. B* **2017**, *5*, 7082–7098. [CrossRef]
101. Labala, S.; Jose, A.; Chawla, S.R.; Khan, M.S.; Bhatnagar, S.; Kulkarni, O.P.; Venuganti, V.V.K. Effective melanoma cancer suppression by iontophoretic co-delivery of STAT3 siRNA and imatinib using gold nanoparticles. *Int. J. Pharm.* **2017**, *525*, 407–417. [CrossRef] [PubMed]

nanomaterials

MDPI

Review

Hyaluronic Acid within Self-Assembling Nanoparticles: Endless Possibilities for Targeted Cancer Therapy

Manuela Curcio [1], Orazio Vittorio [2,3,4], Jessica Lilian Bell [2,3], Francesca Iemma [1,*], Fiore Pasquale Nicoletta [1,*] and Giuseppe Cirillo [1]

1 Department of Pharmacy Health and Nutritional Science, University of Calabria, 87036 Rende, Italy
2 Children's Cancer Institute, Lowy Cancer Research Centre, University of New South Wales, Sidney, NSW 2052, Australia
3 School of Women's and Children's Health, University of New South Wales, Kensington, NSW 2052, Australia
4 ARC Centre of Excellence in Convergent Bio-Nano Science and Technology, Australian Centre for NanoMedicine, University of New South Wales, Kensington, NSW 2052, Australia
* Correspondence: francesca.iemma@unical.it (F.I.); fiore.nicoletta@unical.it (F.P.N.); Tel.: +39-0984-493011 (F.I.); +39-0984-493194 (F.P.N.)

Abstract: Self-assembling nanoparticles (SANPs) based on hyaluronic acid (HA) represent unique tools in cancer therapy because they combine the HA targeting activity towards cancer cells with the advantageous features of the self-assembling nanosystems, i.e., chemical versatility and ease of preparation and scalability. This review describes the key outcomes arising from the combination of HA and SANPs, focusing on nanomaterials where HA and/or HA-derivatives are inserted within the self-assembling nanostructure. We elucidate the different HA derivatization strategies proposed for this scope, as well as the preparation methods used for the fabrication of the delivery device. After showing the biological results in the employed in vivo and in vitro models, we discussed the pros and cons of each nanosystem, opening a discussion on which approach represents the most promising strategy for further investigation and effective therapeutic protocol development.

Keywords: cancer; drug delivery; drug targeting; hyaluronic acid; self-assembling nanoparticles

Citation: Curcio, M.; Vittorio, O.; Bell, J.L.; Iemma, F.; Nicoletta, F.P.; Cirillo, G. Hyaluronic Acid within Self-Assembling Nanoparticles: Endless Possibilities for Targeted Cancer Therapy. *Nanomaterials* **2022**, *12*, 2851. https://doi.org/10.3390/nano12162851

Academic Editor: Horacio Cabral

Received: 14 July 2022
Accepted: 16 August 2022
Published: 18 August 2022

1. Introduction

Over the last decade, the application of nanotechnology gained enormous interest as an interdisciplinary approach for cancer theranostics, with the number of researchers focusing on the development of tumor-targeting nanoparticles growing exponentially [1,2].

The unique properties of the nanoparticle system, including proper size, prolonged serum half-life, and specific cell targeting, together with the peculiar features of the tumor site, i.e., leakage of lymphatic drainage, angiogenesis, and increased vascular permeability, enable enhanced molecule accumulation at the tumor site (Enhanced Permeability and Retention effect—EPR) [3]. This offers solutions for both early-stage diagnosis and efficient delivery of therapeutic agents [4], boosting antitumor effects with reduction or reversal of multidrug resistance [5].

The biological performance of any nanosystem is strictly related to its chemical composition and fabrication method, as well as its architecture [6]. The chemical composition can move from organic to inorganic [7], from polymeric and lipid to hybrid and composite materials [8]. Whereas the fabrication method and the purification steps in particular can affect the matrix-associated toxicity, because of contamination with reaction by-products, residual solvents, and un-reacted species [9]. Finally, considering the architecture, non-spherical (e.g., tubes, cubes, cones) and spherical nanosystems can be distinguished, with further classifications possible regarding micellar, vesicular, solid nanoparticle or pristine, layered, and core-shell structures [10].

Generally, self-assembly refers to a process in which molecular building blocks (small- or macromolecules, nanomaterials) spontaneously organize into ordered structures with a certain geometric arrangement through local non-covalent interactions [11]. Self-assembly nanotechnologies play a pivotal role in nanomedicine since they are inspired by well-known biological processes, including the formation of the DNA double-helix and the arrangement of phospholipids in cell membranes [12]. The easy scalability of self-assembled nanoparticles (SANPs) preparation methods, which often involve green and inexpensive steps, fits well with the requirements needed for approval from regulatory agencies (e.g., FDA and EMA), thus allowing desirable laboratory-to-clinic-to-industry translations [13]. Moreover, the therapeutic outcomes of SANPs benefit from the ability to encapsulate (or co-encapsulate) with high-efficiency drugs with different physicochemical properties (e.g., hydrophilic, hydrophobic, amphiphilic, and ionic) [14,15], and form the possibility of easy modification with site-specific functionalities. This includes stimuli-responsive groups and small or large targeting moieties [16,17], and takes advantage of the peculiar structural and molecular anomalies at the tumor site (e.g., acidic interstitial pH, altered redox state due to increased cellular metabolism, enhanced oxygen perfusion) [18,19], as well as from the presence of overexpressed receptors for molecular components (e.g., growth factor, interleukins, transferrin) assisting tumor development and metastasis [20]. CD44, transmembrane glycoproteins involved in adhesion, aggregation, migration, and signal transduction, are representative biomarkers for cancer early-stage diagnosis and clinical management [21]. These receptors show a higher affinity for different extracellular elements, including hyaluronic acid (HA), a negatively charged, non-sulfated glycosaminoglycan consisting of D-glucuronic acid and N-acetyl-D-glucosamine repeating units bound by beta-linkages [22]. Thus, the insertion of HA moieties within nanoparticle formulations is a successful strategy for cancer targeting [23], although the choice of HA molecular significantly affects the targeting efficiency [24]. HA with different molecular weights, indeed, possess not only diverse biological functions [25], but also different cell uptake tendencies [26]. Several studies involving nanoparticle systems of different nature, from inorganic nanoparticles to liposomes [27], proved that high molecular HA is a more effective targeting element than low molecular weight HA (e.g., 31 kDa HA was better internalized by HeLa cells than 6 kDA HA [28]), although this is not a general statement since there are experimental evidences that in photodynamic therapy (PDT) protocols, 20 kDa HA exerted better efficacy than 50 kDa or 100 kDa HA [29]. Moreover, by virtue of the presence of hydroxyl, carboxylic, and N-acetyl groups, allowing easy chemical derivatizations, HA can be used in either native or modified forms [30].

Within this review, we overview the impact of HA-SANPs in cancer diagnosis and therapy over the last decade, focusing on the key peculiarities of each formulation. By discussing the chemical composition, the preparation methods, and the biological performances, we aim to highlight the pros and cons of the different proposed approaches, offering a multidisciplinary point of discussion for scientists working in cancer-related research areas. Finally, with a glance to the future in the field, we provide a critical analysis concerning the flaws to be considered and solved for effective bench-to-clinic translation.

2. Self-Assembling Nanoparticles Containing Hyaluronic Acid

SANPs are obtained through a process involving the spontaneous organization or aggregation of small molecules, macromolecules, or nanoparticles into stable structures. The interaction forces consist of hydrophobic interactions, π–π aromatic stacking, electrostatic forces, van der Waals forces, and hydrogen bonding [31]. Commonly, hydrophobic interactions drive the formation of SANPs (e.g., micelles or vesicles) composed of amphiphilic molecules where saturated or unsaturated hydrocarbon chain and polar ionic or non-ionic moieties are the lipophilic and hydrophilic counterparts, respectively [32]. In detail, micelles are nanosized particles consisting of a hydrophobic inner portion surrounded by a hydrophilic outer surface [33], while vesicles are hollow structures with an aqueous core surrounded by one or more bilayered membrane [34]. On the other hand, oppositely charged molecules and polymers can self-assembly via electrostatic forces and hydrogen bonding carrying out to nanoplexes and solid nanoparticles [35,36].

The formation of HA-SANPs results from the formation of either electrostatic forces due to their anionic nature or hydrophobic interactions when functionalized with lipophilic moieties. Moreover, supramolecular structures can be obtained when cyclodextrins (CD) and their inclusion counterparts are involved in the self-assembly process [37]. The choice of the suitable preparation technique is driven by the physicochemical properties of the selected HA-based material, as well as by the nature of the interactions between the HA binding blocks and with the loaded therapeutics [38]. Such techniques mainly involved the simple dispersion in water media, with sonication or ultra-sonication methods used as formation co-adjuvants [39], while dialysis processes are used when the HA-derivative needs to be dispersed in organic solvents [40]. The first methodologies can be used when hydrophilic therapeutic agents are used, while hydrophobic molecules should be treated with the second approach [41]. Moreover, typical thin-film hydration or emulsion methods are employed when HA derivatives are organized in liposomal-like structures able to load water-soluble and insoluble bioactive molecules [42,43]. Finally, HA or HA derivatives can be used for the coating (either electrostatic or covalent) of pre-formed SANPs to enhance the targeting behavior [44], but these materials do not fall within the scope of the present review. Here the discussion of the HA-SANPs proposed in the literature for cancer treatment is organized into four main sections, depending on the driving force of the self-assembly process, with further sectioning in native or modified HA.

3. Application of HA-SANPs in Cancer Therapy

As discussed in the previous section, HA-SANPs can be obtained by both electrostatic and hydrophobic interaction forces. In the following paragraphs, we will discuss the main outcome of each approach in cancer theranostics (Figure 1), highlighting the need for HA derivatization to favor the self-assembly process.

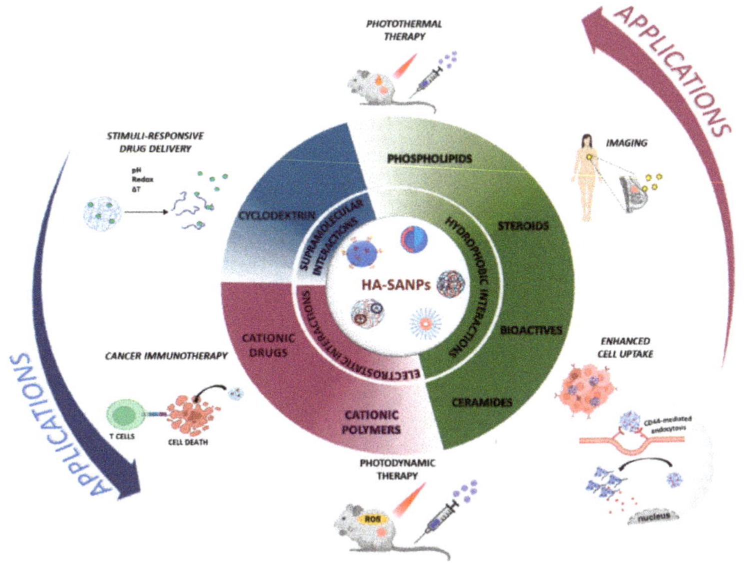

Figure 1. Applications of HA-SANPs for cancer theranostics: indication of the main mechanisms of HA-SANPs formation and the most representative derivatization agents.

3.1. HA-SANPs Obtained by Electrostatic Interactions

Table 1 collects the most relevant examples of HA-SANPs by electrostatic interactions.

Table 1. HA-SANPs obtained by electrostatic interactions.

Composition (*Preparation*)		Bioactive Agent	Performance				Outcome	Ref.
HA-Derivative	Other Components		Cancer Type	In Vitro CD44+	In Vitro CD44−	In Vivo		
HA (*Water dispersion*)	CDDP	CDDP	Lung	LLC	—	LLC Xm	Control Release (pH) Selective Biodistr	[45]
HA (*Water dispersion*)	CDDP/SRF	CDDP/SRF	Liver	HepG2	—	HepG2 Xm	Control Release (pH) Synergism Selective Biodistr	[46]
HA (*Water dispersion*)	CDDP/GFT CDDP/MTX	CDDP/GFT CDDP/MTX	Breast	MDA-MB-231	MCF-7	—	Targeting Multidrug therapy Sustained Release	[47]
HA (*Sonication*)	FCP-Tph	FCP-Tph	Breast	MDA-MB-231 4T1	NIH 3T3	S-D Rats 4T1 Xm	PDT Synergism Selective Biodistr	[48]
HA (*Water dispersion*)	PRTS-miR-34a	miR-34a	Breast	MDA-MB-231	MCF-7	MDA-MB-231 Xm	Control Release (pH) Synergism Selective Biodistr	[49]
HA/TPP (*Ionic crosslinking*)	CS	miR-34a DOX	Breast	MDA-MB-231	—	MDA-MB-231 Xm	Control Release (pH) Synergism	[50]
HA (*Ionic coordination*)	CS SBE-βCD	CUR	Colon	HT-29	I407	—	Targeting Synergism	[51]
HA-SH * (*Water dispersion*)	CS	DOX	Breast	SKBR3	—	—	Control Release (pH/redox)	[52]
HA-SH HS-HA-DA (*Ionic coordination*)	NOCC —	DOX CaP-siRNA	Cervix Ovary	HeLa OVCAR-3/MDR	—	—	Controlled Release (pH/redox) Synergism	[53]
HA-SH #-oDNA * (*K^+-dependent self-assembly*)	—	—	Cervix	HeLa	NIH-3T3	—	Cell Blebbing and Death	[54]
HA (*Ionic crosslinking*)	—	DOX	Bone	K7	—	S-D Rats K7 Xm	Control Release (pH) Synergism Selective Biodistr	[55]
HA-His * (*Ionic crosslinking*)	—	DOX/Ce6/Mn^{2+}	Skin	B16	—	B16 Xm	Control Release (pH/redox) MRI/PDT/Synergism	[56]
HHA (*Desolvation + coordination crosslinking*)	BSA	CDDP/ICG	Liver	HepG2	L929	HepG2 Xm	Control Release (redox) PTT/Synergism Selective Biodistr	[57]
HA (*Dialysis*)	MPL/QS21/ R837	MPL/QS21/ R837/OVA	— Lymphatic system	BMDCs —	RAW 264.7 —	C57BL/6 BALB/c mice EG7-OVA Xm	Selective Biodistr Immunotherapy (OVA antigen)	[58]

* Carbodiimide chemistry; # $NaBH_3CN$ + DTT; BSA: Bovine serum albumin; CaP: Calcium phosphate; CDDP: Cisplatin; Ce6: Chlorin e6; CS: Chitosan; CUR: Curcumin; Cys: Cystamine; DA: Dopamine; DOX: Doxorubicin; FCP: Ferrocene cyclopalladated compound; GFT: Gefitinib; HA: Hyaluronic acid; HHA: Hydrazided HA; His: Histidine; ICG: Indocyanine green; MPL: 3-O-desacyl-4′-monophosphoryl lipid A; MRI: Magnetic Resonance Imaging; MTX: Methotrexate; NOCC: N,O-Carboxymethyl chitosan; oDNA: DNA oligonucleotide; OVA: Ovalbumin; PDT: Photodynamic therapy; PRTS: Protamine sulfate; R837: Imiquimod; S-D: Sprague Dawley; SBE: Sulphobutyl-ether; SRF: Sorafenib; Tph: 5,10,15,20-Tetrakis(4-aminophenyl)-porphin; TPP: Tripolyphosphate; Xm: Xenograft mice.

The negative charge of HA can be exploited for the formation of nanoparticle structures with cationic drugs such as cisplatin (CDDP) acting as ionic crosslinker [45], with the further insertion of therapeutic agents such as Sorafenib (SRF) [46], Gefitinib (GFT) and Methotrexate (MTX) [47] found to be a valuable strategy for an effective multidrug therapy.

The formation of hydrogen bonds (SRF) or π-π stacking (GFT and MTX) interactions, indeed, enhances the stability of the nanoformulation, improving the in vitro and in vivo pharmacological outcomes by virtue of both a pH-controlled release and a selective CD44 targeting and biodistribution. Following a similar approach, micelle nanocarriers were developed by complexing ferrocene cyclopalladated compound (FCP) with HA in the presence of 5,10,15,20-Tetrakis(4-aminophenyl)-porphin (Tph) as a photosensitizer [48]. The obtained HA-SANPs were successfully employed in a photodynamic therapy (PDT) protocol for the treatment of breast cancer models in vivo.

HA-SANPs containing native HA were also proposed for the vectorization of genic materials with high efficiency. In these formulations, cationic macromolecules such as Protamine sulfate (PRTS) [49] or Chitosan (CS) [50] were inserted as complexing agents for a miR-34a mimic to improve the loading efficiency, while the presence of HA guaranteed the targeting of CD44 positive cells. Moreover, when CS was used, nanogel systems can be obtained by adding Tripolyphosphate (TPP) in the reaction feed exploiting the well-known ability of such polyanion to act as a crosslinker upon interaction with the NH_3^+ groups on CS side chains [50]. The resulting nanosystem was found to be suitable for dual therapy where Doxorubicin (DOX) was used as a conventional cytotoxic agent in combination with a miR-34a mimic (Figure 2).

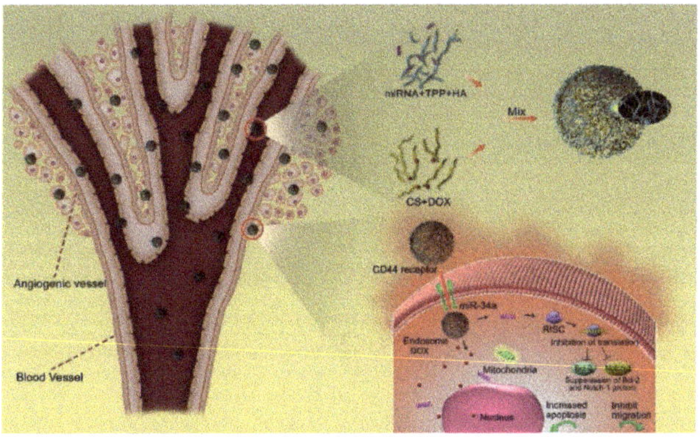

Figure 2. Schematic representation of HA-SANPs obtained from electrostatic interactions between HA and CS for DOX and miR-34a co-delivery. Reprinted with permission from Ref. [50]. 2014, Elsevier Ltd.

SANPs can be also obtained by the direct ionic interaction between HA and CS, with further stabilization of the nanoparticle structure being achieved by oxidation of thiol groups inserted on HA side chains and the formation of disulfide bridges.

HA/CS complexation was used for the vectorization of Curcumin (CUR) to colon cancer cells upon inclusion in a cyclodextrin derivative [51], while disulfide stabilization was proposed by Xia et al. [52] to prepare dual responsive (pH and redox) DOX delivery systems for the treatment of breast cancer cells. In the latter case, the key advantage of the proposed nanosystem is that, by selecting a proper HA to CS ratio, negative or positive surface charges can be obtained in order to optimize the interaction with ionic drugs. Moreover, targeted chimeric nanocarriers for the co-delivery of DOX and siRNA were constructed by conjugating two different HA-SANPs through redox-sensitive thiol–disulfide bonds. HA-SH was combined to N,O-Carboxymethyl chitosan (NOCC) for DOX loading, while the Calcium Phosphate siRNA complex was encapsulated within Dopamine (DA)-HA-SH based SANPs [53].

HA derivatization with SH groups was also proposed for the formation of an oligoDNA complex able to self-assemble in a K^+-dependent manner in a G-quadruplex causing selective cancer cell blebbing and death [54].

Metal chelation is another methodology useful for promoting the self-assembly of HA nanoparticles, with calcium ions being widely explored as pH-responsive crosslinking agents [55]. On the other hand, Mn^{2+} ions are capable of both stabilizing SANPs by forming crosslinks and acting as magnetic imaging agents. Moreover, by competitive coordination, manganese ions are able to decrease the glutathione (GSH) intracellular concentration with a beneficial effect on PDT protocols. These findings were recorded in a work by Pan et al., where multifunctional HA-SANPs for combined chemo-photodynamic therapy were developed taking advantage of the loading of DOX as an antineoplastic agent, and Chlorin e6 as PDT agent, as well as from the HA derivatization with Histidine (His) residues to enhance the affinity towards Mn^{2+} ions [56]. Pt ions within CDDP molecules can also act as metal crosslinkers by ligand exchange between the NH_2 and the hydrazide groups of HA-3,3′-dithiobis(propionohydrazide) derivative (HHA) within HHA/BSA nanoparticles. Moreover, the simultaneous coordination between CDDP and the sulfonic groups of ICG allowed the formation of SANPs for dual chemo-photothermal therapy [57].

The ionic nature of HA can be responsible for the formation of strong hydrogen bonding with different species, including water-insoluble compounds of biological interest. In this regard, HA was proposed as a targeting dispersant agent for the immunostimulatory monophosphoryl lipid A (MPL) in combination with the extract from the bark of the *Quillaja saponaria Molina* tree (QS-21) or Imiquimod (R837). The resulting complexes were found to enhance both humoral and cellular immunity and thus can be used as a vaccine system (Ovalbumin—OVA as model antigen) to induce prophylactic anticancer immune response preventing tumor recurrence and growth in vivo [58].

3.2. HA-SANPs Obtained by Hydrophobic Interactions

The HA functionalization with lipophilic moieties was proved as a valuable strategy to confer amphiphilic properties allowing the organization of HA-derivative in stable nanoparticle systems. Different molecular specimens can be used as lipophilic moieties, which are here classified in three main classes, namely the steroid-, lipid-, and phenyl-based structures.

3.2.1. Steroid Modified HA in the Formation of HA-SANPs

HA backbone was hydrophobically modified by conjugation with cholesterol (CHL) moieties via carbodiimide chemistry to form either nanoparticle structures for drug and gene delivery or liposomes when inserted in a proper mixture of phospholipids (Table 2).

Table 2. HA-SANPs obtained by hydrophobic interactions of steroid-modified HA.

Composition (*Preparation*)		Bioactive Agent	Performance					
			Cancer Type	In Vitro		In Vivo	Outcome	[Ref]
HA-Derivative	Other Components			CD44+	CD44−			
HA-CHL * (*Sonication*)	—	2b/SiRNA	Skin	B16-F10	RAW264.7	—	Targeting Control Release (pH)	[59]
GE11-HA-cys-CHL * (*Sonication*)	—	DOX	Breast	MCF-7 MDA-MB-231	—	MDA-MB-231 Xm	Dual Targeting Control Release (Redox) Synergism	[60]
HA-cys-CHL * (*Dialysis*)	—	IR780	Breast	MDA-MB-231	—	MDA-MB-231 Xm	PTT/PDT Selective Biodistr Synergism	[61]
HA-CHL * (*Embedding*)	HSCP	DOX/PTX	Breast Liver	MCF-7 —	L929 HepG2	—	Control Release (pH) Synergism	[62]
KLVFF-pA [§]- HA-CHL * (*Thin-film hydration*)	LipoidS100/ CHL/ DSPE-mPEG	KLVFF DOX	Breast	4T1	HUVEC	Balb/c mice 4T1 Xm	Synergism Metastasis Inhibition	[63]
HA-TST * (*Dialysis*)	—	CPT/DOX	Breast	MCF-7	—	—	Control Release (pH) Synergism	[64]
HA-5βCA-Cy7.5 * (*Water dispersion*)	—	—	Breast Prostate	MDA-MB 231 PC-3	—	—	Targeting Control Release (HAase)	[65]
HA-5βCA-Cy5.5 * (*Water dispersion*)	—	—	Squamous	SCC7	CV-1	SCC7 Xm	Selective Biodistr	[66]
HA-5βCA * (*Sonication*)	—	PTX	Squamous	SCC7	NIH-3T3	SCC7 Xm	Targeting Synergism Selective Biodistr	[67]
HA-5βCA * (*High-pressure homogenization*)	—	PTX	Colon Lung Breast Liver Skin	HT29 A549 MDA-MB 231 HepG2 MDA-MB-435	NIH-3T3	— — — — MDA-MB-435 Xm	Targeting Control Release (HAase) Selective Biodistr	[68]
HA-5βCA * (*O/W Emulsion*)	—	PFP	Blood Colon	CL —	—	— HT-29 Xm	Echogenic Diagnosis	[69]
PEG-NH_2-HA-5-βCA-Cy5.5 * (*Water dispersion*)	—	—	Squamous Colon Breast	SSC7 HCT116 MDA-MB 231	CV-1	SSC7 Xm — —	Selective Biodistr	[70]
PEG-NH_2-HA-5-βCA * (*Sonication*)	—	DOX CPT	Squamous Colon Breast	SSC7 HCT116 MDA-MB 231	NIH-3T3	SSC7 Xm — —	Control Release (HAase) Selective Biodistr	[71]
PEG-NH_2-HA-5-βCA-Cy5.5 * (*O/W Emulsion*)	—	IRT	Colon	—	—	HT-29 Xm CT-26 Xm	Diagnosis Synergism Selective Biodistr	[72]
HA-DOCA-His * (*Sonication*)	—	PTX	Breast	MCF-7	—	MCF-7 Xm	Control Release (pH) Synergism	[73]

Table 2. *Cont.*

Composition (*Preparation*)			Performance					
				In Vitro				
HA-Derivative	Other Components	Bioactive Agent	Cancer Type	CD44+	CD44−	In Vivo	Outcome	[Ref]
HA-cys-DOCA-His * (*Dialysis*)	—	PTX	Breast	MDA-MB-231	—	MDA-MB-231 Xm	Control Release (Redox) Synergism	[74]
mPEG-HA(DOCA)-NAC * (*Sonication*)	—	PTX	Breast Liver	MCF-7 —	—	— H22 Xm	Control Release (Redox) Synergism Selective Biodistr	[75]
HA-DOCA-His * (*Dialysis*)	PF 127	DOX	Breast	MCF-7 MCF-7/ADR	—	MCF-7/ADR Xm	Control Release (pH) Resistance Reversal	[76]

* Carbodiimide chemistry; § Click chemistry; 2b: 2b RNA-binding protein; 5-βCA: 5-β-Cholanic acid; CHL: Cholesterol; CPT: Camptothecin; Cy: Cyanine; Cys: Cystamine; DOCA: Deoxycholic acid; DOX: Doxorubicin; DSPE: 1,2-Distearoyl-sn-glycero-3-phosphocholine; GE11: targeting peptide; HA: Hyaluronic acid; HAase: Hyaluronidase; His: Histidine; HSCP: Lecithin hydrogenated; IRT: Irinotecan; KLVFF: Lys-Leu-Val-Phe-Phe peptide; mPEG: Poly(ethylene glycol) methyl ether; NAC: N-acethyl cysteine; pA: Propargylamide; PF 127: Pluronic F127; PFP: Perfluoropentane; PTX: Paclitaxel; TST: Testosterone; Xm: Xenograft mice.

Taking into consideration that HA-based nanoparticles cannot directly encapsulate anionic siRNA molecules due to the net negative charge, Choi et al. proposed the incorporation of siRNA/2b protein complexes into HA-CHL nanoparticles. They found that the nanosystem was able to selectively deliver the 2b protein/siRNA complexes to melanoma cells with up-regulated CD44 receptors, release the siRNA within the endocytic compartments due to dissociation of the 2b protein/siRNA at acidic pH, and effectively suppress the expression of the target gene [59]. HA-CHL nanoparticles were also endowed with redox responsivity when a GSH-sensitive linker such as Cystamine (cys) was used to connect HA and CHL molecules. By this strategy, DOX and IR780, as cytotoxic drugs or photosensitizing agents, respectively, were vectorized to breast cancer cells in both in vitro and in vivo models. GE11 was used as a targeting peptide to improve the selectivity of the DOX release [60], while cell death occurred by high ROS generation upon IR780 laser irradiation (PDT step) followed by high increased temperature (photothermal effect—PTT) [61].

The insertion of HA-CHL conjugate in liposome formulations can be performed by either post-insertion in pre-formed vesicular formulation or hydration of the thin layer film with an HA-derivative solution. By the post-insertion method, hydrogenated Lecithin-based liposomes for DOX and Paclitaxel (PTX) combined therapy were prepared, with the HA residues on the outer surface able to discriminate between CD44+ (breast cancer) and CD44− (fibroblast and liver cancer) cells [62]. On the other hand, complex liposomal structures can be obtained when HA-CHL aqueous solution was employed as a hydrating agent. The efficiency of such a system was further enhanced by conjugation with the Lys-Leu-Val-Phe-Phe (KLVFF) peptide, a key sequence involved in the β-sheet fibril formation showing antimetastatic activity [63].

Together with CHL, other steroid structures such as testosterone (TST) [64], 5-β-Cholanic acid (5-βCA), and Deoxycholic acid (DOCA) have been proposed for the lipidization of HA. Cyanine-labeled self-assembled HA-5-βCA nanoparticles (Figure 3), prepared via different techniques, were effectively targeted in different cancer cells and tissues, including breast, prostate, and squamous carcinomas [65,66].

Such nanosystems were proposed for the delivery of PTX [67] with the further possibility to modulate the release by exploiting the hydrolytic activity of Hyaluronidase (HAase) selectively expressed within the tumor cells [68]. Moreover, the encapsulation of Perfluoropentane (PFP) within HA-5-βCA SANPs allowed the obtainment of echogenic materials for the early-stage diagnosis of colon cancer [69].

The effective accumulation of HA-SANPs into the tumor site is a result of a combined EPR and active targeting. Nevertheless, the high affinity of HA towards the HA receptor (HARE) expressed by liver sinusoidal endothelial cells can determine high liver uptake, with a possible reduction of the therapeutic efficiency. The PEGylation of HA-SANPs can be a valuable approach to specifically address this issue: PEG molecules, indeed, form a hydrophilic shell on the nanoparticle outer surface, conferring stealth properties towards the phagocytic cells of the reticuloendothelial system and thus prolonging the blood circulation time. When applied to self-assembled HA-5-βCA nanoparticles, the selective biodistribution of nanocarriers [70] allowed the pharmacokinetics profiles of different chemotherapeutic agents, such as DOX, CPT [71], and Irinotecan [72] to be improved.

The targeting efficiency of HA-SANPs can be enhanced by inserting pH and/or redox responsive functionalities in the nanoparticle structure. For this purpose, the imidazole ring of His (pH responsivity) [73] and the disulfide bridges of cys (GSH responsivity) [74] were conjugated to HA-DOCA derivatives, obtaining PTX vectorization to breast cancer both in vitro and in vivo.

As a further development of these systems, the HA-DOCA derivative was conjugated or co-formulated with PEG and Pluronic (PF 127) species, obtaining nanocarriers able to modulate the release of DOX and PTX in response to the acidic and GSH-rich tumor environment. In detail, the PEGylation processes allowed the enhancement of the biodistribution profiles [75], while the presence of PF 127 was used to improve cellular uptake thus counteracting the insurgence of multidrug resistance processes [76].

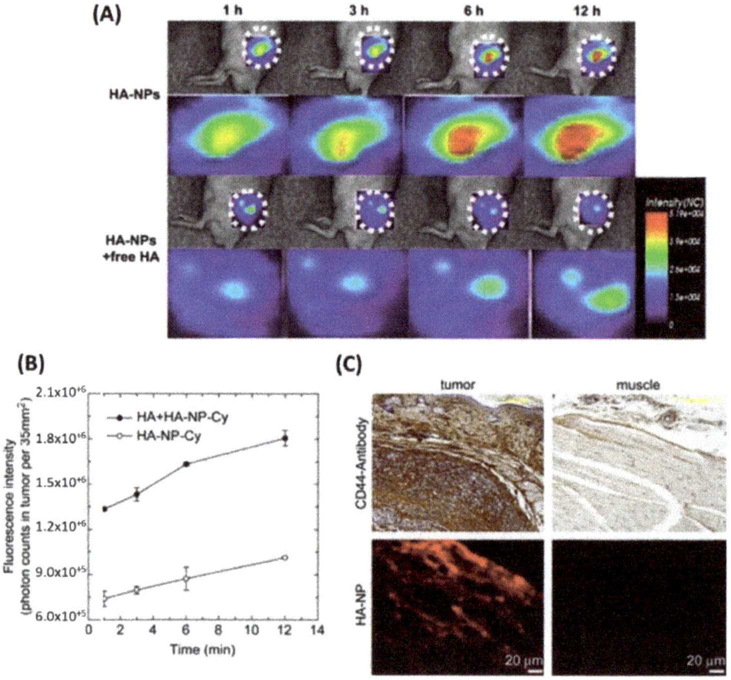

Figure 3. (**A**) In vivo fluorescence images of self-assembled HA-5-βCA nanoparticles and (**B**) their quantification in tumor-bearing mice with and without pre-injection of free-HA. (**C**) Magnified images of tumor and muscle tissues. Adapted with permission from Ref. [66]. 2009, Elsevier Ltd.

3.2.2. Lipid-modified HA in the formation of HA-SANPs

The introduction of lipophilic moieties on the HA backbone can be reached by conjugation with lipid chains, belonging to phospholipids, ceramides, fatty acids, amines, and alcohols (Table 3).

Table 3. HA-SANPs obtained by hydrophobic interactions of lipid-modified HA.

Composition (*Preparation*)		Bioactive Agent	Performance				Outcome	Ref.
			Cancer Type	In Vitro		In Vivo		
HA-Derivative	Other Components			CD44+	CD44−			
HA-DSPE * HA-DMPE * *(Sonication)*	CHL	—	Breast	MCF-7	—	—	Biocompatibility	[77]
HA-DPPE $ *(Thin-film hydration)*	CHL/DPPC/PG	C12GEM	Pancreas	MiaPaCa2	VIT1	MiaPaCa2 Xm	Synergism Selective Biodistr	[78]
HA-PEG-DSPE * *(O/W emulsion)*	Tf-PEG-DSPE * GM/DOTAP/PC	pDNA	Lung	A549	—	A549 Xm	Dual Targeting Sustained Release Enhanced Transfection	[79]
HA-CE £ *(Dialysis)*	—	HB PTX	Lung	A549	—	A549 Xm	Sustained Release Synergism/PDT	[80]
HA-CE £ *(Thin-film hydration)*	DOTAP/DOPE	pDNA	Breast	MDA-MB-231	NIH-3T3	—	Synergism	[81]
HA-CE £ *(Thin-film hydration)*	PC/CHL	DOX/MGV	Breast	MDA-MB-231	—	S-D rats MDA-MB-231 Xm	Control Release (pH) Selective Biodistr Synergism/MR Imaging	[82]
HA-CE £ *(Thin-film hydration)*	P85	DTX	Brain Breast	U87-MG MCF-7 MCF-7/ADR	—	— MCF-7/ADR	Sustained Release Synergism Resistance Reversal	[83]
His-HA-DDA * *(Dialysis)*	—	DOX	Breast	4T1	—	4T1 Xm	Control Release (pH) Selective Biodistr Synergism	[84]
HA-DDA * *(Self-emulsification)*	Miglyol812 Tween80 SolutolHS15 CTAB	DTX	Lung	A549	—	—	Targeting Enhanced Uptake	[85]
HA-HDA * *(Thin lipid film hydration)*	DPPC	IONPs/DTX	Breast	MCF-7	NIH-3T3	—	Synergism PTT Magnetic Targeting	[86]
HA-HDA * *(O/W emulsion)*	PLGA	ZnPHC	Colon Lung Liver	HT-29 A549 —	— — LO2	HT-29 Xm — —	PTT Selective Biodistribution	[87]
HA-DO * *(Thin lipid film hydration)*	CaP	ICG	Lung	A549	—	A549 Xm	Control Release (pH) PTT/PDT	[88]
HA-cys-STA * *(Dialysis)*	—	DOX	Colon	HCT116	HEK293 CT-26	HCT116 Xm CT-26 Xm	Control Release (Redox) Synergism	[89]
HA-AUT * *(Water dispersion)*	—	FITC-DEX NR	Breast	MDA-MB-468	SK-BR-3	—	Control Release (Redox) Targeting	[90]
MPEG-ss-HA-HDO * *(Sonication)*	—	PTX	Breast Liver	MCF-7 —	—	— H22 Xm	Control Release (Redox) Synergism Selective Biodistr	[91]
HA-His-MGK * *(Thin-film hydration)*	—	CUR	Squamous	—	—	SCC7 Xm	Control Release (pH) Selective Biodistr	[92]

Table 3. *Cont.*

Composition (*Preparation*)		Bioactive Agent	Performance				Outcome	Ref.
			Cancer Type	In Vitro		In Vivo		
HA-Derivative	Other Components			CD44+	CD44−			
HA-His-MGK * (*Thin-film hydration*)	PEG-NH_2-CS-K *	CUR	Mesothelioma	HMM-239		HMM-239 Xm	Control Release (pH) Synergism In Vivo	[93]
FA-HA-MGK * (*Dialysis*)	—	CUR	Lung Breast	A549 MCF-7	—	—	Double Targeting Controlled Release (pH)	[94]

* Carbodiimide chemistry; $ reductive amination; £ TBA mediated condensation; AUT: 11-(Aminooxy)-1-undecanethiol; C12GEM: 4-(N)-lauroyl-gemcitabine; CaP: Calcium Phosphate; CE: Ceramide; CHL: Cholesterol; CTAB: Cetyl trimethylammonium bromide; CUR: Curcumin; Cys: Cystamine; DDA: Dodecylamine; DMPE: 1,2-Dimiristoyl-sn-glycerol-3-phosphatidylethanolamine; DO: 1,2-Dioleoyl-3-amino-propane; DOPE: 1,2-Dioleoyl-sn-glycero-3-phoshphoethanolamine; DOTAP: 1,2-dioleoyl-3-trimethylammonium-propane; DOX: Doxorubicin; DPPC: 1,2-dipalmitoyl-sn-glycero-3-phosphocholine; DPPE: 1,2-Dipalmitoyl-sn-glycero-3-phosphoethanolamine; DSPE: 1,2-Distearoyl-sn-glycero-3-phosphocholine; DTX: Docetaxel; FITC-DEX: Fluorescein isothiocyanate-Dextran; GM: Glycerol monostearate; HA: Hyaluronic acid; HB: Hypocrellin B; HDA: Hexadecylamine; HDO: Hexadecanol; His: Histidine; ICG: Indocyanine green; IONPs: Iron oxide nanoparticles; MGK: Menthone 1,2-glycerol ketal; MGV: Magnevist—gadopentetate dimeglumine; MPEG: Poly(ethylene glycol) methyl ether; MR: Magnetic resonance; NR: Nile red; P85: Pluronic P85; PC: Phosphatidylcholine; pDNA: Plasmid DNA; PEG: Poly(ethylene glycol); PG: Phosphatidylglycerol; PHC: Phthalocyanine: PLGA: Poly(lactic-co-glycolic acid); PTX: Paclitaxel; S-D: Sprague Dawley; STA: Stearic acid; Tf: Transferrin; Xm: Xenograft mice.

Phospholipids are highly biocompatible compounds widely employed for the fabrication of different drug delivery systems, such as micelles, liposomes, solid lipid nanoparticles, micro- and nano-emulsions [95]. They can serve as HA lipidizing agents, allowing the obtainment of SANPs with different architectures [77] for the vectorization of cytotoxic drugs and gene to pancreatic [78] and lung [79] carcinomas. In the latter case, the insertion of the transferrin (Tf) motif within the nanoparticle formulation enhanced the targeting behavior and the transfection efficiency which was found to be significantly superior to that of conventional liposomes used as a control.

Ceramides (CE) belong to the sphingolipids group and consist of an acylated long-chain sphingosine base. Although they have a positive net charge, ceramides are used as structural components of nanoformulations by virtue of their ability to easily move across cell membranes [96].

HA-CE conjugates, alone or in combination with phospholipids and pluronics, were properly used as a component of nanoparticle, liposome, and micelle formulations. Pure HA-CE nanoparticles were tested as a vehicle for the vectorization of PTX in combination with Hypocrellin B (HB) as a photosensitizer in synergistic chemo- and photodynamic-treatment of lung cancer [80]. The choice of cationic or neutral phospholipids allowed the insertion of HA-CE in liposomal bilayer for the delivery of plasmid DNA [81] or drug to MDA-MB-231 breast cancer cells, with the possibility to simultaneously load a gadolinium derivative for Magnetic Resonance Imaging (MRI) [82] (Figure 4).

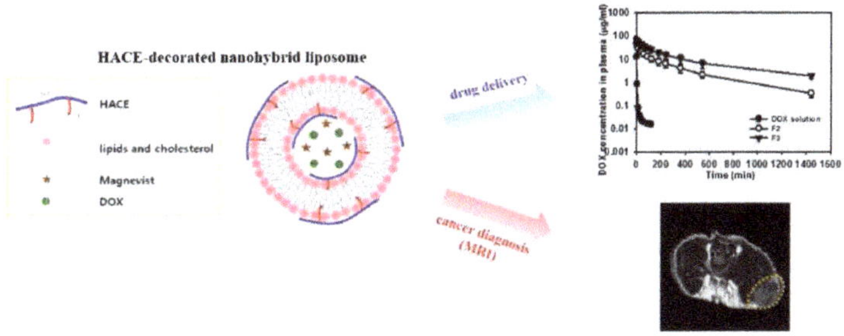

Figure 4. Self-assembled HA-CE nanoparticles for cancer imaging and DOX delivery. Reprinted with permission from Ref. [82]. 2013, Elsevier B.V.

When Pluronic 85 was combined with HA-CE, a micellar drug formulation able to reverse the Docetaxel (DTX) resistance in MCF-7/ADR xenograft mice was obtained [83].

Fatty acid derivatives, including amines and alcohols, represent potentially one of the most useful types of lipidizing agents for macromolecules of biological interest, with the amphiphilic behavior being able not only to promote self-assembling processes in physiological environments but also to module the biological properties of the resulting conjugate [97]. The HA derivatization with Dodecylamine (DDA) residues was used for the obtainment of micelle nanocarriers for the delivery of DOX, and the targeting behavior was further enhanced by inserting pH-responsive His residues [84]. HA-DDA derivative was also co-formulated with different surfactants to improve the intracellular delivery of DTX in lung cancer cells [85], while the insertion of HA-hexadecylamine (HA-HDA) into a liposome architecture was proposed as a strategy to co-encapsulate DTX and Iron Oxide Magnetic Nanoparticles (IONPs) in a combined chemo- and photothermal therapeutic nanoplatform [86]. HA-HDA conjugate was also combined with Poly(lactic-co-glycolic acid) (PLGA) in a O/W emulsion process for the preparation of SANPs for PTT by encapsulation of a Zn-phthalocyanine (PHC) complex [87]. By conjugation of the hydrophobic unit of Dioleic acid (DO) to the carboxyl group of HA by carbodiimide chemistry, Xu and co-workers developed hyalurosomes for the targeted delivery of Indocyanine green (ICG) into lung tumor cells, where it exerts PTT and PDT functions [88]. Moreover, as discussed for steroid-modified HA, HA-SANPs based on HA-fatty acid derivatives were endowed with redox responsivity by insertion of cys [89] or alkanethiol [90] residues and further engineered by PEGylation of the outer surface [91].

Finally, an original approach for the lipidization of HA backbone was proposed by the Chen research group, that synthesized amphiphilic and pH-sensitive HA-acetal-menthone (MGK) derivatives able to self-assembly in micelle systems for the vectorization of CUR to squamous carcinoma [92] and mesothelioma [93], with Folic acid (FA) used as dual targeting element for breast and lung cancer cells [94].

3.2.3. Phenyl Compounds-Modified HA in the Formation of HA-SANPs

Aromatic compounds were also employed as HA lipidizing agents, due to the enhanced loading capacity of hydrophobic bioactive agents via π-π stacking (Table 4).

Aminopropyl-1-pyrenebutanamide (PBA) was found to enhance the loading and selective biodistribution of ICG [98] and Orlistat (ORL) [99], an FDA-approved inhibitor of fatty acid synthase. It was observed an enhanced ORL activity not only in pancreatic cancer cells, the main target of this lipophilic drug but also in breast cancer cells overexpressing CD44 receptors, confirming the high internalization efficiency of HA targeted SANPs.

Another approach involved the HA derivatization with 2,3,5-Triiodobenzoic acid (TIBA), a contrasting agent for X-ray computed tomography [100], allowing the preparation of HA-SANPs that, upon DOX loading, were successfully applied in the treatment of squamous cell carcinoma. The same cancer model was employed to test the in vitro and in vivo efficiency of hybrid nanoparticles consisting of hydrophobic IONPs linked to HA through Dopamine (DA) spacer [101]. Here, Homocamptothecin (HCPT) acted as a cytotoxic agent, while the magnetic properties of IONPs were used for both targeting and imaging applications.

Finally, by virtue of the high efficiency of pH/redox responsive SANPs, cys residues were used as a spacer between HA and the hydrophobic Tetraphenylethylene (TPE) moieties for the preparation of micelles able to selectively vectorize DOX to cervix and ovary cancers [102]. Disulfide-containing HA-SANPs were also obtained by oxidizing an HA-cysteine derivative with 6-Mercaptopurine (MP), with the resulting micelle being destabilized in the tumor micro-environment allowing a selective release of the loaded anticancer drug [103].

3.3. HA-SANPs Obtained by HA Modification with Polymeric Materials

Polymeric SANPs have been widely demonstrated as safe and powerful anticancer nanocarriers due to their high chemical versatility and the possibility to easily tailor the physicochemical properties, the permeability, and thus the kinetics of drug release. HA was used as a targeting motif of the self-assembling polymeric conjugate of both natural and synthetic origin (Table 5).

Table 4. HA-SANPs obtained by hydrophobic interactions of phenyl-modified HA.

Composition (*Preparation*)		Bioactive Agent	Performance					Ref.
			Cancer Type	In Vitro		In Vivo	Outcome	
HA-Derivative	Other Components			CD44+	CD44−			
HA-PBA * (*Dialysis*)	—	ICG	Breast	MDA-MB-231	2000 MS1	MDA-MB-231 Xm	Selective Biodistr	[98]
HA-PBA * (*Dialysis*)	ORL	ORL	Pancreas Breast	PC-3 LNCaP MDA-MB-231	—	—	Synergism	[99]
HA-TIBA * (*Thin-film hydration*)	—	DOX	Squamous	SCC7	NIH-3T3	SCC7 Xm	Control Release (pH) Selective Biodistr Synergism/CT Imaging	[100]
HA-DA *-IONPs (*Water dispersion*)	—	HCPT	Squamous	SCC7	—	SCC7 Xm	Controlled Release (HAase) Synergism Magnetic Targeting MR Imaging	[101]
HA-cys-TPE * (*Dialysis*)	—	DOX	Ovary Cervix	ES2 HeLa	L929	ES2 Xm	Control Release (pH/Redox) Synergism Selective Biodistr	[102]
HA-ss-MP * (*Dialysis*)	—	DOX	Colon	HCT-116	—	BALB/C mice HCT116 Xm	Control Release (pH, redox) Synergism Selective Biodistr	[103]

* Carbodiimide chemistry; Cys: Cystamine; DA: Dopamine; DOX: Doxorubicin; HA: Hyaluronic acid; HAase: Hyaluronidase; HCPT: Homocamptothecin; ICG: Indocyanine green; IONPs: Iron oxide nanoparticles; MP: 6-Mercaptopurine; ORL: Orlistat; PBA: Aminopropyl-1-pyrenebutanamide; TIBA: 2,3,5-Triiodobenzoic acid; TPE: Tetraphenylethylene; Xm: Xenograft mice.

Table 5. HA-SANPs obtained by HA modification with polymeric materials.

Composition (*Preparation*)		Bioactive Agent	Performance					Ref.
			Cancer Type	In Vitro		In Vivo	Outcome	
HA-Derivative	Other Components			CD44+	CD44−			
HA-BSA °° (*Water Dispersion*)	—	PTX IA C-1375	Ovary	SKOV-3	A2780	—	Targeting Synergism	[104]
HA-ss-HSA * (*Water Dispersion*)	—	DOX	Breast	MDA-MB-231	NIH-3T3	—	Control Release (redox) Synergism	[105]
HA-PBLG ££ (*Nanoprecipitation*)	—	—	Breast Brain	MCF-7 —	— U87	S-D rats	Control Release (pH) Synergism	[106]
HA-PBLG ££ (*Nanoprecipitation*)	—	Dy-700	Lung	A549	H322 H358	A549 Xm H358 Xm	Selective Biodistr	[107]
HA-PBLG ££ (*Nanoprecipitation*)	—	GFT VN	Lung	A549	H322 H358	BALB/C mice H358 Xm H322 Xm A549 Xm	Selective Biodistr	[108]
HA-ss-PZLL * (*Dialysis*)	—	DOX IONPs	Liver	HepG2	—	BALB/C mice	Control Release (redox) MR Imaging	[109]

Table 5. *Cont.*

Composition (*Preparation*)		Bioactive Agent	Performance				Outcome	Ref.
HA-Derivative	Other Components		Cancer Type	In Vitro		In Vivo		
				CD44+	CD44−			
HA-PIPASP-Ce6 * (*Dialysis*)	—	DOX Ce6	Colon	HCT-116 CT-26	CV-1	CT-26 Xm	Control Release (photochemical, pH)	[110]
AcHA-PLA (*Dialysis*)	—	DOX	Colon	HCT-116	—	S-D rats	Selective Biodistr	[111]
HA-PLGA * (*Dialysis*)	—	DOX	Colon	HCT-116	—	—	Synergism	[112]
HA-PLGA * (*Dialysis*)	—	DTX	Breast	MDA-MB-231	MCF-7	S-D rats MDA-MB-231 Xm	Targeting Selective Biodistr	[113]
HA-PLGA * (*Dialysis*)	—	PpIX	Lung	A549	—	—	Sustained release PDT Synergism	[114]
HA-prop-PLA (*Nanoprecipitation*)	sPLGA-LA	DTX	Lung	A549	—	A549 Xm	Control Release (redox) Synergism Selective Biodistr	[115]
HA-cys-PLGA (*Sonication*)	TPGS	PTX RTV	Breast	MCF-7 MDA-MB-231	MCF-12A	—	Control Release (pH, redox) Synergism/Targeting Resistance Reversal	[116]
FA-HA-cys-PLGA * (*Dialysis*)	—	DOX	Breast	MCF-7	—	MCF-7 Xm	Control Release (pH, redox) Synergism	[117]
Tf- HA-cys-PLGA * (*Emulsion solvent evaporation*)	PVA	HSP90 AUY922	Brain	U87 P5 P5/TMZ-R	—	U87 Xm	Control Release (redox) Selective Biodistr Synergism Resistance Reversal	[118]
HA-PLGA * (*Endocytosis*)	MSC	PTX	Brain	C6	—	C6 Xm	Sustained Release Synergism Selective Biodistr	[119]
HA-cys-PCL § (*Dialysis*)	—	DOX IONPs	Liver	HepG2	—	—	Control Release (redox) MR Imaging	[120]
HA-PCL (*Dialysis*)	—	I-LIP	Liver	HepG2	CCL-13	—	Targeting Radiotherapy	[121]
PCL-PEG-NH_2-HA * (*O/W emulsion solvent diffusion*)	—	DTX	Breast	MDA-MB-231	NIH-3T3	—	Targeting Synergism	[122]
PDA-HA-prop-PCL § (*O/W emulsion*)	—	DOX	Squamous	SCC7	—	SCC7 Xm	Control Release (redox) Selective Biodistr	[123]
HA-PPDSMA § (*Dialysis*)	—	DOX	Squamous	SCC7	—	SCC7 Xm	Control Release (redox) Selective Biodistr	[124]
HA-P(TMC-DTC) § (*Dialysis*)	—	DTX	Breast	MDA-MB-231	L929	MDA-MB-231 Xm	Control Release (redox) Selective Biodistr	[125]
HA-cys-MA * (*Microfluidics click chemistry*)	HA-tet-GALA *	Sap	Breast Lung Liver	4T1 MDA-MB-231 A549 SMMC-7721	—	MDA-MB-231 Xm — —	Control Release (redox) Synergism Selective Biodistr	[126]

Table 5. *Cont.*

Composition (*Preparation*)		Bioactive Agent	Performance				Outcome	Ref.
			Cancer Type	In Vitro		In Vivo		
HA-Derivative	Other Components			CD44+	CD44−			
HA–ss–PNIPAAm * (*T-triggered self-assembly*)	—	DOX	Lung Breast	A549 —	LO2	— 4T1 Xm	Control Release (redox) Targeting Selective Biodistr	[127]
HA-poly(DEGMA-co-OEGMA) & (*T-triggered self-assembly*)	—	PTX	Ovary	SKOV-3	HCT-8/E11	—	Targeting Synergism	[128]
HA-m-poly(DEGM-co-CMA) & (*T-triggered self-assembly*)	—	PTX	Cervix	HeLa	Vero	HeLa Xm	Control Release (light) Selective Biodistr	[129]
HA-PEI * (*Water Dispersion*)	HA-Cys * PEG-NH_2-HA *	siRNA	Breast Lung Skin Liver	MDA-MB-468 A549/$A549^{DDP}$ B16F10 —	— H69/H69AR — Hep3B	MDA-MB-468 Xm A549/$A549^{DDP}$ Xm H69/H69AR Xm B16F10 Xm —	Selective Biodistr Synergism	[130]
HA-PEI * HA-ODA * (*Water Dispersion*)	PEG-NH_2-HA * PEG-NH_2-HA *	siRNA CDDP	Lung	A549/$A549^{DDP}$	H69/H69AR	A549/$A549^{DDP}$ Xm H69/H69AR Xm	Selective Biodistr Synergism	[131]
HA-BPEI * (*Coordination*)	—	siRNA	Skin	B16F10	HEK-293	—	Targeting	[132]
HA-βCD-OEI $ (*Coordination*)	pDNA	pDNA	Breast	MDA-MB-231	MCF-7	—	Synergism targeting	[133]

* Carbodiimide chemistry, °° Maillard, ££ Huisgen 1,3-dipolar cycloaddition; § Click chemistry; & Reversible addition–fragmentation chain-transfer polymerization; $ reductive amination; Ac: Acetyl; BPEI: Branched polyethylenimine; BSA: Bovine serum albumin; CDDP: Cisplatin; Ce6: Chlorin e6; CMA: 6-Bromo-4-hydroxymethyl-7-coumarinyl methacrylate; cys: Cystamine; DEGM: Di(ethylene glycol)methyl ether methacrylate; DEGMA: Diethyleneglycolmethacrylate; DOX: Doxorubicin; DTX: Docetaxel; Dy-700: near infrared dye 700; FA: Folic acid; GALA: Cell penetrating peptide; GFT: Gefitinib; HA: Hyaluronic acid; HSA: Human serum albumin; IA: Imidazo acridinones; I-LIP: ^{131}I-lipiodol; IONPs: Iron oxide nanoparticles; LA: Lipoic acid; MA: Methacrylic acid; MSC: Mesenchymal stem cells; ODA: Octadecylamine; OEGMA: Oligoethyleneglycolmethacrylate; OEI: Oligoethylenimine; P(TMC-DTC): Poly(trimethylene carbonate-co-dithiolane trimethylene carbonate); PBLG: Poly(γ-benzyl-L-glutamate); PCL: Poly(ε-caprolactone); PDA: 2-(Pyridyldithio)-ethylamine; pDNA: Plasmid DNA; PEG: Poly(ethylene glycol); PEI: Poly(ethylenimine); PIPASP: Poly(diisopropylaminoethyl) aspartamide; PLA: Poly(L-lactic acid); PLGA: Poly(lactic-co-glycolic acid); PNIPAAm: Poly(N-isopropylacrylamide); PPDSMA: Poly(pyridyl disulfide methacrylate); PpIX: Protoporphyrin IX; prop: Propargylamine; PTX: Paclitaxel; PVA: Poly(vinyl alcohol); PZLL: Poly(N-ε-carbobenzyloxy-L-lysine); RTV: Ritonavir; Sap: Saporin; S-D: Sprague Dawley; sPLGA: star PLGA; T: Temperature; Tet: Lysine-tetrazole; Tf: Transferrin; TPGS: D-alpha-tocopheryl poly(ethylene glycol) succinate; VNS: Vorinostat; Xm: Xenograft mice; β-CD: β-Cyclodextrin.

HA was conjugated to serum albumins because of their intrinsic ability to bind and transport biomolecules through the blood circulation [134], allowing the preparation SANPs for PTX, Imidazoacridinones (IA) [104], and DOX [105] vectorization. In the case of DOX vehicles, a cys linker was also inserted between HA and protein for conferring GSH responsivity (Figure 5).

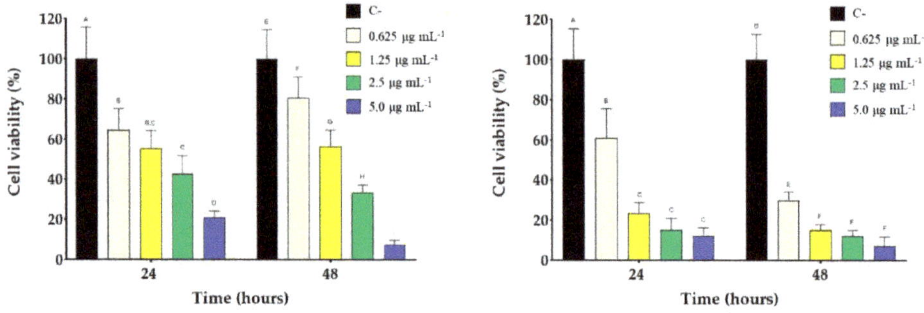

(A) (B)

Figure 5. Cell viability of BALB/3T3 (**A**) and MDA-MB231 cells (**B**) treated with DOX@HA-SANPs (drug concentration from 0 to 5 μg/mL) after 24 and 48 h. Within each group, different letters denote statistical differences for $p < 0.05$, n = 5. Reprinted from Ref. [105].

HA-polypeptides, including Poly(γ-benzyl-L-glutamate) (PBLG), Poly(N-ε-carbobenzyloxy-L-lysine) (PZLL), and Poly(diisopropylaminoethyl) aspartamide (PIPASP) were used as building blocks of HA-SANPs. PBLG and PZLL allow the self-assembly by virtue of their highly ordered α-helix secondary structure [106–108], and hydrophobic behavior [109], respectively, while PDIPASP acts as a pH-responsive moiety [110].

Owing to their high biocompatibility and biodegradability, Poly(lactic acid) (PLA), Poly(glycolic acid) (PGA), and their copolymers (PLGA) have been extensively investigated for the preparation of highly engineered nanocarriers [135]. PLA/PLGA moieties were conjugated to HA to serve as hydrophobic counterparts needed to form robust self-assembling nanostructures in aqueous media, and effectively vectorize chemotherapeutics such as DOX [111,112], DTX [113], and PDT agents [114] to colon, breast, and lung cancers. As discussed for other typologies of HA-SANPs, also for the HA-PLA/PLGA conjugates the redox responsive approach was widely explored in order to obtain a targeted release of the therapeutic agent in the intracellular space of cancer cells. For this context, Wang et al. developed disulfide-crosslinked HA-SANPs consisting of star PLGA-Lipoic acid (sPLGA-LA) conjugate self-assembled in the presence of HA-PLA conjugate. As a post-formulation crosslinking strategy, LA residues were oxidized by Dithiothreitol (DTT) to ensure the selective vectorization of DTX to lung cancer both in vitro and in vivo as a consequence of SANPs destabilization within the tumor environment [115]. Moreover, HA-PLGA conjugates can be organized in redox-responsive nanoparticle structures by inserting GSH-responsive linkers between HA and PLGA counterparts [116], with the possibility to further enhance the site-specificity of the drug release by the insertion of other targeting elements such as FA [117] and Tf [118].

In a more innovative approach, mesenchymal stem cells (MSC)-based "Trojan horse" micelles were proposed as a more selective nanocarrier to overcome non-specific distribution often attributed to the wide expression of CD44 within tissues and organs. In detail, PTX-loaded HA-PLGA SANPs were shielded by endocytosis within MSC micelles for an effective orthotopic glioma therapy [119].

Poly(ε-caprolactone) (PCL) is another key polymer widely used in biomedical fields for the preparation of delivery vehicles due to its ability to control the drug release kinetics and not significantly lower the environmental pH upon degradation [136]. SANPs based on HA-PCL conjugate were, indeed, successfully used for the vectorization of chemo- [120] and radio- [121] therapeutics. Moreover, the further insertion of PEG moieties was found to improve the blood circulation time [122], while the derivatization with 2-(Pyridyldithio)-ethylamine (PDA) conferred the possibility to perform a post-crosslinking step in the presence of DTT for enhanced GSH responsivity [123].

The targeting properties of HA were combined to the high biocompatibility and chemical versatility of acrylic-based polymers [137], and the resulting polymer conjugate was suitable for the preparation of SANPs with a variety of architecture, including micelles and nanogels. In this regard, redox-responsive micelles for the treatment of squamous [124] and breast carcinomas [125] were developed by self-assembly and post-disulfide crosslinking either in the presence or absence of DTT, while microfluidics and catalyst-free photo-click crosslinking allowed the preparation of nanogels with dual targeting efficiency [126]. Different research groups proposed the synthesis of HA derivatives able to organize in nanogel structures upon reaching a critical assembling temperature, with SANPs for the vectorization of DOX and PTX to breast [127] and ovarian [128] cancers being some key examples of this approach. Moreover, the insertion of photocleavable coumarin moieties allowed the possibility to trigger the PTX release in response to the application of a light stimulus [129].

Finally, when poly(ethyleneimines) (PEI) were used as HA derivatizing agents, targeted gene-delivering SANPs were obtained. Ganesh et al. performed a screening of different NH_2-containing HA derivatives to assess the siRNA encapsulation efficiency, showing the superior performance of HA-PEI, as well as the possibility to combine these features with redox responsibility and PEG-shielding properties [130]. The same authors proposed a CDDP and siRNA co-therapy for lung cancer treatment: HA-SANPs obtained by the co-assembly of HA-ODA and HA-PEG derivatives were used as CDDP vehicles co-administrated in xenograft mice in combination with siRNA-loaded HA-PEI/HA-PEG nanosystem [131]. Genetic materials were also loaded in star HA derivatives consisting of HA-branched PEI [132] and β-CD branched oligoethylenimine (OEI) [133], allowing effective transfection to melanoma and breast cancers, respectively.

3.4. HA-SANPs by Supramolecular Assemblies

CDs are water-soluble, nontoxic, and low-cost cyclic oligosaccharides with six to eight D-glucose units linked by α-1,4-glycosidic bonds, obtained from biodegradation of starch using Glucanotransferase enzyme. They are widely explored for the delivery of bioactive agents by virtue of their ability to selectively host inorganic and/or organic molecules in their hydrophobic cavity. Nevertheless, the use of native or simply-modified CD can be limited due to unfavorable pharmacokinetic profiles [138]. To overcome this disadvantage, multiple CD units were combined in the so-called CD-based supramolecular assemblies, nanoarchitectured materials with several binding sites for substrates mimicking the typical cooperative "multimode, multipoint" binding effect observed in biological systems, thus enhancing the loading efficiency and tailoring the release behavior [139].

HA-SANPs involving the formation of supramolecular CD complexes can be obtained by the derivatization of α- and β-CD with either HA or the other components of the nanosystem. Finally, some examples of dual host–guest interactions are listed (Table 6).

Table 6. HA-SANPs obtained by supramolecular assemblies.

Composition		Bioactive Agent	Performance				Outcome	Ref.
				In Vitro				
HA-Derivative	Other Components		Cancer Type	CD44+	CD44−	In Vivo		
HA-βCD *	CUR-OXPt *	CUR-OXPt	Pancreas Lung	PC-3 A549	LO2	—	Control Release (pH, Ease) Synergism	[140]
HA-PMCD *	Ps-PTX	Ps-PTX	Ovary	SKOV-3	NIH-3T3	—	Control Release (HAase) Targeting/Synergism Imaging	[141]
HA-βCD *	Fc-CA	Fc-CA	Breast	MCF-7 4T1	NIH-3T3	4T1 Xm	Control Release (pH) CDT Selective Biodistr	[142]
HA-αCD *	G-CB[8]	G-CB[8]	Lung	A549	293T	—	PDT Targeting	[143]
HA-αCD *	Trans-G	siRNA	Lung	A549	293T	—	Control Release (UV) Synergism	[144]
HA-βCD *	Ad-Pt	Pt	Breast Ovary	MCF-7 SKOV-3	NIH-3T3	— SKOV-3 Xm	Control Release (HAase) Synergism	[145]
AHA-βCD °	Ad-ss-CPT	CPT	Liver Bone	HepG2 —	—	— S180	Control Release (pH/redox) Synergism	[146]
HA-βCD *	Ad-DOTA-Gd Ad-Cy7	Gd Cy7	Breast Brain	MCF-7 —	— U87-MG	—	Targeting MR Imaging NIR Imaging	[147]
Ad-HA *	AM-βCD	CBL	Lung	A549	—	—	Control Release (HAase) Synergism ATP Depletion	[148]
Ad-HA *	βCD-TPE #	TPE DOX	Breast	MCF-7	NIH-3T3	—	Control Release (pH) Targeting	[149]
Ad-HA *	βCD-CPT *	CPT	Colon	HCT-116	NIH-3T3	—	Targeting	[150]
Ad-HA *	βCD-PEI * pDNA	pDNA	Cervix	HeLa	HeLa NIH-3T3	—	Targeting	[151]
HA-βCD *	DAE-βCD §	adPy-Ru	Lung	A549	293T	—	PDT Targeting	[152]
TPhPh-HA-βCD *	PMCD-SS-CPT * adPs	CPT Ps	Lung	A549	293T	—	Control Release (redox) PDT/Targeting	[153]
HA-CE £-MβCD *	—	—	Breast	MDA-MB-231	NIH-3T3 HUVEC	BALB/c mice MDA-MB-231 Xm	CHL Depletion Enhanced Apoptosis Targeting	[154]
Ad-HA *	MβCD	—	Colon	HCT-116	NIH-3T3	—	CHL Depletion Enhanced Apoptosis Targeting	[155]
Ad-HA *	FA-MβCD *	—	Colon	HCT-116	—	—	CHL Depletion Enhanced Apoptosis Targeting	[156]

* Carbodiimide chemistry; ° Shiff base formation; £ TBA mediated condensation; § Click chemistry; # NaBH3CN + DTT; Ad: Adamantane; Ad-Pt: Adamplatin; adPy-Ru: adamantane-polypyridyl ruthenium; AHA: Aldehyde HA; AM-βCD: hexylimidazolium modified βCD; ATP: Adenosine triphosphate; CBL: Chlorambucil; CD: Cyclodextrin; CDT: Chemodynamic therapy; CE: Ceramide; CHL: Cholesterol; CPT: Camptothecin; CUR: Curcumin; Cy: Cyanine; DAE: Diarylethene; DOTA: Tert-butyloxycarbonyl 1,4,7,10-tetraazacyclododecane-1,4,7,10-tetraacetic acid; DOX: Doxorubicin; Ease: Esterase; FA: Folic acid; Fc-CA: Ferrocene-modified cinnamaldehyde prodrug; G-CB[8]: Cucurbit[8]uril carbazole derivative; HA: Hyaluronic acid; HAase: Hyaluronidase; MR: Magnetic Resonance; Mβ-CD: Methyl-β-cyclodextrin; NIR: Near Infrared; OXPt: Oxoplatin; pDNA: Plasmid DNA; PDT: Photodynamic therapy; PEI: poly(ethylenimine); PMCD: Permethyl-β-CD; Ps: Porphyrin; PTX: Paclitaxel; TPE: Tetraphenylethylene; TPhPh: Triphenylphosphine; Trans-G: Azobenzene-modified diphenylalanine; Xm: Xenograft mice.

CUR/Oxaplatin (OXPt) complex was included in HA-βCD with the formation of supramolecular SANPs for the treatment of pancreatic and lung cancers [140], while ultra-strong host-guest interaction between Permethyl-β-CD (PMCD) and Porphyrin (Ps) was used in the preparation of HA-SANPs for the delivery of PTX-Ps to ovarian cancers cells combining the therapeutic efficiency of PTX and the fluorescence properties of Ps [141]. Moreover, high efficient chemodynamic (CDT) and photodynamic therapy (PDT) protocols for breast and lung cancers were developed when Fc-Cinnamaldehyde (Fc-CA) pH-responsive prodrug and Cucurbit[8]uril photosensitizing derivatives were used as the guest molecule of HA-βCDa and HA-αCD, respectively [142,143]. In a different approach, Liu and co-workers explored the possibility to use HA-αCD derivative as the hosting element for a UV-responsive azobenzene-diphenylalanine compound with a positively charged imidazole group able to coordinate siRNA. UV irradiation triggers the cis-trans isomerization of the azobenzene double bond resulting in an HA-SANPs disassembly and siRNA release [144]. Among the different molecules forming strong inclusion complexes with CD, Adamantine (Ad) was widely used as a derivatizing agent of either guest molecules with improved affinity for HA-CD conjugates, or HA with the aim to confer targeting activity to CD-based SANPs.

Following the first approach, the coordination of Adamaplatin (Ad-Pt) [145] and Ad-CPT redox responsive prodrug [146] were explored as therapeutic tools for the treatment of ovarian cancer and osteosarcoma, respectively, while MRI and near-infrared (NIR) imaging protocols were developed when the host-guest interaction involved diagnostic molecules such as Gadolinium and Cyanine dye derivatives [147]. On the other hand, Ad-HA acted as the guest molecule of CD derivatives for Chlorambucil (CBL) [148] and DOX [149] release, and further improvements were obtained upon conjugation of CD to CPT [150], or PEI [151] to enhance the drug and gene targeting efficiency, respectively.

Double host-guest interactions due to the presence of CD on both HA and guest molecules were involved in the formation of supramolecular SANPs for Ruthenium-based PDT [152] or for combined chemo-PDT protocol upon derivatization of guesting permethyl-β-CD and hosting HA-βCD with redox responsive CPT and Triphenylphosphine moieties, respectively [153].

Finally, the specific CHL-binding affinity of Methyl-βCD (MβCD) can be used to extract CHL from the membrane of cancer cells, thus inducing apoptosis, either by the direct conjugation to HA [154] or as a hosting molecule for Ad-HA [155,156] (Figure 6).

3.5. HA-Prodrug Nanoassemblies

Although small-molecule cytotoxic drugs remain the mainstream tools for cancer treatment, the narrow therapeutic window and unfavorable pharmacokinetic properties, due to quick clearance and lack of selectivity, significantly hinder their long-term employment in clinical practice thus limiting the therapeutic outcomes [157]. Apart from the strategies involving the encapsulation of bioactive within SANPs extensively discussed in the previous sections, another approach involves the covalent conjugation of these molecules to polymeric materials, with the formation of the so-called polymeric prodrugs, specimens with enhanced water solubility, chemical stability and enhanced permeation within the tumor environment [158–160].

Moreover, the insertion of proper stimuli-responsive linkages allows the release of the bioactive element to be finely tuned according to the therapeutic needs [161,162]. Polymer prodrugs show the double advantage of high drug loading with negligible formulation-trigged adverse reaction [163], and superior self-assembly ability due to the balancing between the drug to drug (driving self-assembly) and the drug to water (driving dissolution) intermolecular forces [164].

The organization in prodrug nano-assembly can be exploited for combination protocols where a second therapeutic agent is loaded within the nanostructure [165]. The most relevant examples of HA-prodrug nanoassemblies are collected in Table 7 and discussed below.

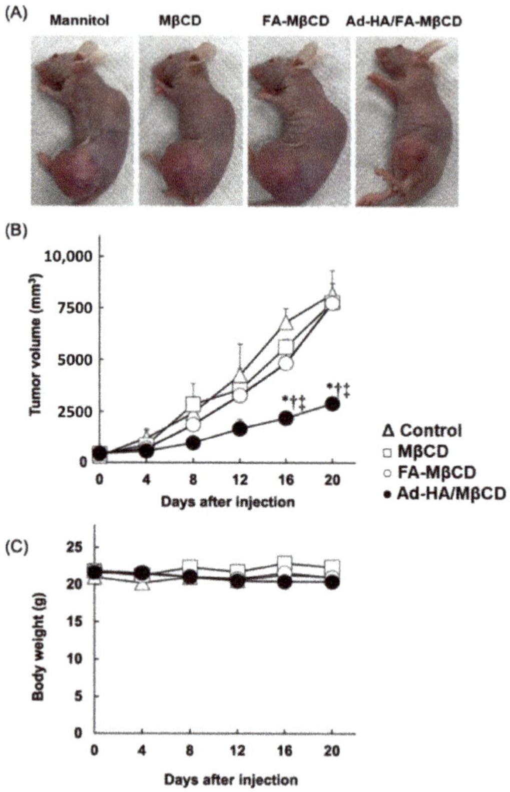

Figure 6. Effects of adamantane-grafted hyaluronic acid/folate-appended methyl-β-cyclodextrin (Ad-HA/FA-MβCD) on tumor growth (**A**,**B**) and body weight (**C**) after an intravenous administration to BALB/c nu/nu mice bearing HCT116 cells. * $p < 0.05$, compared with control (5% mannitol solution). † $p < 0.05$, compared with MβCyD. ‡ $p < 0.05$, compared with FA-MβCyD. Reprinted with permission from Ref. [156]. 2018, Elsevier B.V.

Table 7. HA-prodrug nanoassemblies.

Composition		Bioactive Agent	Performance				Outcome	Ref.
				In Vitro				
HA-Derivative	Other Components		Cancer Type	CD44+	CD44−	In Vivo		
HA–PTX * *(Water dispersion)*	—	PTX	Liver	H22	—	H22 Xm	Targeting Selective Biodistr	[166]
HA-aa-PTX * *(Water dispersion)*	—	PTX	Breast	MCF-7	—	—	Control Release (pH, HAase) Synergism	[167]
HA-prop-dOG-PTX § *(Solvent exchange)*	—	PTX	Breast	MCF-7	—	MCF-7 Xm	Control Release (pH) Targeting Selective Biodistr	[168]
DTX-GFLG-HA-ss-DD * *(Dialysis)*	—	DTX	Breast	MDA-MB-231	MCF-7	MDA-MB-231 Xm	Control Release (pH, redox, protease)	[169]
HA-d-DOX * *(Water dispersion)*	—	DOX	Breast	MDA-MB-231 MDA-MB-468LN	—	S–D rats MDA-MB-468LN Xm	Control Release (pH) Selective Biodistr	[170]
HA-cys-DOX * *(Water dispersion)*	—	DOX	Lung	A549	—	A549 Xm	Control Release (pH, redox) Selective Biodistr	[171]
Gal-PEG-ss-HA-ss-DOX * *(Water dispersion)*	—	DOX	Liver	HepG2	—	—	Control Release (pH, redox) Dual Targeting	[172]
HA-cys *-PMAA-PDMAEMA-P[VHim]NTf2-DOX * *(Dialysis)*		DOX	Breast Colon	4T1 CT-26	L929	4T1 Xm —	Control Release (pH, redox) Synergism	[173]
MTX-HA-ODA * *(Ultrasonication)*	—	MTX/CUR	Cervix Breast	HeLa MCF-7	—	HeLa Xm —	Dual targeting Control Release (pH) Synergism	[174]
HA-cys-MTX * *(Water dispersion)*	—	MTX	Cervix Lung	HeLa A549	NIH-3T3	HeLa Xm —	Control Release (redox) Dual Targeting/Synergism Selective Biodistr	[175]
HA-DTPA-CPT * *(Ultrasonication)*	—	CPT	Breast	4T1	MCF-7	4T1 Xm	Control Release (redox) Synergism Selective Biodistr	[176]
PLA-CDM-HA-DTPA-CPT * *(Electrospun)*	—	CPT	Liver	HepG2	—	H22 Xm	Control Release (pH, redox) Synergism Selective Biodistribution	[177]
HA-DAS * *(Thin-film hydration)*	TPGS	DAS/VES	Nasopharynge	HNE1 HNE1/DDP	—	HNE1 Xm	Control Release (pH) Resistance Reversal Synergism Selective Biodistr	[178]
HA-VES * *(Emulsion solvent evaporation)*	tLyP-1-TPGS*	VES/DTX	Pancreas Breast	PC-3 MDA-MB-231	—	PC-3 Xm —	Sustained Release Synergism Selective Biodistr	[179]
HA-VES * *(Sonication)*	TPGS	VES DOX/CUR	Breast	MCF-7 MCF-7/ADR	—	S-D rats 4T1 Xm	Control Release (pH) Resistance Reversal Synergism Selective Biodistr	[180]

Table 7. *Cont.*

Composition		Bioactive Agent	Performance				Outcome	Ref.
				In Vitro				
HA-Derivative	Other Components		Cancer Type	CD44+	CD44−	In Vivo		
HA-DAS * (*Thin-film hydration*)	TPGS	DAS/ROZ	Breast	MCF-7 MDA-MB-231	—	MDA-MB-231 Xm	Control Release (pH) Synergism Selective Biodistr	[181]
HA-VES * (*Sonication*)	—	VES/DOX	Breast Liver	MCF-7 MCF-7/ADR HepG2	—	4T1 Xm —	Control Release (pH) Resistance Reversal Synergism Selective Biodistr	[182]
HA-VES * (*Sonication*)	—	VES DOX/CUR	Breast Liver	MCF-7 MCF-7/ADR HepG2	—	4T1 Xm —	Control Release (pH) MDR Reversal/Synergism Selective Biodistr	[183]
HA-VES * (*Dialysis*)	—	VES/DTX Anti-PD-L1	Skin	B16	—	B16 Xm	Synergism Immune-chemotherapy	[184]
HA-CUR °° (*Water dispersion*)	—	CUR/DOX	Cervix Kidney Liver	HeLa 786-O —	— 293A HepG2	—	Control Release (pH) Synergism	[185]
HA-QC * (*Dialysis*)	—	QC/DTX	Liver	HepG2	—	HepG2 Xm	Control Release (pH) Synergism Resistance Reversal Selective Biodistribution	[186]
HA-ss-EGCG ££ (*Dialysis*)	—	EGCG CDDP	Ovary Colon	SKOV-3 HCT-116	HEK293T	SKOV-3 Xm	Control Release (HAase) Synergism Selective Biodistr	[187]
HA-Ala-EGCG * (*Water dispersion*)	PEI	EGCG GzmB	Colon Liver	HCT-116 —	— HepG2	—	Synergism	[188]
HA-GCA * (*Dialysis*)	—	GCA PTX	Liver Skin Breast	HepG2 B16-F10 —	HELF	— — MDA-MB-231 Xm	Synergism Selective Biodistr	[189]
HA-GCA * (*Dialysis*)	—	GCA PTX	Liver Skin Breast	HepG2 B16-F10 —	—	— — MDA-MB-231 Xm	Synergism Selective Biodistr	[190]
HA-ATPh-IR780 * (*Water dispersion*)	—	IR780	Bladder	MB-49	—	MB-49 Xm	Control Release (HAase) PTT/Selective Biodistr	[191]
HA-DB * (*Sonication*)	—	DB	Colon	HCT-116	A2780	HCT-116 Xm	Targeting PDT	[192]
HA-Se-Se-Ce6 (*Desolvation*)	BSA	Ce6/CYC	Breast	4T1	—	4T1 Xm	Control Release (redox, $^1O_2$) PDT/Synergism Selective Biodistr	[193]
HA-DNB-DEA/NO ** (*Sonication*)	—	DEA/NO DOX	Liver	SMMC-7721	HL-7702	SMMC-7721 Xm	ROS Generation Control Release (HAase, redox) Synergism	[194]
HA-CHL £- BSAO * (*Sonication*)	—	BSAO	Skin	M14 M14/MDR	—	—	Resistance Reversal Synergism	[195]
HA-PDI £ (*Coordination*)	—	PDI	—	—	—	—	Control Release (HAase) Early Diagnosis	[196]

Table 7. *Cont.*

Composition		Bioactive Agent	Performance				Outcome	Ref.
				In Vitro				
HA-Derivative	Other Components		Cancer Type	CD44+	CD44−	In Vivo		
HA-OPV * (*Coordination*)	PAA/HEP/CHS	OPV	—	—	—	—	Control Release (HAase) Fluorescence Imaging	[197]
HA-OVA $ (*Water dispersion*)	—	OVA	Cervix	TC-1	—	TC-1 Xm	Immunotherapy	[198]
PEG-pep-HA-OVA * (*Dialysis*)	—	OVA	Cervix	TC-1	—	TC-1 Xm	Control Release (MMP9) Immunotherapy	[199]

* Carbodiimide chemistry; °° Radical polymerization; £ TBA mediated condensation; ££ Nucleophilic addition; ** Aromatic Nucleophilic substitution; £ electrostatic interaction; § Click chemistry; $ reductive amination; aa: Aminoacid; Ala: Alanine; anti-PD-L1: programmed cell death ligand 1 (PD-L1) antibodies; ATPh: 4-Aminothiophenol; BSA: Bovine serum albumin; BSAO: bovine serum albumin oxidase; CDDP: Cisplatin; CDM: 2-Propionic-3-methylmaleic anhydride; Ce6: Chlorin e6; CHL: Cholesterol; CHS: Chondroitin 4-sulfate; CPT: Camptothecin; CUR: Curcumin; CYC: Cyclopamine; cys: Cystamine; d: Adipic dihydrazide; DAS: Dasatinib; DB: Diiodostyryl bodipy; DD: Glycodendron; DEA/NO: Diethylamine NONOate; DNB: 2,4-Dinitrobenzene; dOG: Dendritic oligoglycerol block copolymer; DOX: Doxorubicin; DTPA: 3,3′-Dithiodipropionic acid; DTX: Docetaxel; EGCG: Epigallocatechin-3-O-gallate; Gal: Galactosamine; GCA: Glycyrrhetinic acid; GFLG: Cell penetrating tetrapeptide; GzmB: Granzyme B; HA: Hyaluronic acid; HAase: Hyaluronidase; HEP: Heparin; MDR: Multi Drug Resistance; MMP9: Matrix metalloproteinase 9; MTX: Methotrexate; NTf2: Targeting peptide; ODA: Octadecylamine; OPV: Oligophenylenevinylene; OVA: Ovalbumin; P[VHim]: Poly(vinylimidazole); PAA: Poly(acrylic acid); PDI: Perylene diimide derivative; PDMAEMA: Poly(2-(dimethylamino)ethyl methacrylate); PDT: Photodynamic therapy; PEG: Poly(ethylene glycol); PEI: Poly(ethylenimine); Pep: MMP9 sensitive peptide; prop: Propargylamine; PTT: Photothermal therapy; PTX: Paclitaxel; QC: Quercetin; ROZ: Rosiglitazone; tLyP-1: Cell penetrating peptide; TPGS: D-α-tocopheryl poly(ethylene glycol) succinate; VES: α-Tocopheryl succinate; Xm: Xenograft mice.

PTX and DTX, the two most representative members of taxane drugs used in clinical practice, were conjugated to HA with the obtainment of prodrug SANPs for the treatment of liver and breast carcinomas. The conjugation strategies involved the condensation via either carbodiimide [166] or TBA chemistry [167], as well as click chemistry [168], while the insertion of tailored spacers between bioactive and HA counterparts confers responsivity to the acidic and GSH-rich tumor environment. Moreover, the presence of HA [167] and the insertion of peptide spacers [169], susceptible to the hydrolytic activity of HAase and proteases overexpressed within the cancer cells, allowed the enhancement of the targeting efficiency due to enzyme-triggered disassembly. Similarly, DOX was introduced as a bioactive hydrophobic moiety of HA-SANPs, reaching the selective pH-sensitive vectorization of the antineoplastic antibiotic [170], while the insertion of disulfide bridges was used as a dual-stimuli responsive strategy for lung [171], liver [172], and breast [173] cancers.

MTX is another hydrophobic drug used for conferring amphiphilic behavior to HA backbones. Upon conjugation to HA, MTX works as both a cytotoxic agent by inhibiting the Dihydrofolate reductase [174] and targeting moiety because, due to the structural similarity with FA, acts as a ligand for FA receptors overexpressed in many cancer cell types [175].

In the attempt to deliver the therapeutic doses of redox-responsive HA-3,3′-dithiodipropionic acid (DTPA)-CPT micelles into tumors, limiting the high liver accumulation [176], Chen et al. proposed the conjugation of HA-DTPA-CPT conjugate to PLA via acid-labile 2-propionic-3-methylmaleic anhydride (CDM) linkers and the subsequent incorporation of micelles into electrospun fibers [177]. The results confirmed the antitumor performance of fiber fragments, as well as the acidic-triggered release of HA-DTPA-CPT from the fibers and the self-assembly of the prodrug in the tumor tissues.

One of the main drawbacks of conventional chemotherapeutic protocols is the insurgence of multidrug resistance (MDR), a complex biological event involving different pathways, such as increased efflux of drugs, restoration of DNA damages, and development of antiapoptotic mechanisms [200,201]. Different strategies have been developed to address this issue [202–204], mainly based on the inhibition of P-glycoprotein (P-gp), membrane transporters belonging to the ATP binding cassette family, responsible for drug efflux through an ATP-dependent mechanism [205,206]. Among the different P-gp inhibitors proposed in the literature, D-α-tocopheryl poly(ethylene glycol) succinate (TPGS), coupling the intrinsic biological function with the self-assembling properties, was found to be an ideal nanocarrier for MDR reversal [207]. TPGS was successfully combined with different HA prodrugs, obtaining HA-SANPs with superior anticancer performance. Dasatinib (DAS), a second-generation tyrosine kinase inhibitor, was conjugated to HA and together with TPGS, the resulting nanoassemblies were tested as therapeutic agents [178]. Vitamin E succinate could also be conjugated to HA to create carriers for conventional chemotherapeutics such as DTX [179]. The authors also added TPGS conjugated to a cell-penetrating peptide to enhance the internalization efficiency. Moreover, CUR [180] and rosiglitazone (ROZ) [181] were co-loaded as MDR reversing and adipogenesis agents, respectively. HA- SANPs were also obtained with HA-VES and proposed as nanocarriers for DOX treatment in Adriamycin resistant breast cancer cells [182], as well as for the combined DOX/CUR [183] or DTX/programmed cell death ligand 1 (PD-L1) antibodies (anti-PD-L1) [184] protocols exploiting CUR as coadjuvant and the Anti-PD-L1 as an immune checkpoint.

Polyphenols have been explored as anticancer therapeutics acting via multiple mechanisms, mainly related to the pro-apoptotic effect and modulation of the cell redox balance [208–210]. Their application in clinics needs suitable carrier systems to overcome their poor pharmacokinetics, and it was widely accepted that the conjugation to macromolecular systems is a valid approach for improving their stability and bioavailability [211,212]. Polyphenols, such as CUR, Quercetin, and Epigallocatechin-3-O-gallate (EGCG), were used for the synthesis of HA amphiphiles with biological activity. HA-CUR and HA-QC conjugates were explored as functional nanocarriers for the pH-responsive delivery of DOX [185] and DTX [186], respectively, while EGCG was used for enhancing the ability of

HA-SANPs to complex CDDP molecules [187] and cytotoxic proteins such as Granzyme B (GzmB) [188]. HA-SANPs for the delivery of PTX to multiple solid tumors were obtained by the self-assembly of HA-Glycyrrhetinic acid (GCA), taking advantage of the GCA anti-inflammatory and immuno-modulating properties, as well as its ability to reverse MDR [189,190].

Bioactive molecules for PTT and PDT such as IR780 [191] and Diiodostyryl bodipy (DB) [192], were also conjugated to HA for the fabrication of HA-SANPs suitable for the treatment of bladder and colon cancers both in vitro and in vivo. A different PDT protocol, developed by Feng et al., is based on the delivery of Ce6 by a nanoplatform consisting of BSA, Cyclopamine (CYC), and HA-SeSe-Ce6 amphiphile [193]. The anticancer efficiency is the result of the synergistic contribution of each component: BA is the base material for tumor residence, CYC disrupts the extracellular matrix (ECM) barrier thus allowing HA-SeSe-Ce6 penetration, HA-SeSe-Ce6 is the PDT agent selectively releasing Ce6 within the tumor cells in response to the GSH concentrations.

In recent years, Nitric Oxide (NO) donors are emerging as effective anticancer therapeutics able to release NO at the tumor site where it causes tumor regression and metastasis inhibition. Since NO exerts such anticancer activity only at high concentrations, while acting as a pro-carcinogenic agent at low concentrations, it is of key importance to ensure high NO levels at the desired place in the body [213]. HA-SANPs able to generate NO redox reactions catalyzed by intracellular glutathione S-transferase π and to encapsulate DOX in the hydrophobic inner core were developed by derivatizing HA with Diethylamine NONOate (DEA/NO). The resulting material was found to greatly enhance the DOX anticancer efficiency in the treatment of highly aggressive hepatoma cells [194]. An alternative anticancer therapy based on the use of Bovine serum amine oxidase (BSAO) as a bioactive agent was proposed by Montanari et al. [195]. BSAO catalyzes the oxidative deamination of primary amines, such as spermine and spermidine, carrying out the formation of highly cytotoxic aldehyde and hydrogen peroxide. Injectable hydrogels were developed by the self-assembly of HA-CHL-BSAO with the aim to maximize the selectivity of enzymatic activity to melanoma cells. Further extensions of the HA-prodrug nanoassemblies concern the use of Perylene diimide (PDI) [196] and Oligophenylenevinylene (OPV) [197] derivatives as diagnostic tools for the early detection of solid tumors.

Finally, HA-OVA conjugates were proposed as targeted delivery systems for pathogen-derived foreign antigens (OVA), determining a robust CD8+ T cell response upon recognition of tumor cells presenting non-self foreign antigens by the host immune system [198]. As a further upgrade of this concept, Shin et al. proposed the use of a matrix metalloproteinase 9 (MMP9) cleavable linker to attach PEG moieties to HA-OVA conjugate [199]. Within the tumor site, the hydrolytic activity of MMP9 allowed the removal of the PEG shell, with the site-specific HA exposure and the subsequent cellular uptake via CD44-mediated endocytosis. As a result, cancer cells were labeled with antigenic peptides presented by surface major histocompatibility complex class I molecules thus favoring elimination by CD8+ cytotoxic T lymphocytes (Figure 7).

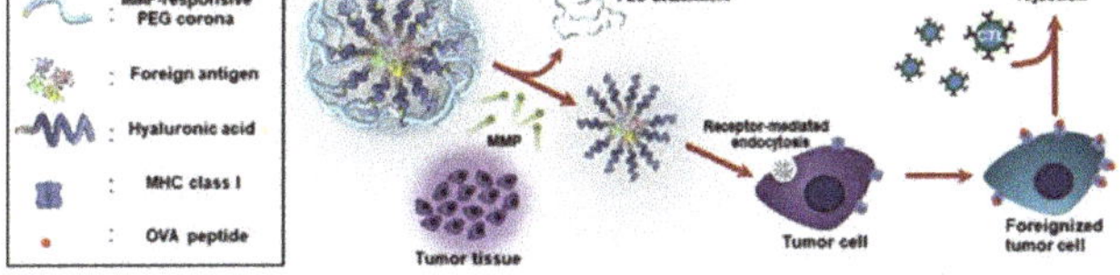

Figure 7. Representation of MMP9 responsive PEGylated HA-OVA targeted cancer immunotherapy. Reprinted with permission from Ref. [199]. 2017, Elsevier B.V.

4. Conclusions and Perspectives

The past two decades have seen great efforts from the scientific community in the development of effective antitumor regimes able to face the highly heterogeneous nature of cancers at the cellular and sub-cellular levels, expanding the concept of personalized medicine from the discovery of new biological targets to the optimization of the vectorization of therapeutics within the body to reduce unfavorable cross-toxicity to healthy organs and tissues.

In this paper, we highlighted the role of HA as a targeting element as nanoparticle systems for the selective delivery of bioactive agents to cancer cells, showing the promising outcomes of using HA-SANPs. Although some promising results in both in vitro and in vivo investigations, severe limitations still hinder an effective bench to clinics translation of the proposed nanocarriers.

Thus, for a more comprehensive analysis of such limitations, and to hypothesize key solutions to these issues, the literature data discussed in this review and shown in Tables 1–7 are summarized in the following table (Table 8).

Table 8. Outcomes of HA-SANPs for cancer therapy expressed as (%) of the reviewed studies.

HA-Derivative	Other Components	Preparation	Cancer Type	Bioactive Agent	Stimuli	In Vitro/In Vivo Success
HA (11) *	Bioactive (53) ** Polymer (27) ** Other (7) **	Water Disp (47) ** Coordination (47) ** Dialysis (6) **	Breast (38) **/Cervix (13) ** Liver (13) **/Bone (6) ** Colon (6) **/Lung (6) ** Lymphatic (6) ** Ovary (6) **/Skin (6) **	Drug (67) ** Gene (27) ** PTT/PDT (20) ** Imaging (7) ** Immuno (7) **	pH (53) ** Redox (27) **	(100) **/(60) **
HA-LIPOID (30) * βCA (19) ** FAD (19) ** CE (10) ** CHL (11) ** DOCA (10) ** PPL (7) ** Other (24) **	PPL (17) ** Polymer (10) ** Bioactive (2) ** Other (7) **	Thin Film (40) ** Dialysis (29) ** Water Disp (17) ** Emulsion (14) **	Breast (40) **/Colon (13) ** Lung (12) **/Squamous (12) ** Liver (5) **/Pancreas (3) ** Skin (3) **/Blood (2) ** Brain (2) **/Cervix (2) ** Mesothelioma (2) ** Ovary (2) **/Prostate (2) **	Drug (62) ** PTT/PDT (12) ** Imaging (10) ** Gene (7) **	pH (33) ** Redox (19) ** HAase (10) **	(92) **/(53) **
HA-POLYMER (22) * PLA/PLGA (30) ** PPEP (24) ** PACRY (20) ** PCL (13) ** PEI (13) **	Polymer (10) ** Bioactive (3) ** Other (3) **	Dialysis (37) ** Emulsion (10) ** Water Disp (17) ** Coordination (7) ** Temperature (10) ** Precipitation (13) ** Other (6) **	Breast (31) **/Lung (22) ** Liver (11) **/Brain (8) ** Colon (8) **/Ovary (6) ** Skin (6) **/Squamous (6) ** Cervix (2) **	Drug (77) ** Gene (13) ** Imaging (7) ** PTT/PDT (7) ** Radio (3) **	pH (13) ** Redox (40) ** Light (3) **	(94) **/(50) **
HA-CD (12) * HA-Ad (35) **	Bioactive (47) ** CD (41) ** Other (12) **	Host–Guest (100) **	Lung (30) **/Breast (25) ** Colon (15) **/Ovary (10) ** Bone (5) **/Cervix (5) ** Liver (5) **/Pancreas (5) **	Drug (47) ** PTT/PDT (24) ** Gene (6) ** Imaging (12) **	pH (24) ** Redox (12) ** HAase (18) ** Enzyme (6) ** Light (6) **	(95) **/(20)**
HA-Prodrug (25) *	Polymer (21) **	Water Disp (53) ** Dialysis (23) ** Coordination (6) ** Thin Film (6) ** Other (12) **	Breast (35) **/Liver (21) ** Cervix (11) **/Colon (9) ** Skin (9) **/Lung (5) ** Bladder (2) **/Kidney (2) ** Nasopharynge (2) ** Ovary (2) **/Pancreas (2) **	Drug (79) ** PTT/PDT (9) ** Immuno (9) ** Imaging (6) **	pH (47) ** Redox (26) ** HAase (15) ** Enzyme (3) **	(95) **/(62) **

* Incidence (%) to total reviewed studies; ** Incidence (%) within each group; 5-βCA: 5-β-Cholanic acid; Ad: Adamantane; CD: Cyclodextrins; CE: Ceramide; CHL: Cholesterol; DOCA: Deoxycholic acid; FAD: Fatty acid derivatives; HA: Hyaluronic acid; PRE: Precipitation; PACRY: Acrylic polymers; PCL: Poly(ε-caprolactone); PDT: Photodynamic therapy; PEI: Poly(ethylenimine); PLA: Poly(L-lactic acid); PLGA: Poly(lactic-co-glycolic acid); PPEP: Polypeptide; PPL: Phospholipids; PTT: Photothermal therapy.

Here, the overviewed research was initially classified into five groups based on the adopted HA derivatization route, and then the incidence in the use of each HA-derivative group for a specific cancer type was calculated as a percentage of total studies, considering that a single paper can cover more than a single cancer cell line and/or in vivo model at once. Moreover, as an indication of the complexity of the fabrication strategy, both preparation methods and the presence of a co-reactant within the nanoformulation were quantified in terms of value (%) within each group. Similarly, the presence (%) of in vitro or in vivo validation of the proposed HA-SANPs was assessed to show the progress of the research, while the stimuli responsivity, together with the choice of the loaded therapeutic, classified in terms of cytotoxic, MDR reversal, PTT, PDT, and imaging agents, allowed the potential application of each system to be quantitatively determined.

From the analysis of report data in Table 8, it is evident that most of the HA derivatization used for the preparation of HA-SANPs involved the coupling with lipidizing (30%) and polymeric (22%) materials. Lipidized materials, indeed, are able to spontaneously reorganize in self-assembling structures in water media, thus allowing easy fabrication methods such as thin-film hydration (40%), dialysis (29%), and simple dispersion in water media (17%). On the other hand, polymeric materials offer high chemical versatility allowing for the insertion of stimuli-responsive functionalities, including redox (40%), pH (13%), and light (3%), as well as the possibility to reach a direct conjugation with the bioactive molecule in prodrug systems (25%). Prodrug HA-SANPs were widely explored as tools for improving the pharmacokinetics profile of conventional cytotoxic drugs (29%) and, more interestingly, of PDT/PTT and immunostimulatory agents. Finally, the formation of supramolecular assemblies was also reported (12%), particularly for obtaining HAase-responsive delivery vehicles (18%). Pristine HA (11%) was also useful to prepare HA-SANPs by virtue of electrostatic interactions with cationic polymers or biologically active molecules such as drugs and genes.

As far as the investigated tumor types, breast cancers are the most studied in almost all groups, followed by lung and colon, due to both the overexpression of CD44+ receptors and the high incidence between populations. Most of the studies are well supported by investigations in in vivo models, and this can facilitate the translation to the clinics, but some key issues should be addressed.

At first, it should be considered that not all the HA-SANPs preparation routes match the requirements of clinical applications. As extensively discussed by Foulkes et al., there is currently very little regulatory guidance in the area of nanomaterials for biomedical applications, with the manufacturing process often being hit or miss for nanomaterial stability [214]. Although the self-assembly process is not the limiting step, since it is mainly based on the spontaneous insurgence of weak intermolecular forces (e.g., electrostatic attraction, hydrogen bonding, and hydrophobic modification) reducing the possibility of any toxic cross-reactivity, the multiple reaction steps often required for the synthesis of the tailored HA-derivative, cannot be easily scaled at the industrial level, and require significant modification to fit with the good manufacturing procedures rules [215]. From a therapeutic point of view, despite the key advantages of high and reproducible drug loading, site-specific vectorization, and the ability to bypass some MDR pathways (e.g., drug efflux transporters), tailoring the physicochemical properties for optimal therapeutic efficacy is still challenging, especially in the case of prodrug HA-SANPs. The conjugation of bioactive molecules to the polymeric backbone, indeed, has two opposite effects. The solubility, circulation through the bloodstream, and permeability are greatly enhanced, but the chemistry of the conjugation can compromise the particle-target interaction [216].

Moreover, recent trends point toward the fabrication of multifunctional HA-SANPs, where a dual targeting element and/or a penetration enhancer moiety are anchored. The different functionalities within multifunctional nanoparticles, indeed, can act synergistically to achieve maximal anti-tumoral activity [217].

In our opinion, only the synergistic combination of different approaches, including active targeting and stimuli responsivity, as well as the co-loading of multiple therapeutics (e.g., conventional cytotoxic drugs and PDT/PTT agents) can lead to some significant results. HA-SANPs well address these needs, and promising results were also obtained at the border between chemo- and immune-therapy, which is the new and more promising approach for cancer eradication. Deep integration between basic and industrial research is required, together with multidisciplinary synergistic expertise exchange, which can make the applicability to HA-SANPs not a chimera but an eye-catching future.

Funding: This work was supported by MIUR Excellence Department Project funds (L.232/2016), awarded to the Department of Pharmacy, Health and Nutritional Sciences, University of Calabria, Italy; PON R&I 2014–2020—ARS01_00568—SI.F.I.PA.CRO.DE.—Sviluppo e industrializzazione farmaci innovativi per terapia molecolare personalizzata PA.CRO.DE. M.C. is funded by PON R&I 2014–2020 Azione IV.6—"Contratti di ricerca su tematiche Green".

Informed Consent Statement: Not applicable.

Conflicts of Interest: The authors declare no conflict of interest.

References

1. Ma, Y.F.; Huang, J.; Song, S.J.; Chen, H.B.; Zhang, Z.J. Cancer-Targeted Nanotheranostics: Recent Advances and Perspectives. *Small* **2016**, *12*, 4936–4954. [CrossRef] [PubMed]
2. Zhao, C.Y.; Cheng, R.; Yang, Z.; Tian, Z.M. Nanotechnology for Cancer Therapy Based on Chemotherapy. *Molecules* **2018**, *23*, 826. [CrossRef] [PubMed]
3. Golombek, S.K.; May, J.N.; Theek, B.; Appold, L.; Drude, N.; Kiessling, F.; Lammers, T. Tumor Targeting Via Epr: Strategies to Enhance Patient Responses. *Adv. Drug Deliv. Rev.* **2018**, *130*, 17–38. [CrossRef] [PubMed]
4. Overchuk, M.; Zheng, G. Overcoming Obstacles in the Tumor Microenvironment: Recent Advancements in Nanoparticle Delivery for Cancer Theranostics. *Biomaterials* **2018**, *156*, 217–237. [CrossRef]
5. Yao, Y.H.; Zhou, Y.X.; Liu, L.H.; Xu, Y.Y.; Chen, Q.; Wang, Y.L.; Wu, S.J.; Deng, Y.C.; Zhang, J.M.; Shao, A.W. Nanoparticle-Based Drug Delivery in Cancer Therapy and Its Role in Overcoming Drug Resistance. *Front. Mol. Biosci.* **2020**, *7*, 193. [CrossRef]
6. Banik, B.L.; Fattahi, P.; Brown, J.L. Polymeric Nanoparticles: The Future of Nanomedicine. *Wiley Interdiscip. Rev. Nanomed. Nanobiotechnology* **2016**, *8*, 271–299. [CrossRef]
7. Sailor, M.J.; Park, J.H. Hybrid Nanoparticles for Detection and Treatment of Cancer. *Adv. Mater.* **2012**, *24*, 3779–3802. [CrossRef]
8. Mukherjee, A.; Waters, A.K.; Kalyan, P.; Achrol, A.S.; Kesari, S.; Yenugonda, V.M. Lipid-Polymer Hybrid Nanoparticles as a Next-Generation Drug Delivery Platform: State of the Art, Emerging Technologies, and Perspectives. *Int. J. Nanomed.* **2019**, *14*, 1937–1952. [CrossRef]
9. Jahangirian, H.; Lemraski, E.G.; Webster, T.J.; Rafiee-Moghaddam, R.; Abdollahi, Y. A Review of Drug Delivery Systems Based on Nanotechnology and Green Chemistry: Green Nanomedicine. *Int. J. Nanomed.* **2017**, *12*, 2957–2977. [CrossRef]
10. Truong, N.P.; Whittaker, M.R.; Mak, C.W.; Davis, T.P. The Importance of Nanoparticle Shape in Cancer Drug Delivery. *Expert Opin. Drug Deliv.* **2015**, *12*, 129–142. [CrossRef]
11. Vincent, M.P.; Navidzadeh, J.O.; Bobbala, S.; Scott, E.A. Leveraging Self-Assembled Nanobiomaterials for Improved Cancer Immunotherapy. *Cancer Cell* **2022**, *40*, 255–276. [CrossRef] [PubMed]
12. Araste, F.; Aliabadi, A.; Abnous, K.; Taghdisi, S.M.; Ramezani, M.; Alibolandi, M. Self-Assembled Polymeric Vesicles: Focus on Polymersomes in Cancer Treatment. *J. Control. Release* **2021**, *330*, 502–528. [CrossRef] [PubMed]
13. Ge, W.; Wang, L.; Zhang, J.Y.; Ou, C.J.; Si, W.L.; Wang, W.J.; Zhang, Q.M.; Dong, X.C. Self-Assembled Nanoparticles as Cancer Therapeutic Agents. *Adv. Mater. Interfaces* **2021**, *8*, 2001602. [CrossRef]
14. Yang, L.; Tang, J.; Yin, H.; Yang, J.; Xu, B.; Liu, Y.K.; Hu, Z.; Yu, B.T.; Xia, F.F.; Zou, G.W. Self-Assembled Nanoparticles for Tumor-Triggered Targeting Dual-Mode Nirf/Mr Imaging and Photodynamic Therapy Applications. *Acs Biomater. Sci. Eng.* **2022**, *8*, 880–892. [CrossRef] [PubMed]
15. Curcio, M.; Brindisi, M.; Cirillo, G.; Frattaruolo, L.; Leggio, A.; Rago, V.; Nicoletta, F.P.; Cappello, A.R.; Iemma, F. Smart Lipid-Polysaccharide Nanoparticles for Targeted Delivery of Doxorubicin to Breast Cancer Cells. *Int. J. Mol. Sci.* **2022**, *23*, 2386. [CrossRef]
16. Li, Q.; Fu, D.S.; Zhang, J.; Yan, H.; Wang, H.F.; Niu, B.L.; Guo, R.J.; Liu, Y.M. Dual Stimuli-Responsive Polypeptide-Calcium Phosphate Hybrid Nanoparticles for Co-Delivery of Multiple Drugs in Cancer Therapy. *Colloids Surf. B Biointerfaces* **2021**, *200*, 111586. [CrossRef]
17. Bevacqua, E.; Curcio, M.; Saletta, F.; Vittorio, O.; Cirillo, G.; Tucci, P. Dextran-Curcumin Nanosystems Inhibit Cell Growth and Migration Regulating the Epithelial to Mesenchymal Transition in Prostate Cancer Cells. *Int. J. Mol. Sci.* **2021**, *22*, 7013. [CrossRef]
18. Yuan, Y.; Liu, J.; Yu, X.N.; Liu, X.X.; Cheng, Y.N.; Zhou, C.; Li, M.Y.; Shi, L.; Deng, Y.; Liu, H.; et al. Tumor-Targeting Ph/Redox Dual-Responsive Nanosystem Epigenetically Reverses Cancer Drug Resistance by Co-Delivering Doxorubicin and Gcn5 Sirna. *Acta Biomater.* **2021**, *135*, 556–566. [CrossRef]
19. Curcio, M.; Paoli, A.; Cirillo, G.; di Pietro, S.; Forestiero, M.; Giordano, F.; Mauro, L.; Amantea, D.; di Bussolo, V.; Nicoletta, F.P.; et al. Combining Dextran Conjugates with Stimuli-Responsive and Folate-Targeting Activity: A New Class of Multifunctional Nanoparticles for Cancer Therapy. *Nanomaterials* **2021**, *11*, 1108. [CrossRef]
20. Vyas, D.; Patel, M.; Wairkar, S. Strategies for Active Tumor Targeting-an Update. *Eur. J. Pharmacol.* **2022**, *915*, 174512. [CrossRef]
21. Kazemi, Y.; Dehghani, S.; Nosrati, R.; Taghdisi, S.M.; Abnous, K.; Alibolandi, M.; Ramezani, M. Recent Progress in the Early Detection of Cancer Based on Cd44 Biomarker; Nano-Biosensing Approaches. *Life Sci.* **2022**, *300*, 120593. [CrossRef] [PubMed]
22. Platt, V.M.; Szoka, F.C. Anticancer Therapeutics: Targeting Macromolecules and Nanocarriers to Hyaluronan or Cd44, a Hyaluronan Receptor. *Mol. Pharm.* **2008**, *5*, 474–486. [CrossRef] [PubMed]
23. Ma, W.; Chen, Q.L.; Xu, W.G.; Yu, M.; Yang, Y.Y.; Zou, B.H.; Zhang, Y.S.; Ding, J.X.; Yu, Z.Q. Self-Targeting Visualizable Hyaluronate Nanogel for Synchronized Intracellular Release of Doxorubicin and Cisplatin in Combating Multidrug-Resistant Breast Cancer. *Nano Res.* **2021**, *14*, 846–857. [CrossRef]

24. Zhong, L.; Liu, Y.Y.; Xu, L.; Li, Q.S.; Zhao, D.Y.; Li, Z.B.; Zhang, H.C.; Zhang, H.T.; Kan, Q.M.; Sun, J.; et al. Exploring the Relationship of Hyaluronic Acid Molecular Weight and Active Targeting Efficiency for Designing Hyaluronic Acid-Modified Nanoparticles. *Asian J. Pharm. Sci.* **2019**, *14*, 521–530. [CrossRef]
25. Misra, S.; Hascall, V.C.; Markwald, R.R.; Ghatak, S. Interactions between Hyaluronan and Its Receptors (Cd44, Rhamm) Regulate the Activities of Inflammation and Cancer. *Front. Immunol.* **2015**, *6*, 201. [CrossRef]
26. Arpicco, S.; Lerda, C.; Pozza, E.D.; Costanzo, C.; Tsapis, N.; Stella, B.; Donadelli, M.; Dando, I.; Fattal, E.; Cattel, L.; et al. Hyaluronic Acid-Coated Liposomes for Active Targeting of Gemcitabine. *Eur. J. Pharm. Biopharm.* **2013**, *85*, 373–380. [CrossRef]
27. Arpicco, S.; Bartkowski, M.; Barge, A.; Zonari, D.; Serpe, L.; Milla, P.; Dosio, F.; Stella, B.; Giordani, S. Effects of the Molecular Weight of Hyaluronic Acid in a Carbon Nanotube Drug Delivery Conjugate. *Front. Chem.* **2020**, *8*, 578008. [CrossRef]
28. Li, J.C.; He, Y.; Sun, W.J.; Luo, Y.; Cai, H.D.; Pan, Y.Q.; Shen, M.W.; Xia, J.D.; Shi, X.Y. Hyaluronic Acid-Modified Hydrothermally Synthesized Iron Oxide Nanoparticles for Targeted Tumor Mr Imaging. *Biomaterials* **2014**, *35*, 3666–3677. [CrossRef]
29. Li, W.J.; Zheng, C.F.; Pan, Z.Y.; Chen, C.; Hu, D.H.; Gao, G.H.; Kang, S.D.; Cui, H.D.; Gong, P.; Cai, L.T. Smart Hyaluronidase-Actived Theranostic Micelles for Dual-Modal Imaging Guided Photodynamic Therapy. *Biomaterials* **2016**, *101*, 10–19. [CrossRef]
30. Tiwari, S.; Bahadur, P. Modified Hyaluronic Acid Based Materials for Biomedical Applications. *Int. J. Biol. Macromol.* **2019**, *121*, 556–571. [CrossRef]
31. Abbas, M.; Zou, Q.L.; Li, S.K.; Yan, X.H. Self-Assembled Peptide- and Protein-Based Nanomaterials for Antitumor Photodynamic and Photothermal Therapy. *Adv. Mater.* **2017**, *29*, 1605021. [CrossRef] [PubMed]
32. Yang, C.B.; Lin, Z.I.; Chen, J.A.; Xu, Z.R.; Gu, J.Y.; Law, W.C.; Yang, J.H.C.; Chen, C.K. Organic/Inorganic Self-Assembled Hybrid Nano-Architectures for Cancer Therapy Applications. *Macromol. Biosci.* **2022**, *22*, 2100349. [CrossRef] [PubMed]
33. Qiu, L.Y.; Zheng, C.; Jin, Y.; Zhu, K.J.E. Polymeric Micelles as Nanocarriers for Drug Delivery. *Expert Opin. Ther. Pat.* **2007**, *17*, 819–830. [CrossRef]
34. Antonietti, M.; Forster, S. Vesicles and Liposomes: A Self-Assembly Principle Beyond Lipids. *Adv. Mater.* **2003**, *15*, 1323–1333. [CrossRef]
35. Cheow, W.S.; Hadinoto, K. Self-Assembled Amorphous Drug-Polyelectrolyte Nanoparticle Complex with Enhanced Dissolution Rate and Saturation Solubility. *J. Colloid Interface Sci.* **2012**, *367*, 518–526. [CrossRef]
36. Qiu, F.; Becker, K.W.; Knight, F.C.; Baljon, J.J.; Sevimli, S.; Shae, D.; Gilchuk, P.; Joyce, S.; Wilson, J.T. Poly(Propylacrylic Acid)-Peptide Nanoplexes as a Platform for Enhancing the Immunogenicity of Neoantigen Cancer Vaccines. *Biomaterials* **2018**, *182*, 82–91. [CrossRef]
37. Yi, S.H.; Liao, R.Q.; Zhao, W.; Huang, Y.S.; He, Y. Multifunctional Co-Transport Carriers Based on Cyclodextrin Assembly for Cancer Synergistic Therapy. *Theranostics* **2022**, *12*, 2560–2579. [CrossRef]
38. Yang, J.H.; Jia, C.Y.; Yang, J.S. Designing Nanoparticle-Based Drug Delivery Systems for Precision Medicine. *Int. J. Med. Sci.* **2021**, *18*, 2943–2949. [CrossRef]
39. Grzelczak, M.; Vermant, J.; Furst, E.M.; Liz-Marzan, L.M. Directed Self-Assembly of Nanoparticles. *Acs. Nano* **2010**, *4*, 3591–3605. [CrossRef]
40. Barros, C.H.N.; Hiebner, D.W.; Fulaz, S.; Vitale, S.; Quinn, L.; Casey, E. Synthesis and Self-Assembly of Curcumin-Modified Amphiphilic Polymeric Micelles with Antibacterial Activity. *J. Nanobiotechnol.* **2021**, *19*, 1–15. [CrossRef]
41. Mahalingam, M.; Krishnamoorthy, K. Selection of a Suitable Method for the Preparation of Polymeric Nanoparticles: Multi-Criteria Decision Making Approach. *Adv. Pharm. Bull.* **2015**, *5*, 57–67.
42. Kim, Y.J.; Lee, K.P.; Lee, D.Y.; Kim, Y.T.; Koh, D.; Lim, Y.; Yoon, M.S. Anticancer Activity of a New Chalcone Derivative-Loaded Polymeric Micelle. *Macromol. Res.* **2019**, *27*, 48–54. [CrossRef]
43. Park, J.E.; Hickey, D.R.; Jun, S.; Kang, S.; Hu, X.; Chen, X.J.; Park, S.J. Surfactant-Assisted Emulsion Self-Assembly of Nanoparticles into Hollow Vesicle-Like Structures and 2d Plates. *Adv. Funct. Mater.* **2016**, *26*, 7791–7798. [CrossRef]
44. Mauro, N.; Utzeri, M.A.; Drago, S.E.; Nicosia, A.; Costa, S.; Cavallaro, G.; Giammona, G. Hyaluronic Acid Dressing of Hydrophobic Carbon Nanodots: A Self-Assembling Strategy of Hybrid Nanocomposites with Theranostic Potential. *Carbohydr. Polym.* **2021**, *267*, 118213. [CrossRef]
45. Fan, X.H.; Zhao, X.S.; Qu, X.K.; Fang, J. Ph Sensitive Polymeric Complex of Cisplatin with Hyaluronic Acid Exhibits Tumor-Targeted Delivery and Improved in Vivo Antitumor Effect. *Int. J. Pharm.* **2015**, *496*, 644–653. [CrossRef] [PubMed]
46. Zhang, W.; Cai, J.Z.; Wu, B.; Shen, Z.Y. Ph-Responsive Hyaluronic Acid Nanoparticles Coloaded with Sorafenib and Cisplatin for Treatment of Hepatocellular Carcinoma. *J. Biomater. Appl.* **2019**, *34*, 219–228. [CrossRef]
47. Zhang, W.Q.; Tung, C.H. Cisplatin Cross-Linked Multifunctional Nanodrugplexes for Combination Therapy. *Acs Appl. Mater. Interfaces* **2017**, *9*, 8547–8555. [CrossRef]
48. Gong, G.D.; Pan, J.Z.; He, Y.X.; Shang, J.J.; Wang, X.L.; Zhang, Y.Y.; Zhang, G.L.; Wang, F.; Zhao, G.; Guo, J.L. Self-Assembly of Nanomicelles with Rationally Designed Multifunctional Building Blocks for Synergistic Chemo-Photodynamic Therapy. *Theranostics* **2022**, *12*, 2028–2040. [CrossRef]
49. Wang, S.H.; Cao, M.J.; Deng, X.W.; Xiao, X.Q.; Yin, Z.X.; Hu, Q.; Zhou, Z.X.; Zhang, F.; Zhang, R.R.; Wu, Y.; et al. Degradable Hyaluronic Acid/Protamine Sulfate Interpolyelectrolyte Complexes as Mirna-Delivery Nanocapsules for Triple-Negative Breast Cancer Therapy. *Adv. Healthc. Mater.* **2015**, *4*, 281–290. [CrossRef]

50. Deng, X.W.; Cao, M.J.; Zhang, J.K.; Hu, K.L.; Yin, Z.X.; Zhou, Z.X.; Xiao, X.Q.; Yang, Y.S.; Sheng, W.; Wu, Y.; et al. Hyaluronic Acid-Chitosan Nanoparticles for Co-Delivery of M1r-34a and Doxorubicin in Therapy against Triple Negative Breast Cancer. *Biomaterials* **2014**, *35*, 4333–4344. [CrossRef]
51. Abruzzo, A.; Zuccheri, G.; Belluti, F.; Provenzano, S.; Verardi, L.; Bigucci, F.; Cerchiara, T.; Luppi, B.; Calonghi, N. Chitosan Nanoparticles for Lipophilic Anticancer Drug Delivery: Development, Characterization and in Vitro Studies on Ht29 Cancer Cells. *Colloids Surf. B-Biointerfaces* **2016**, *145*, 362–372. [CrossRef] [PubMed]
52. Xia, D.D.; Wang, F.L.; Pan, S.; Yuan, S.P.; Liu, Y.S.; Xu, Y.X. Redox/Ph-Responsive Biodegradable Thiol-Hyaluronic Acid/Chitosan Charge-Reversal Nanocarriers for Triggered Drug Release. *Polymers* **2021**, *13*, 3785. [CrossRef] [PubMed]
53. Wu, H.C.; Kuo, W.T. Redox/Ph-Responsive 2-in-1 Chimeric Nanoparticles for the Co-Delivery of Doxorubicin and Sirna. *Polymers* **2021**, *13*, 4362. [CrossRef]
54. Beals, N.; Model, M.A.; Worden, M.; Hegmann, T.; Basu, S. Intermolecular G-Quadruplex Induces Hyaluronic Acid-DNA Superpolymers Causing Cancer Cell Swelling, Blebbing, and Death. *Acs Appl. Mater. Interfaces* **2018**, *10*, 6869–6878. [CrossRef] [PubMed]
55. Zhang, Y.; Cai, L.L.; Li, D.; Lao, Y.H.; Liu, D.Z.; Li, M.Q.; Ding, J.X.; Chen, X.S. Tumor Microenvironment-Responsive Hyaluronate-Calcium Carbonate Hybrid Nanoparticle Enables Effective Chemotherapy for Primary and Advanced Osteosarcomas. *Nano Res.* **2018**, *11*, 4806–4822. [CrossRef]
56. Pan, Y.T.; Ding, Y.F.; Han, Z.H.; Yuwen, L.H.; Ye, Z.; Mok, G.S.P.; Li, S.K.; Wang, L.H. Hyaluronic Acid-Based Nanogels Derived from Multicomponent Self-Assembly for Imaging-Guided Chemo-Photodynamic Cancer Therapy. *Carbohydr. Polym.* **2021**, *268*, 118257s. [CrossRef]
57. Hou, G.H.; Qian, J.M.; Guo, M.; Xu, W.J.; Wang, J.L.; Wang, Y.P.; Suo, A.L. Hydrazided Hyaluronan/Cisplatin/Indocyanine Green Coordination Nanoprodrug for Photodynamic Chemotherapy in Liver Cancer. *Carbohydr. Polym.* **2022**, *276*, 118810. [CrossRef]
58. Shin, W.J.; Noh, H.J.; Noh, Y.W.; Kim, S.; Um, S.H.; Lim, Y.T. Hyaluronic Acid-Supported Combination of Water Insoluble Immunostimulatory Compounds for Anti-Cancer Immunotherapy. *Carbohydr. Polym.* **2017**, *155*, 1–10. [CrossRef]
59. Choi, K.M.; Jang, M.; Kim, J.H.; Ahn, H.J. Tumor-Specific Delivery of Sirna Using Supramolecular Assembly of Hyaluronic Acid Nanoparticles and 2b Rna-Binding Protein/Sirna Complexes. *Biomaterials* **2014**, *35*, 7121–7132. [CrossRef]
60. Hu, D.R.; Mezghrani, O.; Zhang, L.; Chen, Y.; Ke, X.; Ci, T.Y. Ge11 Peptide Modified and Reduction-Responsive Hyaluronic Acid-Based Nanoparticles Induced Higher Efficacy of Doxorubicin for Breast Carcinoma Therapy. *Int. J. Nanomed.* **2016**, *11*, 5125–5147. [CrossRef]
61. Li, X.L.; Wang, X.; Zhao, C.Y.; Shao, L.H.; Lu, J.Q.; Tong, Y.J.; Chen, L.; Cui, X.Y.; Sun, H.L.; Liu, J.X.; et al. From One to All: Self-Assembled Theranostic Nanoparticles for Tumor-Targeted Imaging and Programmed Photoactive Therapy. *J. Nanobiotechnol.* **2019**, *17*, 1–12. [CrossRef]
62. Song, M.J.; Liang, Y.; Li, K.K.; Zhang, J.; Zhang, N.; Tian, B.C.; Han, J.T. Hyaluronic Acid Modified Liposomes for Targeted Delivery of Doxorubicin and Paclitaxel to Cd44 Overexpressing Tumor Cells with Improved Dual-Drugs Synergistic Effect. *J. Drug Deliv. Sci. Technol.* **2019**, *53*, 101179. [CrossRef]
63. Luo, S.; Feng, J.X.; Xiao, L.Y.; Guo, L.; Deng, L.; Du, Z.W.; Xue, Y.; Song, X.; Sun, X.; Zhang, Z.R.; et al. Targeting Self-Assembly Peptide for Inhibiting Breast Tumor Progression and Metastasis. *Biomaterials* **2020**, *249*, 120055. [CrossRef] [PubMed]
64. Quinones, J.P.; Jokinen, J.; Keinanen, S.; Covas, C.P.; Bruggemann, O.; Ossipov, D. Self-Assembled Hyaluronic Acid-Testosterone Nanocarriers for Delivery of Anticancer Drugs. *Eur. Polym. J.* **2018**, *99*, 384–393. [CrossRef]
65. Kelkar, S.S.; Hill, T.K.; Marini, F.C.; Mohs, A.M. Near Infrared Fluorescent Nanoparticles Based on Hyaluronic Acid: Self-Assembly, Optical Properties, and Cell Interaction. *Acta Biomater.* **2016**, *36*, 112–121. [CrossRef] [PubMed]
66. Choi, K.Y.; Chung, H.; Min, K.H.; Yoon, H.Y.; Kim, K.; Park, J.H.; Kwon, I.C.; Jeong, S.Y. Self-Assembled Hyaluronic Acid Nanoparticles for Active Tumor Targeting. *Biomaterials* **2010**, *31*, 106–114. [CrossRef]
67. Thomas, R.G.; Moon, M.; Lee, S.; Jeong, Y.Y. Paclitaxel Loaded Hyaluronic Acid Nanoparticles for Targeted Cancer Therapy: In Vitro and in Vivo Analysis. *Int. J. Biol. Macromol.* **2015**, *72*, 510–518. [CrossRef]
68. Tang, Y.T.; Chen, M.L.; Xie, Q.; Li, L.; Zhu, L.; Ma, Q.J.; Gao, S. Construction and Evaluation of Hyaluronic Acid-Based Copolymers as a Targeted Chemotherapy Drug Carrier for Cancer Therapy. *Nanotechnology* **2020**, *31*, 305702. [CrossRef]
69. Min, H.S.; Son, S.; Lee, T.W.; Koo, H.; Yoon, H.Y.; Na, J.H.; Choi, Y.; Park, J.H.; Lee, J.; Han, M.H.; et al. Liver-Specific and Echogenic Hyaluronic Acid Nanoparticles Facilitating Liver Cancer Discrimination. *Adv. Funct. Mater.* **2013**, *23*, 5518–5529. [CrossRef]
70. Choi, K.Y.; KMin, H.; Yoon, H.Y.; Kim, K.; Park, J.H.; Kwon, I.C.; Choi, K.; Jeong, S.Y. Pegylation of Hyaluronic Acid Nanoparticles Improves Tumor Targetability in Vivo. *Biomaterials* **2011**, *32*, 1880–1889. [CrossRef]
71. Choi, K.Y.; Yoon, H.Y.; Kim, J.H.; Bae, S.M.; Park, R.W.; Kang, Y.M.; Kim, I.S.; Kwon, I.C.; Choi, K.; Jeong, S.Y.; et al. Smart Nanocarrier Based on Pegylated Hyaluronic Acid for Cancer Therapy. *Acs Nano* **2011**, *5*, 8591–8599. [CrossRef] [PubMed]
72. Choi, K.Y.; Jeon, E.J.; Yoon, H.Y.; Lee, B.S.; Na, J.H.; Min, K.H.; Kim, S.Y.; Myung, S.J.; Lee, S.; Chen, X.Y.; et al. Theranostic Nanoparticles Based on Pegylated Hyaluronic Acid for the Diagnosis, Therapy and Monitoring of Colon Cancer. *Biomaterials* **2012**, *33*, 6186–6193. [CrossRef] [PubMed]
73. Liu, Y.H.; Zhou, C.M.; Wang, W.P.; Yang, J.H.; Wang, H.; Hong, W.; Huang, Y. Cd44 Receptor Targeting and Endosomal Ph-Sensitive Dual Functional Hyaluronic Acid Micelles for Intracellular Paclitaxel Delivery. *Mol. Pharm.* **2016**, *13*, 4209–4221. [CrossRef] [PubMed]

74. Li, J.; Huo, M.R.; Wang, J.; Zhou, J.P.; Mohammad, J.M.; Zhang, Y.L.; Zhu, Q.N.; Waddad, A.Y.; Zhang, Q. Redox-Sensitive Micelles Self-Assembled from Amphiphilic Hyaluronic Acid-Deoxycholic Acid Conjugates for Targeted Intracellular Delivery of Paclitaxel. *Biomaterials* **2012**, *33*, 2310–2320. [CrossRef]
75. Li, S.P.; Zhao, W.; Liang, N.; Xu, Y.X.; Kawashima, Y.; Sun, S.P. Multifunctional Micelles Self-Assembled from Hyaluronic Acid Conjugate for Enhancing Anti-Tumor Effect of Paclitaxel. *React. Funct. Polym.* **2020**, *152*, 104608. [CrossRef]
76. Cao, A.C.; Ma, P.Q.; Yang, T.; Lan, Y.; Yu, S.Y.; Liu, L.; Sun, Y.; Liu, Y.H. Multifunctionalized Micelles Facilitate Intracellular Doxorubicin Delivery for Reversing Multidrug Resistance of Breast Cancer. *Mol. Pharm.* **2019**, *16*, 2502–2510. [CrossRef]
77. Saadat, E.; Amini, M.; Dinarvand, R.; Dorkoosh, F.A. Polymeric Micelles Based on Hyaluronic Acid and Phospholipids: Design, Characterization, and Cytotoxicity. *J. Appl. Polym. Sci.* **2014**, *131*, 40944. [CrossRef]
78. Dalla Pozza, E.; Lerda, C.; Costanzo, C.; Donadelli, M.; Dando, I.; Zoratti, E.; Scupoli, M.T.; Beghelli, S.; Scarpa, A.; Fattal, E.; et al. Targeting Gemcitabine Containing Liposomes to Cd44 Expressing Pancreatic Adenocarcinoma Cells Causes an Increase in the Antitumoral Activity. *Biochim. Et Biophys. Acta-Biomembr.* **2013**, *1828*, 1396–1404. [CrossRef]
79. Zhang, B.; Zhang, Y.Y.; Yu, D.M. Lung Cancer Gene Therapy: Transferrin and Hyaluronic Acid Dual Ligand-Decorated Novel Lipid Carriers for Targeted Gene Delivery. *Oncol. Rep.* **2017**, *37*, 937–944. [CrossRef]
80. Chang, J.E.; Cho, H.J.; Yi, E.; Kim, D.D.; Jheon, S. Hypocrellin B and Paclitaxel-Encapsulated Hyaluronic Acid-Ceramide Nanoparticles for Targeted Photodynamic Therapy in Lung Cancer. *J. Photochem. Photobiol. B-Biol.* **2016**, *158*, 113–121. [CrossRef]
81. Mallick, S.; Park, J.H.; Cho, H.J.; Kim, D.D.; Choi, J.S. Hyaluronic Acid-Ceramide-Based Liposomes for Targeted Gene Delivery to Cd44-Positive Cancer Cells. *Bull. Korean Chem. Soc.* **2015**, *36*, 874–881.
82. Park, J.H.; Cho, H.J.; Yoon, H.Y.; Yoon, I.S.; Ko, S.H.; Shim, J.S.; Cho, J.H.; Park, J.H.; Kim, K.; Kwon, I.C.; et al. Hyaluronic Acid Derivative-Coated Nanohybrid Liposomes for Cancer Imaging and Drug Delivery. *J. Control. Release* **2014**, *174*, 98–108. [CrossRef] [PubMed]
83. Cho, H.J.; Yoon, H.Y.; Koo, H.; Ko, S.H.; Shim, J.S.; Lee, J.H.; Kim, K.; Kwon, I.C.; Kim, D.D. Self-Assembled Nanoparticles Based on Hyaluronic Acid-Ceramide (Ha-Ce) and Pluronic (R) for Tumor-Targeted Delivery of Docetaxel. *Biomaterials* **2011**, *32*, 7181–7190. [CrossRef] [PubMed]
84. Yang, S.D.; Chen, W.L.; Li, W.; Song, J.C.; Gao, Y.; Si, W.H.; Li, X.P.; Cui, B.W.; Yu, T.T. Cd44-Targeted Ph-Responsive Micelles for Enhanced Cellular Internalization and Intracellular on-Demand Release of Doxorubicin. *Artif. Cells Nanomed. Biotechnol.* **2021**, *49*, 173–184. [CrossRef] [PubMed]
85. Cadete, A.; Olivera, A.; Besev, M.; Dhal, P.K.; Goncalves, L.; Almeida, A.J.; Bastiat, G.; Benoit, J.P.; de la Fuente, M.; Garcia-Fuentes, M.; et al. Self-Assembled Hyaluronan Nanocapsules for the Intracellular Delivery of Anticancer Drugs. *Sci. Rep.* **2019**, *9*, 1–11. [CrossRef] [PubMed]
86. Nguyen, V.D.; Zheng, S.; Han, J.; Le, V.H.; Park, J.O.; Park, S. Nanohybrid Magnetic Liposome Functionalized with Hyaluronic Acid for Enhanced Cellular Uptake and near-Infrared-Triggered Drug Release. *Colloids Surf. B-Biointerfaces* **2017**, *154*, 104–114. [CrossRef]
87. Gao, D.; Wong, R.C.H.; Wang, Y.; Guo, X.Q.; Yang, Z.; Lo, P.C. Shifting the Absorption to the near-Infrared Region and Inducing a Strong Photothermal Effect by Encapsulating Zinc(Ii) Phthalocyanine in Poly(Lactic-Co-Glycolic Acid)-Hyaluronic Acid Nanoparticles. *Acta Biomater.* **2020**, *116*, 329–343. [CrossRef]
88. Xu, H.P.; Dong, L.; Bin, Z.; Huo, Y.S.; Lin, S.F.; Chang, L.; Chen, C.; Wang, C.L. Supramolecular Self-Assembly of a Hybrid 'Hyalurosome' for Targeted Photothermal Therapy in Non-Small Cell Lung Cancer. *Drug Deliv.* **2020**, *27*, 378–386. [CrossRef]
89. Jeong, G.W.; Jeong, Y.I.; Nah, J.W. Triggered Doxorubicin Release Using Redox-Sensitive Hyaluronic Acid-G-Stearic Acid Micelles for Targeted Cancer Therapy. *Carbohydr. Polym.* **2019**, *209*, 161–171. [CrossRef]
90. Carvalho, A.M.; Teixeira, R.; Novoa-Carballal, R.; Pires, R.A.; Reis, R.L.; Pashkuleva, I. Redox-Responsive Micellar Nanoparticles from Glycosaminoglycans for Cd44 Targeted Drug Delivery. *Biomacromolecules* **2018**, *19*, 2991–2999. [CrossRef]
91. Liu, J.Y.; Liang, N.; Li, S.P.; Han, Y.; Yan, P.F.; Kawashima, Y.; Cui, F.D.; Sun, S.P. Tumor-Targeting and Redox-Sensitive Micelles Based on Hyaluronic Acid Conjugate for Delivery of Paclitaxel. *J. Biomater. Appl.* **2020**, *34*, 1458–1469. [CrossRef] [PubMed]
92. Chen, D.Q.; Lian, S.N.; Sun, J.F.; Liu, Z.L.; Zhao, F.; Jiang, Y.T.; Gao, M.M.; Sun, K.X.; Liu, W.H.; Fu, F.H. Design of Novel Multifunctional Targeting Nano-Carrier Drug Delivery System Based on Cd44 Receptor and Tumor Microenvironment Ph Condition. *Drug Deliv.* **2016**, *23*, 798–803. [CrossRef] [PubMed]
93. Chen, D.Q.; Sun, J.F.; Lian, S.N.; Liu, Z.L.; Sun, K.X.; Liu, W.H.; Wu, Z.M.; Zhang, Q. Dual Ph-Responsive and Cd44 Receptor Targeted Multifunctional Nanoparticles for Anticancer Intracellular Delivery. *J. Nanoparticle Res.* **2014**, *16*, 1–7. [CrossRef]
94. Chen, D.Q.; XSong, Y.; Wang, K.L.; Guo, C.J.; Yu, Y.M.; Fan, H.Y.; Zhao, F. Design and Evaluation of Dual Cd44 Receptor and Folate Receptor-Targeting Double-Smart Ph-Response Multifunctional Nanocarrier. *J. Nanoparticle Res.* **2017**, *19*, 1–11. [CrossRef]
95. Drescher, S.; van Hoogevest, P. The Phospholipid Research Center: Current Research in Phospholipids and Their Use in Drug Delivery. *Pharmaceutics* **2020**, *12*, 1235. [CrossRef]
96. Alrbyawi, H.; Poudel, I.; Dash, R.P.; Srinivas, N.R.; Tiwari, A.K.; Arnold, R.D.; Babu, R.J. Role of Ceramides in Drug Delivery. *Aaps Pharmscitech* **2019**, *20*, 1–14. [CrossRef]
97. Ekrami, H.M.; Kennedy, A.R.; Shen, W.C. Water-Soluble Fatty-Acid Derivatives as Acylating Agents for Reversible Lipidization of Polypeptides. *Febs Lett.* **1995**, *371*, 283–286.
98. Hill, T.K.; Abdulahad, A.; Kelkar, S.S.; Marini, F.C.; Long, T.E.; Provenzale, J.M.; Mohs, A.M. Indocyanine Green-Loaded Nanoparticles for Image-Guided Tumor Surgery. *Bioconjugate Chem.* **2015**, *26*, 294–303. [CrossRef]

99. Hill, T.K.; Davis, A.L.; Wheeler, F.B.; Kelkar, S.S.; Freund, E.C.; Lowther, W.T.; Kridel, S.J.; Mohs, A.M. Development of a Self-Assembled Nanoparticle Formulation of Orlistat, Nano-Orl, with Increased Cytotoxicity against Human Tumor Cell Lines. *Mol. Pharm.* **2016**, *13*, 720–728. [CrossRef]
100. Lee, J.Y.; Chung, S.J.; Cho, H.J.; Kim, D.D. Iodinated Hyaluronic Acid Oligomer-Based Nanoassemblies for Tumor-Targeted Drug Delivery and Cancer Imaging. *Biomaterials* **2016**, *85*, 218–231. [CrossRef]
101. Wang, G.H.; Gao, S.; Tian, R.; Miller-Kleinhenz, J.; Qin, Z.N.; Liu, T.J.; Li, L.; Zhang, F.; Ma, Q.J.; Zhu, L. Theranostic Hyaluronic Acid-Iron Micellar Nanoparticles for Magnetic-Field-Enhanced Invivo Cancer Chemotherapy. *Chemmedchem* **2018**, *13*, 78–86. [CrossRef] [PubMed]
102. Yan, K.; Feng, Y.C.; Gao, K.; Shi, X.J.; Zhao, X.B. Fabrication of Hyaluronic Acid-Based Micelles with Glutathione-Responsiveness for Targeted Anticancer Drug Delivery. *J. Colloid Interface Sci.* **2022**, *606*, 1586–1596. [CrossRef]
103. Debele, T.A.; Yu, L.Y.; Yang, C.S.; Shen, Y.A.; Lo, C.L. Ph- and Gsh-Sensitive Hyaluronic Acid-Mp Conjugate Micelles for Intracellular Delivery of Doxorubicin to Colon Cancer Cells and Cancer Stem Cells. *Biomacromolecules* **2018**, *19*, 3725–3737. [CrossRef] [PubMed]
104. Edelman, R.; Assaraf, Y.G.; Levitzky, I.; Shahar, T.; Livney, Y.D. Hyaluronic Acid-Serum Albumin Conjugate-Based Nanoparticles for Targeted Cancer Therapy. *Oncotarget* **2017**, *8*, 24337–24353. [CrossRef] [PubMed]
105. Curcio, M.; Diaz-Gomez, L.; Cirillo, G.; Nicoletta, F.P.; Leggio, A.; Iemma, F. Dual-Targeted Hyaluronic Acid/Albumin Micelle-Like Nanoparticles for the Vectorization of Doxorubicin. *Pharmaceutics* **2021**, *13*, 304. [CrossRef] [PubMed]
106. Upadhyay, K.K.; Bhatt, A.N.; Mishra, A.K.; Dwarakanath, B.S.; Jain, S.; Schatz, C.; le Meins, J.F.; Farooque, A.; Chandraiah, G.; Jain, A.K.; et al. The Intracellular Drug Delivery and Anti Tumor Activity of Doxorubicin Loaded Poly(Gamma-Benzyl L-Glutamate)-B-Hyaluronan Polymersomes. *Biomaterials* **2010**, *31*, 2882–2892. [CrossRef]
107. Jeannot, V.; Mazzaferro, S.; Lavaud, J.; Vanwonterghem, L.; Henry, M.; Arboleas, M.; Vollaire, J.; Josserand, V.; Coll, J.L.; Lecommandoux, S.; et al. Targeting Cd44 Receptor-Positive Lung Tumors Using Polysaccharide-Based Nanocarriers: Influence of Nanoparticle Size and Administration Route. *Nanomed. Nanotechnol. Biol. Med.* **2016**, *12*, 921–932. [CrossRef]
108. Jeannot, V.; Gauche, C.; Mazzaferro, S.; Couvet, M.; Vanwonterghem, L.; Henry, M.; Didier, C.; Vollaire, J.; Josserand, V.; Coll, J.L.; et al. Anti-Tumor Efficacy of Hyaluronan-Based Nanoparticles for the Co-Delivery of Drugs in Lung Cancer. *J. Control. Release* **2018**, *275*, 117–128. [CrossRef]
109. Yang, H.K.; Miao, Y.L.; Chen, L.P.; Li, Z.R.; Yang, R.M.; Xu, X.D.; Liu, Z.S.; Zhang, L.M.; Jiang, X.Q. Redox-Responsive Nanoparticles from Disulfide Bond-Linked Poly-(N-Epsilon-Carbobenzyloxy-L-Lysine)-Grafted Hyaluronan Copolymers as Theranostic Nanoparticles for Tumor-Targeted Mri and Chemotherapy. *Int. J. Biol. Macromol.* **2020**, *148*, 483–492. [CrossRef]
110. Lee, C.S.; Na, K. Photochemically Triggered Cytosolic Drug Delivery Using Ph-Responsive Hyaluronic Acid Nanoparticles for Light-Induced Cancer Therapy. *Biomacromolecules* **2014**, *15*, 4228–4238. [CrossRef]
111. Jo, Y.U.; Lee, C.B.; Bae, S.K.; Na, K. Acetylated Hyaluronic Acid-Poly(L-Lactic Acid) Conjugate Nanoparticles for Inhibition of Doxorubicinol Production from Doxorubicin. *Macromol. Res.* **2020**, *28*, 67–73. [CrossRef]
112. Jeong, Y.I.; Kim, D.H.; Chung, C.W.; Yoo, J.J.; Choi, K.H.; Kim, C.H.; Ha, S.H.; Kang, D.H. Self-Assembled Nanoparticles of Hyaluronic Acid/Poly(Dl-Lactide-Co-Glycolide) Block Copolymer. *Colloids Surf. B-Biointerfaces* **2012**, *90*, 28–35. [CrossRef]
113. Huang, J.B.; Zhang, H.; Yu, Y.; Chen, Y.; Wang, D.; Zhang, G.Q.; Zhou, G.C.; Liu, J.J.; Sun, Z.G.; Sun, D.X.; et al. Biodegradable Self-Assembled Nanoparticles of Poly (D,L-Lactide-Co-Glycolide)/Hyaluronic Acid Block Copolymers for Target Delivery of Docetaxel to Breast Cancer. *Biomaterials* **2014**, *35*, 550–566. [CrossRef] [PubMed]
114. Wang, X.L.; Wang, J.B.; Li, J.J.; Huang, H.X.; Sun, X.Y.; Lv, Y.Y. Development and Evaluation of Hyaluronic Acid-Based Polymeric Micelles for Targeted Delivery of Photosensitizer for Photodynamic Therapy in Vitro. *J. Drug Deliv. Sci. Technol.* **2018**, *48*, 414–421. [CrossRef]
115. Wang, X.X.; Cheng, R.; Zhong, Z.Y. Facile Fabrication of Robust, Hyaluronic Acid-Surfaced and Disulfide-Crosslinked Plga Nanoparticles for Tumor-Targeted and Reduction-Triggered Release of Docetaxel. *Acta Biomater.* **2021**, *125*, 280–289. [CrossRef] [PubMed]
116. Gote, V.; Sharma, A.D.; Pal, D. Hyaluronic Acid-Targeted Stimuli-Sensitive Nanomicelles Co-Encapsulating Paclitaxel and Ritonavir to Overcome Multi-Drug Resistance in Metastatic Breast Cancer and Triple-Negative Breast Cancer Cells. *Int. J. Mol. Sci.* **2021**, *22*, 1257. [CrossRef]
117. Zhang, R.; Jiang, Y.Y.; Hao, L.K.; Yang, Y.; Gao, Y.; Zhang, N.N.; Zhang, X.C.; Song, Y.M. Cd44/Folate Dual Targeting Receptor Reductive Response Plga-Based Micelles for Cancer Therapy. *Front. Pharmacol.* **2022**, *13*, 829590. [CrossRef]
118. Debele, T.A.; Wu, P.C.; Wei, Y.F.; Chuang, J.Y.; Chang, K.Y.; Tsai, J.H.; Su, W.P. Transferrin Modified Gsh Sensitive Hyaluronic Acid Derivative Micelle to Deliver Hsp90 Inhibitors to Enhance the Therapeutic Efficacy of Brain Cancers. *Cancers* **2021**, *13*, 2375. [CrossRef]
119. Wang, X.L.; Zhao, W.Z.; Fan, J.Z.; Jia, L.C.; Lu, Y.N.; Zeng, L.H.; Lv, Y.Y.; Sun, X.Y. Tumor Tropic Delivery of Hyaluronic Acid-Poly (D,L-Lactide-Co-Glycolide) Polymeric Micelles Using Mesenchymal Stem Cells for Glioma Therapy. *Molecules* **2022**, *27*, 2419. [CrossRef]
120. Yang, H.K.; Wang, N.H.; Mo, L.; Wu, M.; Yang, R.M.; Xu, X.D.; Huang, Y.G.; Lin, J.T.; Zhang, L.M.; Jiang, X.Q. Reduction Sensitive Hyaluronan-Ss-Poly(E-Caprolactone) Block Copolymers as Theranostic Nanocarriers for Tumor Diagnosis and Treatment. *Mater. Sci. Eng. C-Mater. Biol. Appl.* **2019**, *98*, 9–18. [CrossRef]

121. Chen, S.C.; Yang, M.H.; Chung, T.W.; Jhuang, T.S.; Yang, J.D.; Chen, K.C.; Chen, W.J.; Huang, Y.F.; Jong, S.B.; Tsai, W.C.; et al. Preparation and Characterization of Hyaluronic Acid-Polycaprolactone Copolymer Micelles for the Drug Delivery of Radioactive Iodine-131 Labeled Lipiodol. *Biomed. Res. Int.* **2017**, *2017*, 4051763. [CrossRef] [PubMed]
122. Youm, I.; Agrahari, V.; Murowchick, J.B.; Youan, B.B.C. Uptake and Cytotoxicity of Docetaxel-Loaded Hyaluronic Acid-Grafted Oily Core Nanocapsules in Mda-Mb 231 Cancer Cells. *Pharm. Res.* **2014**, *31*, 2439–2452. [CrossRef] [PubMed]
123. Han, H.S.; Thambi, T.; Choi, K.Y.; Son, S.; Ko, H.; Lee, M.C.; Jo, D.G.; Chae, Y.S.; Kang, Y.M.; Lee, J.Y.; et al. Bioreducible Shell-Cross-Linked Hyaluronic Acid Nanoparticles for Tumor-Targeted Drug Delivery. *Biomacromolecules* **2015**, *16*, 447–456. [CrossRef]
124. Han, H.S.; Choi, K.Y.; Ko, H.; Jeon, J.; Saravanakumar, G.; Suh, Y.D.; Lee, D.S.; Park, J.H. Bioreducible Core-Crosslinked Hyaluronic Acid Micelle for Targeted Cancer Therapy. *J. Control. Release* **2015**, *200*, 158–166. [CrossRef]
125. Zhu, Y.Q.; Zhang, J.; Meng, F.H.; Cheng, L.; Feijen, J.; Zhong, Z.Y. Reduction-Responsive Core-Crosslinked Hyaluronic Acid-B-Poly(Trimethylene Carbonate-Co-Dithiolane Trimethylene Carbonate) Micelles: Synthesis and Cd44-Mediated Potent Delivery of Docetaxel to Triple Negative Breast Tumor in Vivo. *J. Mater. Chem. B* **2018**, *6*, 3040–3047. [CrossRef]
126. Huang, K.; He, Y.H.; Zhu, Z.H.; Guo, J.K.; Wang, G.L.; Deng, C.; Zhong, Z.Y. Small, Traceable, Endosome-Disrupting, and Bioresponsive Click Nanogels Fabricated Via Microfluidics for Cd44-Targeted Cytoplasmic Delivery of Therapeutic Proteins. *Acs Appl. Mater. Interfaces* **2019**, *11*, 22171–22180. [CrossRef] [PubMed]
127. Chen, J.J.; Wu, M.; Veroniaina, H.; Mukhopadhyay, S.; Li, J.Q.; Wu, Z.H.; Wu, Z.H.; Qi, X.L. Poly(N-Isopropylacrylamide) Derived Nanogels Demonstrated Thermosensitive Self-Assembly and Gsh-Triggered Drug Release for Efficient Tumor Therapy. *Polym. Chem.* **2019**, *10*, 4031–4041. [CrossRef]
128. Jing, J.; Alaimo, D.; de Vlieghere, E.; Jerome, C.; de Wever, O.; de Geest, B.G.; Auzely-Velty, R. Tunable Self-Assembled Nanogels Composed of Well-Defined Thermoresponsive Hyaluronic Acid-Polymer Conjugates. *J. Mater. Chem. B* **2013**, *1*, 3883–3887. [CrossRef]
129. Stefanello, T.F.; Couturaud, B.; Szarpak-Jankowska, A.; Fournier, D.; Louage, B.; Garcia, F.P.; Nakamura, C.V.; de Geest, B.G.; Woisel, P.; van der Sanden, B.; et al. Coumarin-Containing Thermoresponsive Hyaluronic Acid-Based Nanogels as Delivery Systems for Anticancer Chemotherapy. *Nanoscale* **2017**, *9*, 12150–12162. [CrossRef]
130. Ganesh, S.; Iyer, A.K.; Morrissey, D.V.; Amiji, M.M. Hyaluronic Acid Based Self-Assembling Nanosystems for Cd44 Target Mediated Sirna Delivery to Solid Tumors. *Biomaterials* **2013**, *34*, 3489–3502. [CrossRef]
131. Ganesh, S.; Iyer, A.K.; Gattacceca, F.; Morrissey, D.V.; Amiji, M.M. In Vivo Biodistribution of Sirna and Cisplatin Administered Using Cd44-Targeted Hyaluronic Acid Nanoparticles. *J. Control. Release* **2013**, *172*, 699–706. [CrossRef] [PubMed]
132. Jiang, G.; Park, K.; Kim, J.; Kim, K.S.; Oh, E.J.; Kang, H.G.; Han, S.E.; Oh, Y.K.; Park, T.G.; Hahn, S.K. Hyaluronic Acid-Polyethyleneimine Conjugate for Target Specific Intracellular Delivery of Sirna. *Biopolymers* **2008**, *89*, 635–642. [CrossRef] [PubMed]
133. Yin, H.; Zhao, F.; Zhang, D.H.; Li, J. Hyaluronic Acid Conjugated Beta-Cyclodextrin-Oligoethylenimine Star Polymer for Cd44-Targeted Gene Delivery. *Int. J. Pharm.* **2015**, *483*, 169–179. [CrossRef] [PubMed]
134. Rahimizadeh, P.; Yang, S.; Lim, S.I. Albumin: An Emerging Opportunity in Drug Delivery. *Biotechnol. Bioprocess Eng.* **2020**, *25*, 985–995. [CrossRef]
135. Shen, X.; Li, T.T.; Xie, X.X.; Feng, Y.; Chen, Z.Y.; Yang, H.; Wu, C.H.; Deng, S.Q.; Liu, Y.Y. Plga-Based Drug Delivery Systems for Remotely Triggered Cancer Therapeutic and Diagnostic Applications. *Front. Bioeng. Biotechnol.* **2020**, *8*, 381. [CrossRef]
136. Rai, A.; Senapati, S.; Saraf, S.K.; Maiti, P. Biodegradable Poly(Epsilon-Caprolactone) as a Controlled Drug Delivery Vehicle of Vancomycin for the Treatment of Mrsa Infection. *J. Mater. Chem. B* **2016**, *4*, 5151–5160. [CrossRef]
137. Sharma, A.; Raghunathan, K.; Solhaug, H.; Antony, J.; Stenvik, J.; Nilsen, A.M.; Einarsrud, M.A.; Bandyopadhyay, S. Modulating Acrylic Acid Content of Nanogels for Drug Delivery & Biocompatibility Studies. *J. Colloid Interface Sci.* **2022**, *607*, 76–88.
138. Chen, Y.; Liu, Y. Cyclodextrin-Based Bioactive Supramolecular Assemblies. *Chem. Soc. Rev.* **2010**, *39*, 495–505. [CrossRef]
139. Chen, Y.; Zhang, Y.M.; Liu, Y. Multidimensional Nanoarchitectures Based on Cyclodextrins. *Chem. Commun.* **2010**, *46*, 5622–5633. [CrossRef]
140. Bai, Y.; Liu, C.P.; Chen, D.; Liu, C.F.; Zhuo, L.H.; Li, H.; Wang, C.; Bu, H.T.; Tian, W. Beta-Cyclodextrin-Modified Hyaluronic Acid-Based Supramolecular Self-Assemblies for Ph- and Esterase-Dual-Responsive Drug Delivery. *Carbohydr. Polym.* **2020**, *246*, 116654. [CrossRef]
141. Yang, Y.; Zhang, Y.M.; Chen, Y.; Chen, J.T.; Liu, Y. Polysaccharide-Based Noncovalent Assembly for Targeted Delivery of Taxol. *Sci. Rep.* **2016**, *6*, 1–10. [CrossRef] [PubMed]
142. Xu, X.Y.; ZZeng, S.; Chen, J.; Huang, B.Y.; Guan, Z.L.; Huang, Y.J.; Huang, Z.Q.; Zhao, C.S. Tumor-Targeted Supramolecular Catalytic Nanoreactor for Synergistic Chemo/Chemodynamic Therapy Via Oxidative Stress Amplification and Cascaded Fenton Reaction. *Chem. Eng. J.* **2020**, *390*, 124628. [CrossRef]
143. Wu, X.; Chen, Y.; Yu, Q.L.; Li, F.Q.; Liu, Y. A Cucurbituril/ Polysaccharide/ Carbazole Ternary Supramolecular Assembly for Targeted Cell Imaging. *Chem. Commun.* **2019**, *55*, 4343–4346. [CrossRef] [PubMed]
144. Li, F.Q.; Yu, Q.L.; Liu, Y.H.; Yu, H.J.; Chen, Y.; Liu, Y. Highly Efficient Photocontrolled Targeted Delivery of Sirna by a Cyclodextrin-Based Supramolecular Nanoassembly. *Chem. Commun.* **2020**, *56*, 3907–3910. [CrossRef]
145. Yang, Y.; Zhang, Y.M.; Chen, Y.; Chen, J.T.; Liu, Y. Targeted Polysaccharide Nanoparticle for Adamplatin Prodrug Delivery. *J. Med. Chem.* **2013**, *56*, 9725–9736. [CrossRef] [PubMed]

146. Kang, Y.; Ju, X.; Wang, L.; Ding, L.S.; Liu, G.T.; Zhang, S.; Li, B.J. Ph and Glutathione Dual-Triggered Supramolecular Assemblies as Synergistic and Controlled Drug Release Carriers. *Polym. Chem.* **2017**, *8*, 7260–7270. [CrossRef]
147. Wang, H.H.; Sun, D.S.; Liao, H.; Wang, Y.F.; Zhao, S.; Zhang, Y.; Lv, G.J.; Ma, X.J.; Liu, Y.; Sun, G.W. Synthesis and Characterization of a Bimodal Nanoparticle Based on the Host-Guest Self-Assembly for Targeted Cellular Imaging. *Talanta* **2017**, *171*, 8–15. [CrossRef]
148. Chen, C.H.; Chen, Y.; Dai, X.Y.; Li, J.J.; Jia, S.S.; Wang, S.P.; Liu, Y. Multicharge Beta-Cyclodextrin Supramolecular Assembly for Atp Capture and Drug Release. *Chem. Commun.* **2021**, *57*, 2812–2815. [CrossRef]
149. Zhao, Q.; Chen, Y.; Sun, M.; Wu, X.J.; Liu, Y. Construction and Drug Delivery of a Fluorescent Tpe-Bridged Cyclodextrin/Hyaluronic Acid Supramolecular Assembly. *Rsc Adv.* **2016**, *6*, 50673–50679. [CrossRef]
150. Yang, Y.; Zhang, Y.M.; Li, D.Z.; Sun, H.L.; Fan, H.X.; Liu, Y. Camptothecin-Polysaccharide Co-Assembly and Its Controlled Release. *Bioconjugate Chem.* **2016**, *27*, 2834–2838.
151. Badwaik, V.; Liu, L.J.; Gunasekera, D.; Kulkarni, A.; Thompson, D.H. Mechanistic Insight into Receptor-Mediated Delivery of Cationic-Beta-Cyclodextrin:Hyaluronic Acid-Adamantamethamidyl Host:Guest Pdna Nanoparticles to Cd44(+) Cells. *Mol. Pharm.* **2016**, *13*, 1176–1184. [CrossRef] [PubMed]
152. Dai, X.Y.; Dong, X.Y.; Liu, Z.X.; Liu, G.X.; Liu, Y. Controllable Singlet Oxygen Generation in Water Based on Cyclodextrin Secondary Assembly for Targeted Photodynamic Therapy. *Biomacromolecules* **2020**, *21*, 5369–5379. [CrossRef] [PubMed]
153. Dai, X.Y.; Zhang, B.; Zhou, W.L.; Liu, Y. High-Efficiency Synergistic Effect of Supramolecular Nanoparticles Based on Cyclodextrin Prodrug on Cancer Therapy. *Biomacromolecules* **2020**, *21*, 4998–5007. [CrossRef] [PubMed]
154. Lee, S.Y.; Ko, S.H.; Shim, J.S.; Kim, D.D.; Cho, H.J. Tumor Targeting and Lipid Rafts Disrupting Hyaluronic Acid-Cyclodextrin-Based Nanoassembled Structure for Cancer Therapy. *Acs Appl. Mater. Interfaces* **2018**, *10*, 36628–36640. [CrossRef] [PubMed]
155. Elamin, K.M.; Yamashita, Y.; Higashi, T.; Motoyama, K.; Arima, H. Supramolecular Complex of Methyl-Beta-Cyclodextrin with Adamantane-Grafted Hyaluronic Acid as a Novel Antitumor Agent. *Chem. Pharm. Bull.* **2018**, *66*, 277–285. [CrossRef]
156. Elamin, K.M.; Motoyama; Higashi, T.; Yamashita, Y.; Tokuda, A.; Arima, H. Dual Targeting System by Supramolecular Complex of Folate-Conjugated Methyl-Beta-Cyclodextrin with Adamantane-Grafted Hyaluronic Acid for the Treatment of Colorectal Cancer. *Int. J. Biol. Macromol.* **2018**, *113*, 386–394. [CrossRef]
157. Yang, K.K.; Yang, Z.Q.; Yu, G.C.; Nie, Z.H.; Wang, R.B.; Chen, X.Y. Polyprodrug Nanomedicines: An Emerging Paradigm for Cancer Therapy. *Adv. Mater.* **2022**, *34*, 2107434. [CrossRef]
158. Duncan, R. Polymer Conjugates as Anticancer Nanomedicines. *Nat. Rev. Cancer* **2006**, *6*, 688–701. [CrossRef]
159. Haag, R.; Kratz, F. Polymer Therapeutics: Concepts and Applications. *Angew. Chem. Int. Ed.* **2006**, *45*, 1198–1215. [CrossRef]
160. Hoste, K.; de Winne, K.; Schacht, E. Polymeric Prodrugs. *Int. J. Pharm.* **2004**, *277*, 119–131. [CrossRef]
161. Hu, C.; WZhuang, H.; Yu, T.; Chen, L.; Liang, Z.; Li, G.C.; Wang, Y.B. Multi-Stimuli Responsive Polymeric Prodrug Micelles for Combined Chemotherapy and Photodynamic Therapy. *J. Mater. Chem. B* **2020**, *8*, 5267–5279. [CrossRef] [PubMed]
162. Luo, L.; Xu, F.S.; Peng, H.L.; Luo, Y.H.; Tian, X.H.; Battaglia, G.; Zhang, H.; Gong, Q.Y.; Gu, Z.W.; Luo, K. Stimuli-Responsive Polymeric Prodrug-Based Nanomedicine Delivering Nifuroxazide and Doxorubicin against Primary Breast Cancer and Pulmonary Metastasis. *J. Control. Release* **2020**, *318*, 124–135. [CrossRef]
163. Luo, X.L.; Wang, S.C.; Xu, S.S.; Lang, M.D. Relevance of the Polymeric Prodrug and Its Drug Loading Efficiency: Comparison between Computer Simulation and Experiment. *Macromol. Theory Simul.* **2019**, *28*, 1900026. [CrossRef]
164. Fu, S.W.; Li, G.T.; Zang, W.L.; Zhou, X.Y.; Shi, K.X.; Zhai, Y.L. Pure Drug Nano-Assemblies: A Facile Carrier-Free Nanoplatform for Efficient Cancer Therapy. *Acta Pharm. Sin. B* **2022**, *12*, 92–106. [CrossRef] [PubMed]
165. Wang, Z.R.; Chen, J.W.; Little, N.; Lu, J.Q. Self-Assembling Prodrug Nanotherapeutics for Synergistic Tumor Targeted Drug Delivery. *Acta Biomater.* **2020**, *111*, 20–28. [CrossRef] [PubMed]
166. Xu, C.R.; He, W.; Lv, Y.Q.; Qin, C.; Shen, L.J.; Yin, L.F. Self-Assembled Nanoparticles from Hyaluronic Acid-Paclitaxel Prodrugs for Direct Cytosolic Delivery and Enhanced Antitumor Activity. *Int. J. Pharm.* **2015**, *493*, 172–181. [CrossRef]
167. Xin, D.C.; Wang, Y.; Xiang, J.N. The Use of Amino Acid Linkers in the Conjugation of Paclitaxel with Hyaluronic Acid as Drug Delivery System: Synthesis, Self-Assembled Property, Drug Release, and in Vitro Efficiency. *Pharm. Res.* **2010**, *27*, 380–389. [CrossRef]
168. Zhong, Y.N.; Goltsche, K.; Cheng, L.; Xie, F.; Meng, F.H.; Deng, C.; Zhong, Z.Y.; Haag, R. Hyaluronic Acid-Shelled Acid-Activatable Paclitaxel Prodrug Micelles Effectively Target and Treat Cd44-Overexpressing Human Breast Tumor Xenografts in Vivo. *Biomaterials* **2016**, *84*, 250–261. [CrossRef]
169. Wang, W.J.; Zhang, X.Q.; Li, Z.Q.; Pan, D.Y.; Zhu, H.Y.; Gu, Z.W.; Chen, J.; Zhang, H.; Gong, Q.Y.; Luo, K. Dendronized Hyaluronic Acid-Docetaxel Conjugate as a Stimuli-Responsive Nano-Agent for Breast Cancer Therapy. *Carbohydr. Polym.* **2021**, *267*, 118160. [CrossRef]
170. Cai, S.A.; Thati, S.; Bagby, T.R.; Diab, H.M.; Davies, N.M.; Cohen, M.S.; Forrest, M.L. Localized Doxorubicin Chemotherapy with a Biopolymeric Nanocarrier Improves Survival and Reduces Toxicity in Xenografts of Human Breast Cancer. *J. Control. Release* **2010**, *146*, 212–218. [CrossRef]
171. Yin, T.J.; Wang, Y.Y.; Chu, X.X.; Fu, Y.; Wang, L.; Zhou, J.P.; Tang, X.M.; Liu, J.Y.; Huo, M.R. Free Adriamycin-Loaded Ph/Reduction Dual-Responsive Hyaluronic Acid-Adriamycin Prodrug Micelles for Efficient Cancer Therapy. *Acs Appl. Mater. Interfaces* **2018**, *10*, 35693–35704. [CrossRef] [PubMed]

172. Xu, M.H.; Qian, J.M.; Suo, A.L.; Wang, H.J.; Yong, X.Q.; Liu, X.F.; Liu, R.R. Reduction/Ph Dual-Sensitive Pegylated Hyaluronan Nanoparticles for Targeted Doxorubicin Delivery. *Carbohydr. Polym.* **2013**, *98*, 181–188. [CrossRef] [PubMed]
173. Lu, B.B.; Xiao, F.; Wang, Z.Y.; Wang, B.S.; Pan, Z.C.; Zhao, W.W.; Zhu, Z.Y.; Zhang, J.H. Redox-Sensitive Hyaluronic Acid Polymer Prodrug Nanoparticles for Enhancing Intracellular Drug Self-Delivery and Targeted Cancer Therapy. *Acs Biomater. Sci. Eng.* **2020**, *6*, 4106–4115. [CrossRef]
174. Song, L.; Pan, Z.; Zhang, H.B.; Li, Y.X.; Zhang, Y.Y.; Lin, J.Y.; Su, G.H.; Ye, S.F.; Xie, L.Y.; Li, Y.; et al. Dually Folate/Cd44 Receptor-Targeted Self-Assembled Hyaluronic Acid Nanoparticles for Dual-Drug Delivery and Combination Cancer Therapy. *J. Mater. Chem. B* **2017**, *5*, 6835–6846. [CrossRef]
175. Zhang, Y.B.; Li, Y.; Tian, H.N.; Zhu, Q.X.; Wang, F.F.; Fan, Z.X.; Zhou, S.; Wang, X.W.; Xie, L.Y.; Hou, Z.Q. Redox-Responsive and Dual-Targeting Hyaluronic Acid-Methotrexate Prodrug Self-Assembling Nanoparticles for Enhancing Intracellular Drug Self-Delivery. *Mol. Pharm.* **2019**, *16*, 3133–3144. [CrossRef] [PubMed]
176. Chen, Z.J.; He, N.; Chen, M.H.; Zhao, L.; Li, X.H. Tunable Conjugation Densities of Camptothecin on Hyaluronic Acid for Tumor Targeting and Reduction-Triggered Release. *Acta Biomater.* **2016**, *43*, 195–207. [CrossRef]
177. Chen, Z.J.; Liu, W.P.; Zhao, L.; Xie, S.Z.; Chen, M.H.; Wang, T.; Li, X.H. Acid-Labile Degradation of Injectable Fiber Fragments to Release Bioreducible Micelles for Targeted Cancer Therapy. *Biomacromolecules* **2018**, *19*, 1100–1110. [CrossRef]
178. Zhang, Y.W.; Zeng, X.L.; Wang, H.R.; Fan, R.N.; Hu, Y.K.; Hu, X.J.; Li, J. Dasatinib Self-Assembled Nanoparticles Decorated with Hyaluronic Acid for Targeted Treatment of Tumors to Overcome Multidrug Resistance. *Drug Deliv.* **2021**, *28*, 670–679. [CrossRef]
179. Liang, D.S.; Su, H.T.; Liu, Y.J.; Wang, A.T.; Qi, X.R. Tumor-Specific Penetrating Peptides-Functionalized Hyaluronic Acid-D-Alpha-Tocopheryl Succinate Based Nanoparticles for Multi-Task Delivery to Invasive Cancers. *Biomaterials* **2015**, *71*, 11–23. [CrossRef]
180. Wang, J.L.; Li, Y.; Wang, L.F.; Wang, X.H.; Tu, P.F. Comparison of Hyaluronic Acid-Based Micelles and Polyethylene Glycol-Based Micelles on Reversal of Multidrug Resistance and Enhanced Anticancer Efficacy in Vitro and in Vivo. *Drug Deliv.* **2018**, *25*, 330–340. [CrossRef]
181. Wang, H.R.; Zhang, Y.W.; Zeng, X.L.; Pei, W.J.; Fan, R.R.; Wang, Y.S.; Wang, X.; Li, J.C. A Combined Self-Assembled Drug Delivery for Effective Anti-Breast Cancer Therapy. *Int. J. Nanomed.* **2021**, *16*, 2373–2388. [CrossRef] [PubMed]
182. Wang, J.L.; Ma, W.Z.; Guo, Q.; Li, Y.; Hu, Z.D.; Zhu, Z.X.; Wang, X.H.; Zhao, Y.F.; Chai, X.Y.; Tu, P.F. The Effect of Dual-Functional Hyaluronic Acid-Vitamin E Succinate Micelles on Targeting Delivery of Doxorubicin. *Int. J. Nanomed.* **2016**, *11*, 5851–5870. [CrossRef] [PubMed]
183. Ma, W.Z.; Guo, Q.; Li, Y.; Wang, X.H.; Wang, J.L.; Tu, P.F. Co-Assembly of Doxorubicin and Curcumin Targeted Micelles for Synergistic Delivery and Improving Anti-Tumor Efficacy. *Eur. J. Pharm. Biopharm.* **2017**, *112*, 209–223. [CrossRef]
184. Zhou, C.P.; Dong, X.X.; Song, C.X.; Cui, S.; Chen, T.T.; Zhang, D.J.; Zhao, X.L.; Yang, C.R. Rational Design of Hyaluronic Acid-Based Copolymer-Mixed Micelle in Combination Pd-L1 Immune Checkpoint Blockade for Enhanced Chemo-Immunotherapy of Melanoma. *Front. Bioeng. Biotechnol.* **2021**, *9*, 653417. [CrossRef]
185. Bai, F.; Wang, Y.; Han, Q.Q.; Wu, M.L.; Luo, Q.; Zhang, H.M.; Wang, Y.Q. Cross-Linking of Hyaluronic Acid by Curcumin Analogue to Construct Nanomicelles for Delivering Anticancer Drug. *J. Mol. Liq.* **2019**, *288*, 111079. [CrossRef]
186. Xu, C.F.; Ding, Y.; Ni, J.; Yin, L.F.; Zhou, J.P.; Yao, J. Tumor-Targeted Docetaxel-Loaded Hyaluronic Acid-Quercetin Polymeric Micelles with P-Gp Inhibitory Property for Hepatic Cancer Therapy. *Rsc Adv.* **2016**, *6*, 27542–27556. [CrossRef]
187. Bae, K.H.; Tan, S.S.; Yamashita, A.; Ang, W.X.; Gao, S.J.; Wang, S.; Chung, J.E.; Kurisawa, M. Hyaluronic Acid-Green Tea Catechin Micellar Nanocomplexes: Fail-Safe Cisplatin Nanomedicine for the Treatment of Ovarian Cancer without Off-Target Toxicity. *Biomaterials* **2017**, *148*, 41–53. [CrossRef] [PubMed]
188. Liang, K.; Ng, S.; Lee, F.; Lim, J.; Chung, J.E.; Lee, S.S.; Kurisawa, M. Targeted Intracellular Protein Delivery Based on Hyaluronic Acid-Green Tea Catechin Nanogels. *Acta Biomater.* **2016**, *33*, 142–152. [CrossRef]
189. Zhang, L.; Yao, J.; Zhou, J.P.; Wang, T.; Zhang, Q. Glycyrrhetinic Acid-Graft-Hyaluronic Acid Conjugate as a Carrier for Synergistic Targeted Delivery of Antitumor Drugs. *Int. J. Pharm.* **2013**, *441*, 654–664. [CrossRef]
190. Zhang, L.; Zhou, J.P.; Yao, J. Improved Anti-Tumor Activity and Safety Profile of a Paclitaxel-Loaded Glycyrrhetinic Acid-Graft-Hyaluronic Acid Conjugate as a Synergistically Targeted Drug Delivery System. *Chin. J. Nat. Med.* **2015**, *13*, 915–924. [CrossRef]
191. Lin, T.S.; Yuan, A.; Zhao, X.Z.; Lian, H.B.; Zhuang, J.L.; Chen, W.; Zhang, Q.; Liu, G.X.; Zhang, S.W.; Chen, W.; et al. Self-Assembled Tumor-Targeting Hyaluronic Acid Nanoparticles for Photothermal Ablation in Orthotopic Bladder Cancer. *Acta Biomater.* **2017**, *53*, 427–438. [CrossRef] [PubMed]
192. Shi, H.X.; Sun, W.C.; Liu, C.B.; Gu, G.Y.; Ma, B.; Si, W.L.; Fu, N.N.; Zhang, Q.; Huang, W.; Dong, X.C. Tumor-Targeting, Enzyme-Activated Nanoparticles for Simultaneous Cancer Diagnosis and Photodynamic Therapy. *J. Mater. Chem. B* **2016**, *4*, 113–120. [CrossRef] [PubMed]
193. Feng, C.; Chen, L.; Lu, Y.L.; Liu, J.; Liang, S.J.; Lin, Y.; Li, Y.Y.; Dong, C.Y. Programmable Ce6 Delivery Via Cyclopamine Based Tumor Microenvironment Modulating Nano-System for Enhanced Photodynamic Therapy in Breast Cancer. *Front. Chem.* **2019**, *7*, 853. [CrossRef]
194. Hou, L.; Zhang, Y.L.; Yang, X.M.; Tian, C.Y.; Yan, Y.S.; Zhang, H.L.; Shi, J.J.; Zhang, H.J.; Zhang, Z.Z. Intracellular No-Generator Based on Enzyme Trigger for Localized Tumor-Cytoplasm Rapid Drug Release and Synergetic Cancer Therapy. *Acs Appl. Mater. Interfaces* **2019**, *11*, 255–268. [CrossRef]

195. Montanari, E.; Capece, S.; di Meo, C.; Meringolo, M.; Coviello, T.; Agostinelli, E.; Matricardi, P. Hyaluronic Acid Nanohydrogels as a Useful Tool for Bsao Immobilization in the Treatment of Melanoma Cancer Cells. *Macromol. Biosci.* **2013**, *13*, 1185–1194. [CrossRef] [PubMed]
196. Ma, L.; Gao, W.J.; Han, X.; Qu, F.L.; Xia, L.; Kong, R.M. A Label-Free and Fluorescence Turn-on Assay for Sensitive Detection of Hyaluronidase Based on Hyaluronan-Induced Perylene Self-Assembly. *N. J. Chem.* **2019**, *43*, 3383–3389. [CrossRef]
197. Noguchi, T.; Roy, B.; Yoshihara, D.; Sakamoto, J.; Yamamoto, T.; Shinkai, S. Emergent Molecular Recognition through Self-Assembly: Unexpected Selectivity for Hyaluronic Acid among Glycosaminoglycans. *Angew. Chem. Int. Ed.* **2016**, *55*, 5708–5712. [CrossRef]
198. Lee, Y.H.; Yoon, H.Y.; Shin, J.M.; Saravanakumar, G.; Noh, K.H.; Song, K.H.; Jeon, J.H.; Kim, D.W.; Lee, K.M.; Kim, K.; et al. A Polymeric Conjugate Foreignizing Tumor Cells for Targeted Immunotherapy in Vivo. *J. Control. Release* **2015**, *199*, 98–105. [CrossRef]
199. Shin, J.M.; Oh, S.J.; Kwon, S.; Deepagan, V.G.; Lee, M.; Song, S.H.; Lee, H.J.; Kim, S.; Song, K.H.; Kim, T.W.; et al. A Pegylated Hyaluronic Acid Conjugate for Targeted Cancer Immunotherapy. *J. Control. Release* **2017**, *267*, 181–190. [CrossRef]
200. Bukowski, K.; Kciuk, M.; Kontek, R. Mechanisms of Multidrug Resistance in Cancer Chemotherapy. *Int. J. Mol. Sci.* **2020**, *21*, 3233. [CrossRef]
201. Vaidya, F.U.; Chhipa, A.S.; Mishra, V.; Gupta, V.K.; Rawat, S.G.; Kumar, A.; Pathak, C. Molecular and Cellular Paradigms of Multidrug Resistance in Cancer. *Cancer Rep.* **2020**, e1291. [CrossRef] [PubMed]
202. Zhang, M.; Liu, E.G.; Cui, Y.N.; Huang, Y.Z. Nanotechnology-Based Combination Therapy for Overcoming Multidrug-Resistant Cancer. *Cancer Biol. Med.* **2017**, *14*, 212–227. [CrossRef] [PubMed]
203. Zhang, C.Y.; Zhou, X.M.; Zhang, H.Y.; Han, X.L.; Li, B.J.; Yang, R.; Zhou, X. Recent Progress of Novel Nanotechnology Challenging the Multidrug Resistance of Cancer. *Front. Pharmacol.* **2022**, *13*, 122. [CrossRef] [PubMed]
204. Curcio, M.; Farfalla, A.; Saletta, F.; Valli, E.; Pantuso, E.; Nicoletta, F.P.; Iemma, F.; Vittorio, O.; Cirillo, G. Functionalized Carbon Nanostructures Versus Drug Resistance: Promising Scenarios in Cancer Treatment. *Molecules* **2020**, *25*, 2102. [CrossRef]
205. Xue, X.; Liang, X.J. Overcoming Drug Efflux-Based Multidrug Resistance in Cancer with Nanotechnology. *Chin. J. Cancer* **2012**, *31*, 100–109. [CrossRef]
206. Wang, C.D.; Li, F.S.; Zhang, T.A.; Yu, M.; Sun, Y. Recent Advances in Anti-Multidrug Resistance for Nano-Drug Delivery System. *Drug Deliv.* **2022**, *29*, 1684–1697. [CrossRef]
207. Yan, H.X.; Du, X.Y.; Wang, R.J.; Zhai, G.X. Progress in the Study of D-Alpha-Tocopherol Polyethylene Glycol 1000 Succinate (Tpgs) Reversing Multidrug Resistance. *Colloids Surf. B Biointerfaces* **2021**, *205*, 111914. [CrossRef]
208. Patra, S.; Pradhan, B.; Nayak, R.; Behera, C.; Das, S.; Patra, S.K.; Efferth, T.; Jena, M.; Bhutia, S.K. Dietary Polyphenols in Chemoprevention and Synergistic Effect in Cancer: Clinical Evidences and Molecular Mechanisms of Action. *Phytomedicine* **2021**, *90*, 153554. [CrossRef]
209. Vittorio, O.; Brandl, M.; Cirillo, G.; Kimpton, K.; Hinde, E.; Gaus, K.; Yee, E.; Kumar, N.; Duong, H.; Fleming, C.; et al. Dextran-Catechin: An Anticancer Chemically-Modified Natural Compound Targeting Copper That Attenuates Neuroblastoma Growth. *Oncotarget* **2016**, *7*, 47479–47493. [CrossRef]
210. Tatullo, M.; Simone, G.M.; Tarullo, F.; Irlandese, G.; de Vito, D.; Marrelli, M.; Santacroce, L.; Cocco, T.; Ballini, A.; Scacco, S. Antioxidant and Antitumor Activity of a Bioactive Polyphenolic Fraction Isolated from the Brewing Process. *Sci. Rep.* **2016**, *6*, 1–7. [CrossRef]
211. Dai, Q.; Geng, H.M.; Yu, Q.; Hao, J.C.; Cui, J.W. Polyphenol-Based Particles for Theranostics. *Theranostics* **2019**, *9*, 3170–3190. [CrossRef] [PubMed]
212. Vittorio, O.; Curcio, M.; Cojoc, M.; Goya, G.F.; Hampel, S.; Iemma, F.; Dubrovska, A.; Cirillo, G. Polyphenols Delivery by Polymeric Materials: Challenges in Cancer Treatment. *Drug Deliv.* **2017**, *24*, 162–180. [CrossRef] [PubMed]
213. Huang, Z.J.; Fu, J.J.; Zhang, Y.H. Nitric Oxide Donor-Based Cancer Therapy: Advances and Prospects. *J. Med. Chem.* **2017**, *60*, 7617–7635. [CrossRef] [PubMed]
214. Foulkes, R.; Man, E.; Thind, J.; Yeung, S.; Joy, A.; Hoskins, C. The Regulation of Nanomaterials and Nanomedicines for Clinical Application: Current and Future Perspectives. *Biomater. Sci.* **2020**, *8*, 4653–4664. [CrossRef] [PubMed]
215. Gessner, I. Optimizing Nanoparticle Design and Surface Modification toward Clinical Translation. *Mrs Bull.* **2021**, *46*, 643–649. [CrossRef] [PubMed]
216. Yoo, J.W.; Chambers, E.; Mitragotri, S. Factors That Control the Circulation Time of Nanoparticles in Blood: Challenges, Solutions and Future Prospects. *Curr. Pharm. Des.* **2010**, *16*, 2298–2307. [CrossRef]
217. Srinivasan, M.; Rajabi, M.; Mousa, S.A. Multifunctional Nanomaterials and Their Applications in Drug Delivery and Cancer Therapy. *Nanomaterials* **2015**, *5*, 1690–1703. [CrossRef]

nanomaterials

MDPI

Article

Anti-Fn14-Conjugated Prussian Blue Nanoparticles as a Targeted Photothermal Therapy Agent for Glioblastoma

Nicole F. Bonan [1,2,†], Debbie K. Ledezma [1,2,†], Matthew A. Tovar [1,3,†], Preethi B. Balakrishnan [1,†] and Rohan Fernandes [1,2,3,4,*]

[1] George Washington Cancer Center, George Washington University, Washington, DC 20052, USA; nbonan@gwmail.gwu.edu (N.F.B.); dkledezma@gwmail.gwu.edu (D.K.L.); mtovar@gwmail.gwu.edu (M.A.T.); preethibala18@gmail.com (P.B.B.)
[2] Institute for Biomedical Sciences, George Washington University, Washington, DC 20052, USA
[3] School of Medicine and Health Sciences, George Washington University, Washington, DC 20052, USA
[4] Department of Medicine, George Washington University, Washington, DC 20052, USA
* Correspondence: rfernandes@gwu.edu
† These authors contributed equally to this work.

Abstract: Prussian blue nanoparticles (PBNPs) are effective photothermal therapy (PTT) agents: they absorb near-infrared radiation and reemit it as heat via phonon-phonon relaxations that, in the presence of tumors, can induce thermal and immunogenic cell death. However, in the context of central nervous system (CNS) tumors, the off-target effects of PTT have the potential to result in injury to healthy CNS tissue. Motivated by this need for targeted PTT agents for CNS tumors, we present a PBNP formulation that targets fibroblast growth factor-inducible 14 (Fn14)-expressing glioblastoma cell lines. We conjugated an antibody targeting Fn14, a receptor abundantly expressed on many glioblastomas but near absent on healthy CNS tissue, to PBNPs (aFn14-PBNPs). We measured the attachment efficiency of aFn14 onto PBNPs, the size and stability of aFn14-PBNPs, and the ability of aFn14-PBNPs to induce thermal and immunogenic cell death and target and treat glioblastoma tumor cells in vitro. aFn14 remained stably conjugated to the PBNPs for at least 21 days. Further, PTT with aFn14-PBNPs induced thermal and immunogenic cell death in glioblastoma tumor cells. However, in a targeted treatment assay, PTT was only effective in killing glioblastoma tumor cells when using aFn14-PBNPs, not when using PBNPs alone. Our methodology is novel in its targeting moiety, tumor application, and combination with PTT. To the best of our knowledge, PBNPs have not been investigated as a targeted PTT agent in glioblastoma via conjugation to aFn14. Our results demonstrate a novel and effective method for delivering targeted PTT to aFn14-expressing tumor cells via aFn14 conjugation to PBNPs.

Keywords: photothermal therapy; prussian blue nanoparticles; aFn14 antibody; glioblastoma; targeted therapy; thermal therapy; immunogenic cell death

Citation: Bonan, N.F.; Ledezma, D.K.; Tovar, M.A.; Balakrishnan, P.B.; Fernandes, R. Anti-Fn14-Conjugated Prussian Blue Nanoparticles as a Targeted Photothermal Therapy Agent for Glioblastoma. *Nanomaterials* **2022**, *12*, 2645. https://doi.org/10.3390/nano12152645

Academic Editor: Bing Yan

Received: 17 June 2022
Accepted: 28 July 2022
Published: 1 August 2022

1. Introduction

Tumors of the central nervous system (CNS) are the most common tumor diagnosed in patients under the age of 14 and the 8th most common tumor diagnosed in patients over 40. CNS tumors account for a disproportionately higher rate of cancer-related mortality in the United States [1,2]. Glioblastoma (GBM) is histologically defined as a grade IV astrocytoma exhibiting incredible resistance to conventional treatment methods and an overall poor prognosis. The current point-of-care for GBM includes surgery, often followed by high-dose radiation therapy and administration of temozolomide (TMZ), a small molecule DNA-alkylating agent. GBM has a median survival length of 10–14 months even with the most aggressive treatment options [3–8]. Therefore, there is an urgent need for novel treatment options for GBM patients.

Photothermal therapy (PTT) mediated by light-absorbing and biocompatible nanoparticles that are activated by a laser is a promising treatment strategy for GBM. PTT has been extensively described by other groups and us as a means to achieve tumor control through both direct heat-based cytotoxicity, recruitment of the body's endogenous immune system, and pro-immunogenic modulation of the tumor microenvironment [9–13]. Further, in the clinical setting, thermal therapies such as laser interstitial thermal therapy are safe and well tolerated in patients with non-accessible or recurrent GBM tumors [14]. This precedent indicates the potential application and enhanced efficacy of nanoparticle-based, targeted PTT for GBM.

In the context of GBM, several groups have described the use of PTT in vitro and in vivo [15]. The nanomaterials and nanoparticles investigated in these studies include metal-based, carbon-based, hybrid, and "other" nanoparticles that do not fit the aforementioned classifications [16–22]. The nanoparticles were either actively or passively targeted to GBM tumor cells or tumors to administer PTT. Overall, PTT induced significant cell death in both in vitro and in vivo models. The efficacy of these studies is summarized in a review by Bastiancich et. al. [15]. However, when tumor cells are heated using non-targeted nanoparticle-based PTT agents, the heating can be non-selective, with the potential for heating non-malignant tissue. To translate PTT to CNS tumors, non-selective heating should be limited to minimize the risk of off-target toxicities. Thus, a major step towards unlocking the potential of nanoparticle-based PTT for malignant CNS tumors is to engineer a robust, stable, and biocompatible tumor-specific nanotherapeutic modality with high on-target specificity without sacrificing direct tumor cytotoxicity.

To target nanoparticle-based PTT specifically to GBM tumor cells, we target the highly upregulated and GBM-specific receptor fibroblast growth factor-inducible 14 (Fn14). Fn14 is overexpressed on GBM tumors and is a cognate receptor for the tumor necrosis factor weak inducer of apoptosis (TWEAK) [23]. GBM is an aggressive tumor, often invading deep into the normal brain parenchyma, making it difficult for surgical resection and a major reason for tumor recurrence in treated patients. Fn14 receptor and its binding ligands have been shown to be significantly upregulated in poorly prognostic glioma compared to healthy neuronal or glial tissue and thus it represents a favorable candidate for developing targeted nanotherapeutics against GBM, mitigating off-target complications [22–27]. Fn14 also serves as a promising target not only for its abundance in GBM, but also its localization. Fn14 is expressed within both the tumor core and the invasive outer rim region of GBM, making it a potent target to treat all regions of this highly invasive tumor. [24,25] Furthermore, Fn14-TWEAK engagement leads to poor prognostic events, including proliferation, migration, and further invasion of GBM, which is often correlated with GBM's resistance to chemotherapeutics [24,26–28]. Hence, engaging or blocking Fn14 expressed on the GBM tumor cells could make it unavailable for TWEAK binding and could be utilized to elicit antitumor effects. This suggests Fn14 could serve as a therapeutic target for targeted nanoparticle-based PTT.

Few groups have investigated the use of anti-Fn14 conjugation to nanoparticles. One group demonstrated that anti-Fn14 conjugation to gold nanoparticles acted as a TWEAK agonist, which would suggest activation of a more aggressive tumor cell phenotype [29]. This effect presents a strong argument for using PTT to ablate the tumor once the particles are delivered and bound to the surface of the GBM tumor cells. Another group demonstrated that aFn14 conjugation to carboxylate-modified polystyrene nanoparticles enhanced particle retention in on the surface of and within Fn14-expressing tumor cells [30], providing further rationale for Fn14-based targeting. However, while these studies show promise, the feasibility of translating them to clinic would be restricted by the limited clinical studies into the potential toxicities of gold nanoparticles and polystyrene nanoparticles [31,32]. Therefore, we used PBNPs conjugated with aFn14 as a nanotherapeutic with a potentially higher translational feasibility.

Prussian blue is an FDA-approved material used to treat radioactive poisoning when orally consumed as Radiogardase® [33,34]. Additionally, PBNPs can be easily synthesized at large scales using low cost starting materials and is therefore amenable to clinical translation. We have previously described that, when irradiated by an 808-nm near infrared (NIR) laser, PBNPs facilitate photothermal energy conversion [35]. This energy conversion, mediated by electron-phonon coupling, followed by phonon-phonon relaxations in the setting of NIR radiation incident on the PBNP crystal lattice structure [9,36], has the capacity to produce tumor microenvironmental temperatures upwards of 80 °C, depending on the nanoparticle concentration and incident laser power, [9,10] triggering thermally induced cell death. Tumor cell death via this method is one of the most critical effects of PTT, as demonstrated by our group and others [37].

In addition to the safety offered by PBNPs, these particles also offer an immunological advantage. PTT can induce the release of endogenous immunoadjuvants such as damage-associated molecular patterns (DAMPs) from dying cells, indicative of a process known as immunogenic cell death (ICD) [38–40]. ICD can convert an immunologically "cold" tumor to a "hot" one, where released antigens can activate an immune response against the tumor. Our group and others have previously shown that PBNP-mediated PTT can induce ICD in pediatric neuroblastoma [9,10,41,42], melanoma [43], and triple-negative breast adenocarcinoma preclinical models, to name a few, and is being investigated in this study as a novel treatment option for GBM. ICD is not always induced by nanoparticle-mediated PTT, but the consistency with which our lab has observed PBNP-PTT-induced ICD makes PBNPs an attractive material.

The overall hypothesis in this study is that covalent conjugation of the Fn14 targeting antibody to the surface of the PBNP results in a stable nanotherapeutic (aFn14-PBNP) that can be utilized for targeted PTT without sacrificing the ability of the PBNPs to generate on-target heat-based cytotoxicity and ICD. We test this hypothesis by optimizing the synthesizing scheme for generating aFn14-PBNPs and assessing the critical quality attributes of the resulting nanotherapeutic including the attachment efficiency of aFn14 on PBNPs, the size distributions, surface charge, and UV-Vis-NIR spectra of aFn14-PBNPs. We then test the ability of the aFn14-PBNPs to elicit thermal and immunogenic cell death as well as effectively target and be retained on multiple human GBM tumor lines in vitro. Finally, in a targeted treatment assay, we test the ability of aFn14-PBNPs to target and treat GBM tumor lines with differential Fn14 expression compared to non-targeted PBNPs in vitro. Through our results, we seek to demonstrate that this novel nanoparticle-antibody conjugate provides a benefit over nanoparticles alone by retaining particles on aFn14-expressing tumor cells, thus enabling targeted PTT. These targeting studies with our aFn14-PBNP nanoformulation provide important proof-of-concept data to proceed with evaluating aFn14-PBNPs for PTT of GBM in vivo.

2. Materials and Methods

2.1. Synthesis of PBNPs

PBNPs were synthesized using a one-pot scheme as previously described [9,35,41,44]. Briefly, 20 mL of 10 mM aqueous $FeCl_3 \bullet 6H_2O$ (54.06 mg) (Millipore Sigma, Darmstadt, Germany) containing 5 mmol of citric acid (961 mg) (Millipore Sigma, Darmstadt, Germany) was added to an equal volume of 10 mM aqueous $K_4[Fe(CN)_6] \bullet 3H_2O$ (84.5 mg) (Millipore Sigma, Darmstadt, Germany) containing 5 mmol of citric acid, under vigorous stirring at 60 °C using a magnetic stirring hot plate. After heating and stirring for 60 s, the solution was allowed to cool to room temperature (RT) for 5 min under constant stirring. The 40 mL solution was then divided equally between two 50 mL tubes and PBNPs were then collected by adding equal volume of acetone (Millipore Sigma, Darmstadt, Germany) and 5 mL 5 M NaCl (Millipore Sigma, Darmstadt, Germany). The solutions were then centrifuged at 10,000 rpm for 15 min at RT. The centrifugation step with 20 mL MilliQ water, 20 mL acetone, and 5 mL 5M NaCl was repeated twice. The final PBNP pellet was sonicated (at 40% amplitude for 30 s using a microtip probe) in 10 mL MilliQ water to achieve colloidal resuspension. The PBNPs were stored under ambient conditions in DI water prior to further bioconjugation.

2.2. Synthesis of Bioconjugated aFn14-PBNP

Covalent synthesis of aFn14-PBNPs was carried out using 1-ethyl-3-(3-dimethylaminopropyl) carbodiimide (EDC) chemistry. To begin, 21.5 µL of the as synthesized PBNPs (23.21 mg/mL) was combined with 100 µL of EDC solution (2.2 mg/mL; Thermo Fisher Scientific, Waltham, MA, USA) and 100 µL of Sulfo-NHS solution (8.0 mg/mL; Millipore Sigma, Darmstadt, Germany) in 1 mL of MES buffer (0.1 M MES, 0.5 M NaCl, pH = 5; ACROS Organics, Geel, Belgium). The first crosslinking reaction occurred for 15 min at RT and was stopped via addition of 100 µL of 2-mercaptoethanol (8.9 mg/mL; Thermo Fisher Scientific, Waltham, MA, USA). PBNP crosslinked to Sulfo-NHS were then centrifuged at 22,000× *g* for 30 min using a table-top microcentrifuge unit at RT. Particles were resuspended in 1 mL MES buffer and sonicated using a microtip probe at 40% amplitude for 30 s to achieve a homogeneous colloidal solution. FITC-conjugated aFn14 antibody (Santa Cruz Biotechnology, Santa Cruz, CA, USA) was then added to a final concentration of 0.25 µg/mL, corresponding to a 1:2000 mass-to-mass ratio of aFn14:PBNP. The mixture was contacted in the dark at RT for 3 h on an orbital shaker, after which 100 µL of 0.1 M hydroxylamine (Thermo Fisher Scientific, Waltham, MA, USA) was added to quench any remaining primary amine sites. aFn14-PBNP were again centrifuged at 22,000× *g* for 30 min at RT, resuspended in 1 mL DI H_2O, and sonicated. This process was repeated twice for a total of three washes. The particles were then resuspended in the desired volume of sterile DI water and stored at 4 °C, protected from light.

2.3. Attachment Efficiency of aFn14 to PBNPs

The attachment efficiency of aFn14 to PBNPs was calculated based on the amount of aFn14 that remained unbound in the aFn14-PBNP synthesis supernatants. A standard curve of fluorescence intensity (λ_{em} = 490 nm, λ_{ex} = 525 nm) vs. known concentrations of FITC-conjugated aFn14 was generated using a SpectraMax i3x Multimode Microplate Reader (Molecular Devices, LLC, San Jose, CA, USA) (Figure S1). To determine the amount of aFn14 that did not bind to PBNPs after aFn14-PBNP synthesis, supernatants of the syntheses were collected and compared to the standard curve. The concentration and then mass of unbound aFn14 were calculated; the unbound mass was subtracted from the initial mass of aFn14 utilized for the synthesis to determine the final mass of aFn14 attached onto the PBNP collected. This value was then divided by the initial mass of aFn14 and multiplied by 100 to determine the attachment efficiency of the antibody.

2.4. Characterization of aFn14-PBNP

To quantify the size and charge of the aFn14-PBNPs, the hydrodynamic diameter and zeta potential of PNBPs and aFn14-PBNP were measured using dynamic light scattering (DLS) spectroscopy and zeta anemometry on a Zetasizer Nano ZS (Malvern Instruments, Malvern, UK). Optical characteristics of the constructs were measured via UV-Vis-NIR Spectroscopy using a Genesys 10S spectrophotometer and VISIONlite software (Thermo Fisher Scientific, Waltham, MA, USA). Attachment efficiency was measured via fluorescence spectroscopy as described in Section 2.3. To measure nanoparticle stability over time, DLS, zeta-anemometry, and fluorescence spectroscopy was performed at Day 0, +2, +4, +8, +16, and +20 following the initial particle synthesis. These physical characteristics were measured for every subsequent nanoparticle synthesis to assess whether the critical quality attributes of the nanoparticles were within acceptable standards for PBNP and aFn14-PBNP.

2.5. Cell Lines and Culture

Human U87 glioblastoma cells (ATCC, Manassas, VA, USA) were cultured in Eagle's Minimal Essential Medium (EMEM; ATCC, Manassas, VA, USA) containing L-glutamine (Thermo Fisher Scientific, Waltham, MA, USA) supplemented with 10% fetal bovine serum (FBS; Thermo Fisher Scientific) and 1% penicillin/streptomycin antibiotic (Thermo Fisher Scientific, Waltham, MA, USA). Human U251 glioblastoma cells (NCI Developmental Therapeutics Program, Bethesda, MD, USA) were maintained in EMEM containing L-glutamine supplemented with 10% FBS, 1% non-essential amino acids (Thermo Fisher Scientific, Waltham, MA, USA), and 1% penicillin/streptomycin.

2.6. Characterization of the PTT Properties of aFn14-PBNP

In the clinic, PBNPs would be administered to the tumor site and then irradiated to ablate tumor tissue. In our in vitro protocol to mimic this procedure, samples treated with PBNPs are irradiated with an NIR laser at various laser powers, and the temperature and thermal dose are measured over time to characterize the impact of PBNP excitation and relaxation on the surrounding environment. This protocol is a simple and cost-effective way of modeling PTT in a laboratory setting under reproducible conditions.

In this study, 0.5 mL of water or 5×10^6 U87 or U251 in 0.5 mL culture media were treated with 0.15 mg/mL PBNP or aFn14-PBNP. Note that 0.15 mg/mL aFn14-PBNP refers to the concentration of PBNP, not aFn14, to keep the total concentration of PBNPs in aFn14-PBNP consistent with the PBNP only condition. Suspensions were irradiated with an 808 nm NIR continuous wave collimated diode laser (Laserglow Technologies, Toronto, Canada) for 10 min. Temperature was measured using an infrared thermal camera (i7 thermal imaging camera, FLIR, Arlington, VA, USA) at 1 min intervals. The thermal dose administered was modulated by increasing the laser power administered in a stepwise manner. We utilized laser powers of 0.75 W, 1 W, 1.5 W, and 2 W, monitored using a power meter (Thorlabs, Newton, NJ, USA). The thermal dose output was then calculated using the CEM43°C formula, as shown in Equation (1):

$$\text{CEM43°C} = \sum_{i=1}^{n} t_i * R^{(43-T_i)} \tag{1}$$

where T_i is the i-th time interval, R is related to the temperature dependence of the rate of cell death (R(T < 43 °C) = 0.25, R(T > 43 °C) = 0.5) and T is the average temperature during time interval T_i [45–47]. Cyclic heating/cooling studies were also performed on the nanoparticles for 3 cycles with the laser on and off for 10 min each for each cycle at a power of 2 W to elucidate the ability of the nanoparticles to withstand several cycles of laser illumination.

2.7. Elucidation of the Glioblastoma Tumor Cell Phenotype Post-PTT

PTT was administered to 0.5 mL suspensions containing 5×10^6 U87 or U251 cells as described in Section 2.6. The treated cell suspension was then centrifuged, and the cells were suspended in their respective cell culture media and plated in 6 well plates at 37 °C for 24 h. After 24 h of undisturbed incubation, the cell culture media and the cells were harvested from the plates using TrypLE (Thermo Fisher Scientific, Waltham, MA, USA). Cells were then collected with media and transferred to 15 mL conical tubes and collected by centrifugation at 400× g for 5 min. The final cell pellets were then resuspended in PBS, and aliquoted accordingly for subsequent analysis. Viability was measured using a Luna cell counter (Logos Biosystems, Anyang, Korea) and acridine orange (Logos Biosystems, Anyang, Korea).

Intracellular levels of ATP were assessed to indirectly determine the amount of ATP released. Using the CellTiter-Glo Luminescent Viability Assay (Promega, Madison, WI, USA) and following the manufacturer's protocol, 100 μL of cells at a concentration of 2.86×10^6 mL in PBS from every condition were aliquoted into a 96-well opaque bottom plate. Once ATP assay reagents were at RT and mixed together, the ATP reagent was aliquoted at 100 μL per well to the cells. The plate was then covered in foil and placed on an orbital shaker for 2 min. The plate was then incubated at RT for 5–10 min before luminescence was measured via SpectraMax. PBS without cells was used as a control. All samples were performed in triplicate.

For flow cytometry analysis, 1×10^6 cells were first stained with 1 μL Zombie Violet™ fixable viability dye, reconstituted at the manufacturer's recommended concentration, for 20 min in 100 μL PBS (BioLegend, San Diego, CA, USA). After washing cells with flow buffer (PBS + 1% FBS), cells were resuspended in 100 μL flow buffer and blocked with 5 μL/tube Human TruStain FcX™ (Fc Receptor Blocking solution, BioLegend, San Diego, CA, USA) for 10 min at 4 °C. The cells were then separated into three panels for staining, besides the fluorescence minus one (FMO), isotype, and unstained control groups: two cell surface staining panels and one cell surface and intracellular ICD staining panel. For the first panel, the following extracellular stains were added: GD2 (APC, clone 30-F11, 5 μL/tube), CD137L (PE, clone 5F4, 5 μL/tube), HLA-A,B,C (AlexaFluor700, clone W6/32, 5 μL/tube), B7-H3 (PE/Cy7, clone MIH42, 5 μL/tube), and PD-L1 (BV650, clone 29E.2A3, 5 μL/tube). For the second panel, the following extracellular stains were added: Fn14 (PE, clone ITEM-1, 2.5 μL/tube), CD86 (APC, clone BU63, 5 μL/tube), CD80 (BrilliantViolet650, clone 2D10, 5 μL/tube), CD40 (AlexaFluor700, clone 5C3, 5 μL/tube), and HLA-DR (PE/Cy7, clone L243, 5 μL/tube). All antibodies used in panels 1 and 2 were purchased from BioLegend. For the third panel, an antibody against calreticulin was added (PE, clone ab92516, 0.5 μL/tube) (Abcam, Cambridge, UK). All panels were stained for 30 min at 4 °C, fixed, and permeabilized using 250 μL/tube of 1× BD CytoFix/CytoPerm Solution (BD Biosciences, Franklin Lakes, NJ, USA) for 30 min at 4 °C. Panel 3 was then stained for intracellular HMGB1 (AlexaFluor647, clone ab195011, 2 μL/tube) (Abcam, Cambridge, UK) for 30 min at 4 °C. Flow cytometry was performed using the BD Biosciences Celesta Cell Analyzer or BD CytoFLEX (Indianapolis, IN, USA). Flow cytometry gating and analysis was performed using FlowJo software (v10.7.1, Ashland, OR, USA) and plotted using GraphPad Prism (v9.0.0, San Diego CA, USA).

2.8. Determining aFn14 Binding to U87 and U251 Cells via Flow Cytometry

U87 or U251 cells were stained with either the Fn14 antibody used for aFn14-PBNP synthesis (FITC, clone ITEM-4, Santa Cruz Biotechnology Inc., Dalles, TX, USA at manufacturer's recommended concentration) or with the Fn14 antibody used for staining in the ICD panel (PE, clone ITEM-1, Biolegend, at manufacturer's recommended concentration). Percent positive cells were gated on size and single cells.

2.9. Quantifying the Attachment of aFn14-PBNP to U87 Tumor Cells

Inductively Coupled Plasma Mass Spectroscopy (ICP-MS; Neptune Series High Resolution Multicollector ICP-MS; Thermo Fisher Scientific, Waltham, MA, USA) was performed in collaboration with the University of Maryland Plasma Mass Spectroscopy Lab. U87 GBM cells were grown in 6 well plates; aFn14-PBNP was then added to 1×10^6 cells at a final concentration of 0.15 mg/mL for 15, 30, 45, 60, or 120 min. The cells were then rinsed with PBS, harvested using TrypLE, counted, washed, and all cells were resuspended in 1 mL PBS. The cells were then transferred to thick screw-top Teflon beakers and treated with high-purity concentrated nitric acid (OmniTrace Ultra, Supelco, Waltham, MA, USA) for 20 min at RT. The concentrated nitric acid was then diluted to a concentration of 2% (v/v) using DI water and digested further overnight at 60 °C. The samples were then filtered through a 40 μm cell strainer, transferred to a sealed tube and transported to the University of Maryland Plasma Mass Spectroscopy Lab. PBNP attachment was detected by probing for the ^{57}Fe elemental isotope of non-radioactive iron, which is a component of the PBNPs.

2.10. Assessing the Efficacy of Using aFn14-PBNP for Targeted PTT of Glioblastoma Tumor Cells

We developed a protocol to determine whether aFn14-PBNPs provided an advantage over PBNPs in the context of a tumor where external forces, such as the circulatory or lymphatic system, may wash out unbound particles. 1×10^6 U87 or U251 cells were incubated with 5.8 mg/mL PBNP, 5.8 mg/mL aFn14-PBNP, 3 μ/mL FITC-conjugated aFn14 (to match the concentration of aFn14 used in the aFn14-PBNP condition), or a vehicle control (water) for 2 h at 37 °C. Cells were then washed twice to remove unbound particles and/or antibody, resuspended in 500 μL media, and PTT-treated as described in Methods 2.6. Prior to commencing PTT, 50,000 cells from each condition were assessed for FITC expression via flow cytometry, as a measure of aFn14 attachment to the tumor cells. Cells were plated into 6 well plates (one well per sample, in 2 mL total cell culture media) and incubated at 37 °C. Viability was measured 24 h post-PTT using the Luna cell counter and acridine orange.

2.11. Statistical Analysis

Statistical significance for this study was determined using Welch's *t*-test, and values with $p < 0.05$ were considered statistically significant. Descriptive statistics are reported as mean ± standard deviation. All statistical analyses were performed using GraphPad Prism.

3. Results

3.1. aFn14 Can Be Covalently Conjugated on PBNPs to Generate Stable aFn14-PBNPs

The size, charge, and UV-Vis-NIR absorption properties of PBNPs can affect their efficacy as PTT agents in the CNS. Nanoparticles that are too large (>200 nm) cannot effectively cross the blood–brain barrier [48,49] while nanoparticles with a positive surface charge can form nonspecific ionic adhesions to the negatively charged cell membranes and extracellular space of the CNS [50], resulting in unwanted off-target binding and/or toxicity. Additionally, a change in absorption properties of a PTT agent can alter the efficacy of photothermal energy conversion. Therefore, we aimed to synthesize nanoparticles that met the above design attributes. Using the scheme described in Section 2.2, the FITC-conjugated aFn14 antibody was covalently linked to the PBNPs to generate aFn14-PBNPs (Figure 1A). Based on the use of fluorescence spectroscopy to generate a standard curve for aFn14 quantification (Figure S1A,B), we calculated the attachment efficiency on the day of aFn14-PBNP synthesis (Day 0) to be 99.4% (SD 0.88%). Consequently, we estimated that the aFn14-PBNP nanoparticle had 0.497 μg of bound aFn14 per mg of PBNP. While the attachment efficiency of aFn14 on PBNP was initially very high, it decreased to 87.1% (SD 1.48%) by Day 20 post-synthesis, indicating that the conjugation resulted in the retention of the majority of aFn14 on the PBNPs for nearly three weeks. On Day 20, we estimated that aFn14-PBNP had 0.436 μg of bound aFn14 per mg of PBNP (Figure 1B).

Following synthesis, the hydrodynamic diameter of the aFn14-PBNPs increased from a mean of 58.7 nm for PBNPs to 122.4 nm for aFn14-PBNPs with polydispersity indices of 0.42 and 0.37, respectively (Figure 1C). Importantly, the mean hydrodynamic diameter and mean polydispersity index of the aFn14-PBNPs on Day 20 were unchanged compared to the freshly synthesized aFn14-PBNPs. Although the zeta potentials increased from -33.0 ± 1.2 mV for PBNPs alone to -23.1 ± 1.7 mV for aFn14-PBNPs (Figure 1D), indicative of the presence of the covalently attached antibody on the surface of the nanoparticles, the zeta potentials of aFn14-PBNP and PBNPs alone remained stable over 21 days. Pertinent to the PTT properties of PBNPs and aFn14-PBNPs, we measured their UV-Vis-NIR spectra (Figure 1E). Importantly, when matched in terms of PBNP concentrations, the UV-Vis-NIR spectrum of aFn14-PBNPs overlapped with that of PBNPs. Specifically, the aFn14-PBNP spectrum exhibited the characteristic absorption band of PBNP between 650–900 nm, indicating that conjugation of the antibody did not alter the absorbance spectrum attributed to the PBNPs in aFn14-PBNPs. These results indicate that our synthesis scheme yielded aFn14-PBNPs wherein aFn14 was successfully covalently attached onto the PBNPs with high attachment efficiency and that the resulting aFn14-PBNPs had stable size distributions, zeta potentials, and retained the absorption properties of PBNPs.

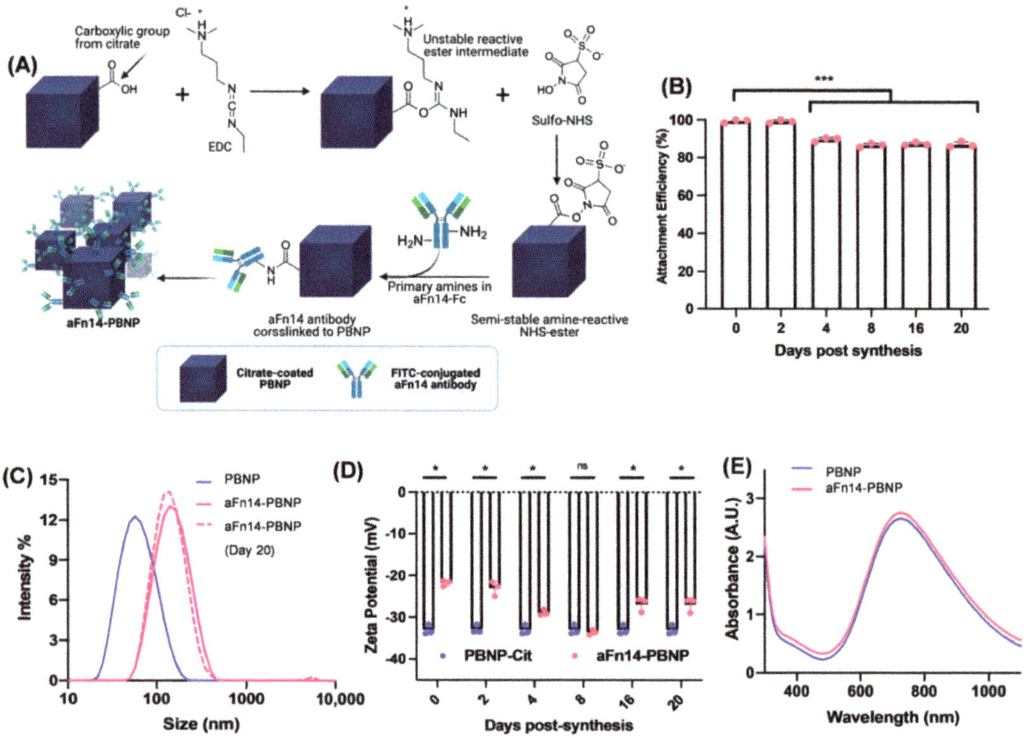

Figure 1. aFn14 can be covalently conjugated to PBNPs to generate stable aFn14-PBNPs. (**A**) Synthesis scheme used to generate aFn14-PBNPs, described in Methods 2.2. EDC-NHS chemistry was used to coat FITC-conjugated aFn14 antibody onto the surface of the PBNPs. The carboxyl groups from citrate on the surface of PBNPs are covalently conjugated to the amine groups on aFn14. (**B**) Attachment efficiency of aFn14 on to PBNPs over time. The initial attachment efficiency on Day 0 (99.4% SD 0.88%) did not significantly change on Day +2, but a statistically significant change was observed after Day +4 to a final attachment efficiency of 87.1% (SD 1.48%) by Day 20 ($p < 0.0001$). (**C**) Size distributions of PBNPs and aFn14-PBNPs as measured by dynamic light scattering. The size shift between PBNP and aFn14-PBNP indicates successful attachment of aFn14. There was no size change of aFn14-PBNP between Day 0 after synthesis (pink line) and Day 20 (dotted pink line), demonstrating long-term stability of aFn14-PBNPs. (**D**) Zeta potentials of PBNPs and aFn14-PBNPs. Statistically significant differences in the surface charge were observed between PBNPs and aFn14-PBNPs at all time points (except Day +8). This suggests the presence of the aFn14 antibody on the surface of aFn14-PBNPs. (**E**) UV-Vis-NIR spectra demonstrating the characteristic absorption peak of PBNP between 650–900 nm, which yields NIR responsiveness at 808 nm. Coating with aFn14 did not affect this absorption property. Welch's *t*-test was utilized to test for statistical significance. * = $p < 0.05$ ** = $p < 0.01$ *** = $p < 0.001$; ns = not significant.

3.2. *aFn14-PBNPs Retain the PTT Properties of PBNPs and Can Be Used to Administer a Range of Thermal Doses to U87 and U251 GBM Tumor Cells*

PBNPs absorb NIR radiation and reemit it as heat via phonon-to-phonon relaxations. This heating makes PBNPs useful as PTT agents to treat tumors, as the heat released can impart ablative thermal dosages onto surrounding tissues. For the aFn14-PBNPs to be functional as effective PTT agents, their photothermal properties must be retained after aFn14 conjugation. We therefore conducted studies assessing the PTT properties of both PBNPs and aFn14-PBNPs to determine whether conjugation to aFn14 alters these properties. Aqueous solutions of PBNPs or aFn14-PBNPs (both containing 0.15 mg/mL PBNPs) in DI water were irradiated with 808 nm NIR laser (Figure 2A) at various laser powers, and the temperatures of the water-nanoparticle systems were recorded every minute for 10 min. The aFn14-PBNP system exhibited laser power-dependent heating. Negligible heating was observed for the control system containing just water and irradiated with the laser at a 2 W ("LASER Alone") and for the control system containing aFn14-PBNP but not irradiated ("aFn14-PBNP Alone"). The maximal temperatures attained for the aFn14-PBNP system were 88.2 °C for 2 W, 70.5 °C for 1.5 W, 66.4 °C for 1 W and 58.3 °C for 0.75 W laser power (Figure 2B). The corresponding thermal doses (log(CEM43)) delivered by the aFn14-PBNPs to the water following irradiation were 11.4 at 2 W, 10.6 at 1.5 W, 7.4 at 1.0 W, and 4.85 at 0.75 W (Figure 2C). The maximal temperatures attained for the PBNP system were 86.8 °C for 2 W, 79.6 °C for 1.5 W, 69.9 °C for 1 W, and 60.7 °C for 0.75 W laser power (Figure 2D). The corresponding thermal doses (log(CEM43)) delivered by the PBNPs to the water following irradiation were 13.65 at 2 W, 11.46 at 1.5 W, 8.46 at 1.0 W, and 5.69 at 0.75 W (Figure 2E). The heating curves and corresponding thermal doses generated with aFn14-PBNPs were comparable to those of PBNPs alone for all laser powers studied (Figure 2D,E). Cyclic heating and cooling studies, where the laser was turned on and off at specific intervals to allow heating and cooling, demonstrated comparable PTT characteristics between PBNP and aFn14-PBNP at equivalent PBNP concentrations as evidenced by the near overlap of their heating/cooling curves over three cycles (Figure 2F). These studies demonstrate that the addition of aFn14 did not alter the PTT properties of the aFn14-PBNPs, indicating the suitability of their use as PTT agents.

Next, we evaluated whether aFn14-PBNPs could heat tumor cells to temperatures consistent with those that are required for thermal ablation. PTT was conducted on 5×10^6 U87 (Figure 2G,H) or U251 (Figure 2I,J) GBM tumor cell lines suspended in cell culture media containing 0.15 mg/mL PBNPs. Similar to the previous study, the heating curves and thermal dose delivered to each cell line by aFn14-PBNPs were assessed as a function of laser power. The maximal temperatures attained for U87 were 77 °C for 2 W, 69 °C for 1.5 W, 59 °C for 1 W and 52 °C for 0.75 W laser power (Figure 2G), with thermal doses of 9.4 at 2 W, 8.2 at 1.5 W, 6.5 at 1.0 W, and 4.9 at 0.75 W (Figure 2H). When U251 were treated with aFn14-PBNPs, similar temperatures and thermal doses were attained: 78 °C for 2 W, 68 °C for 1.5 W, 59 °C for 1 W and 51 °C for 0.75 W (Figure 2I), and thermal doses of 10.1 at 2 W, 8.1 at 1.5 W, 6.6 at 1.0 W, and 4.9 at 0.75 W (Figure 2J). The thermal doses attained during PTT with aFn14-PBNP were lower in the presence of GBM tumor cells for both cell lines as compared to those in water (e.g., 89 °C for aFn14-PBNP alone vs. 77 °C at 2 W for U87). This attenuation can be attributed to components present in the media (e.g., serum proteins) and is consistent with our observations when using uncoated PBNPs for PTT, as well as literature precedent [43,51–53]. However, the cells attained temperatures above those required to generate thermal ablation in treated tumor cells (>45 °C) at all laser powers tested for both GBM lines. Therefore, these results suggest that aFn14-PBNP can function as effective PTT agents for GBM tumor cells and can be used to administer a range of ablative thermal doses to these tumor cells.

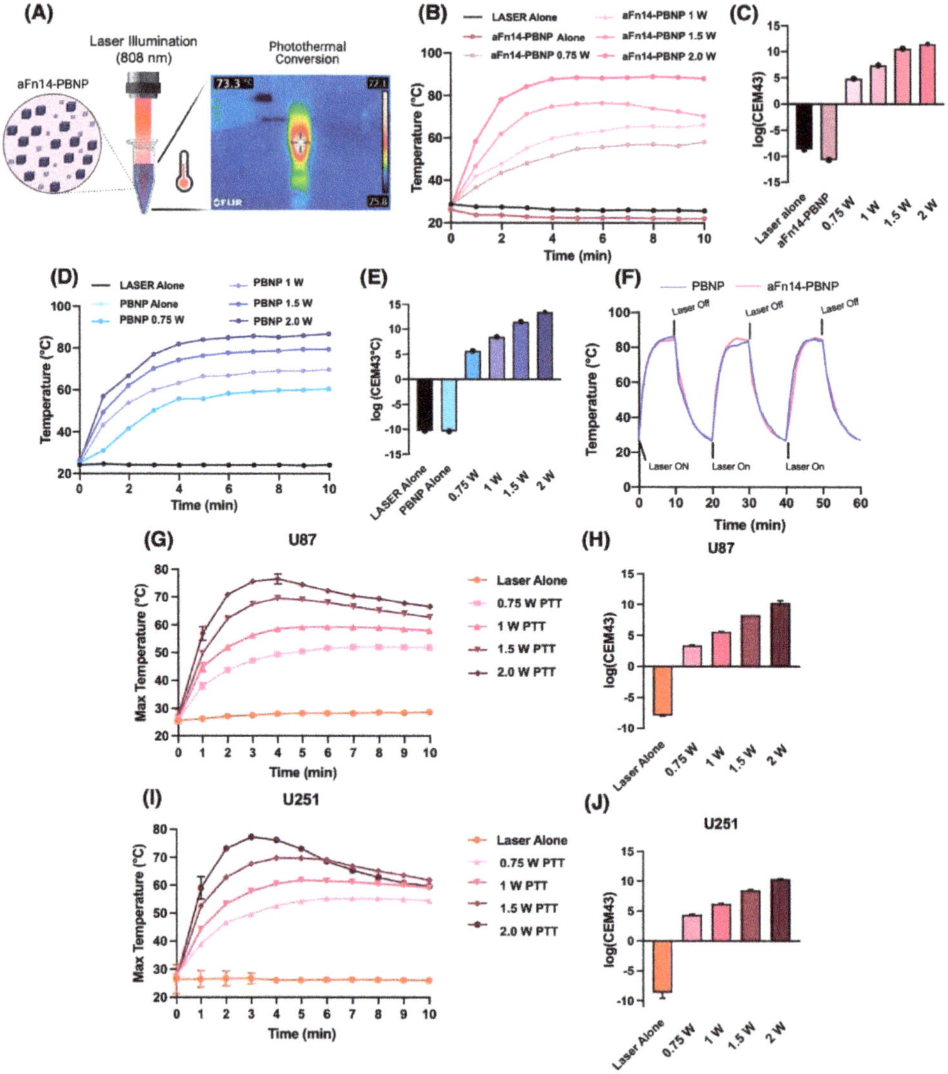

Figure 2. Conjugation to aFn14 does not alter the PTT properties of PBNPs, and aFn14-PBNPs can be used to administer a range of thermal doses to U87 and U251 GBM tumor cells. (**A**) Schematic of the PTT studies as described in Methods 2.6. Water or U87 or U251 cells in media were treated with 0.15 mg/mL PBNP or aF14-PBNP and then irradiated with an 808 nm laser. (**B**) Heating curves of aFn14-PBNP as a function of laser power. (**C**) Thermal doses delivered by aFn14-PBNPs as a function of laser power. (**D**,**E**) Heating curves (**D**) and thermal doses I delivered by PBNPs as a function of laser power. (**F**) Cyclic heating of PBNPs and aFn14-PBNPs over 3 heating-cooling cycles. (**B**–**F**) PTT properties of PBNP and aFn14-PBNP are similar, indicating that conjugation to aFn14 does not negatively affect PBNP PTT properties. (**G**) Heating curves (**H**) and thermal doses delivered to U87 GBM cells by aFn14-PBNPs as a function of laser power. (**I**) Heating curves (**J**) and thermal doses delivered to U251 GBM cells by aFn14-PBNPs as a function of laser power.

3.3. *PTT Using aFn14-PBNP Triggers Thermal and Immunogenic Cell Death in Treated GBM Tumor Lines*

After undergoing PTT using aFn14-PBNPs as described in Figure 2G–J, the GBM tumor cells were plated for 24 h and then harvested for the analysis of PTT-triggered thermal and immunogenic cell death as well as immunophenotypic markers. PTT using aFn14-PBNP elicited thermally induced cell death in both GBM tumor lines as evidenced by a decrease in cellular viability in a laser power-dependent manner (Figure 3A). For U87 cells, the cellular viability decreased from 97.85% for aFn14-PBNP (without the laser) to 42.95% at 0.75 W, with the lowest viability of 18.45% observed at the highest laser power of 2 W. Similarly, for U251 cells, the cellular viability decreased from 95.00% for aFn14-PBNP (without the laser) to 55.25% at 0.75 W, with the lowest viability of 11.19% observed at the highest laser power of 2 W. For both tumor lines, the IC50 was attained at thermal doses of 4.72 for U87 and 4.86 for U251, respectively corresponding to a laser power of 0.75 W. These findings with aFn14-PBNP-based PTT in GBM tumor cells are consistent with our observations when using PBNPs for PTT in diverse tumor cell lines such as neuroblastoma, melanoma, and breast cancer [9,10,42,43]. We did not conduct studies assessing the effects of PBNPs alone (without the laser) as we have extensively observed that the PBNPs have a negligible effect on tumor cellular viability at the concentrations used for this study [9,10,42].

To assess the effect of PTT using aFn14-PBNP on eliciting ICD and immunophenotypic changes in treated GBM tumor cells, flow cytometry analysis of various surface and intracellular targets were conducted. The gating strategy for these analyses is elaborated in Figures S2–S5. PTT using aFn14-PBNPs induced ICD in both cell lines as measured by a decrease in total intracellular ATP levels (Figure 3B), the overexpression of calreticulin at the cell surface (Figure 3C, Figures S2A and S3A), and the release of HMGB1 from the cell as measured by a decrease in intracellular HGMB1 (Figure 3D, Figures S2B and S3B). The expression of all three biochemical correlates of ICD were most prominent for both U87 and U251 at laser powers of 1.5 W and higher.

PTT also induced changes in the tumor cell immunophenotype that may increase T cell immunity against these cells. MHC-I (HLA-A, B, and C) expression was high and retained in both U87 and U251 (Figure 3E, Figures S4A and S5A), and MHC-II expression (HLA-DR) remained unchanged in U87 cells but increased in U251 (Figure 3F, Figures S4B and S5B) as a function of laser power. Regarding expression of tumor-specific antigens, PTT caused a significant decrease in GD2 expression in U251 cells at 1.5 and 2.0 W but not in U87 cells, suggesting the retention of this tumor-specific antigen for this tumor line (Figure 3G, Figures S4E and S5E). In both cell lines, PTT maintained the expression of the immunosuppressive PD-L1 ligand (Figure 3H, Figures S4C and S5C) and no changes in expression of the immunosuppressive B7-H3 ligand were observed (Figure 3I, Figures S4D and S5D). Finally, PTT upregulated various T cell costimulatory markers in both cell lines, including CD137L (Figure 3J, Figures S4F and S5F), CD80 (particularly in U251; (Figure 3K, Figures S4G and S5G), CD86 (Figure 3L, Figures S4H and S5H), and CD40 (Figure 3M, Figures S4I and S5I), all in a dose-dependent manner. Overall, the changes induced by aFn14-PBNP were similar to those induced by PBNP, indicating that antibody coating did not drastically alter thermal or ICD-inducing properties of the PBNPs, except that expression of GD2 and CD137L remained higher in the aFn14-PBNP conditions compared to the PBNP-treated conditions (Figure S6). Together, these results demonstrate that PTT using aFn14-PBNPs generates thermal and immunogenic cell death, as well as favorably alters the surface immunophenotype of GBM tumor cells.

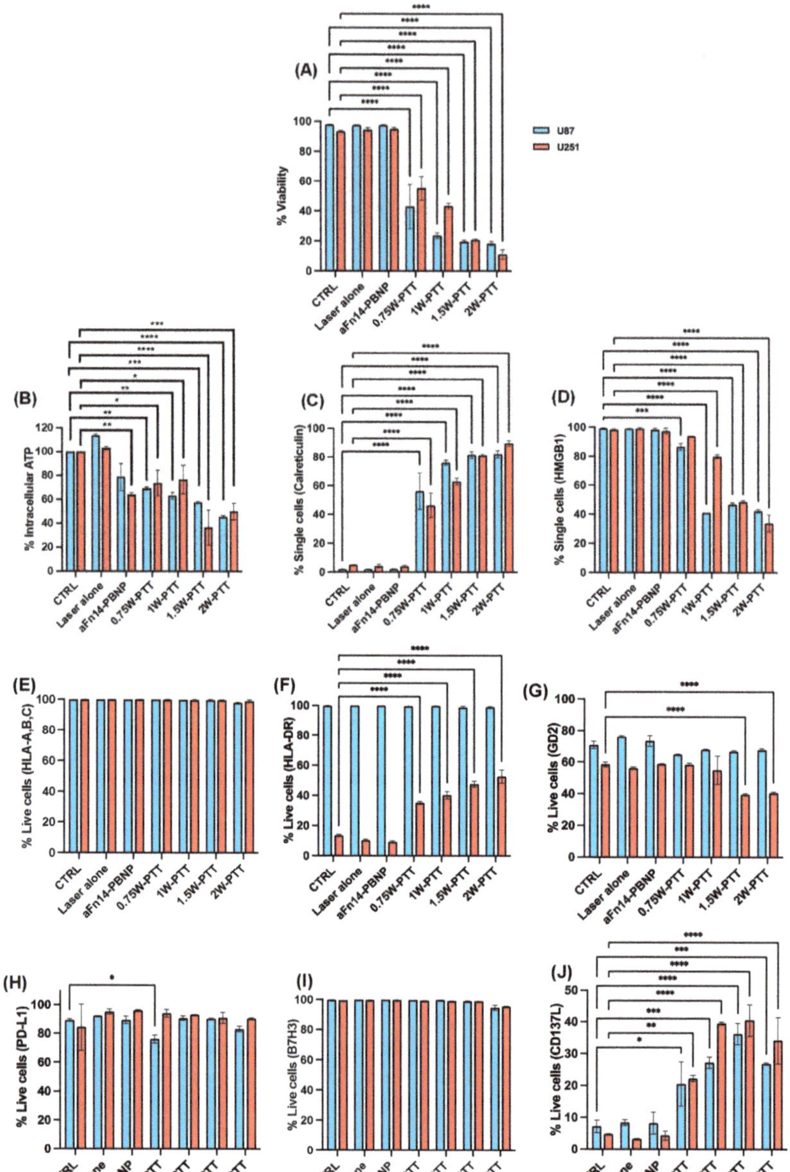

Figure 3. *Cont.*

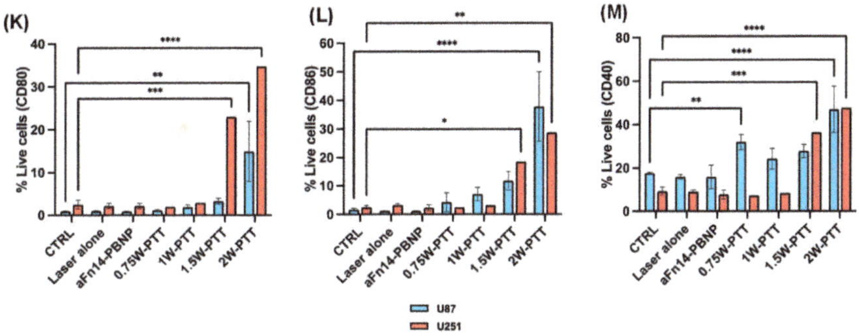

Figure 3. PTT using aFn14-PBNP triggers thermal and immunogenic cell death in treated GBM tumor lines. (**A–D**) U87 and U251 cells undergo thermal (**A**) and immunogenic cell death (**B–D**) after PTT with aFn14-PBNP, as noted by a decrease in cell viability (**A**), decrease in intracellular ATP (**B**), increase in surface calreticulin expression (**C**), and decrease in intracellular HMGB1 (**D**) with increasing laser powers (0.75–2 W). (**E–M**) PTT-induced changes in immunophenotype in GBM tumor lines, including MHC expression (**E**,**F**), tumor specific antigen expression (**G**), immune checkpoint inhibitor expression (**H**,**I**), and T cell costimulatory markers (**J–M**). CTRL = untreated cells not irradiated; laser alone = untreated cells irradiated with laser; aFn14-PBNP = aFn14-PBNP-treated cells not irradiated; remaining conditions are aFn14-PBNP-treated cells irradiated at indicated laser powers. * = $p < 0.05$; ** = $p < 0.01$; *** = $p < 0.001$; **** = $p < 0.0001$.

3.4. GBM Tumor Cells Differentially Express Fn14 That Can Be Successfully Targeted by aFn14-PBNPs

We conducted studies to assess the expression levels of Fn14 on the GBM tumor cells as described in Methods 2.8 (Figure 4). When probed with a PE-conjugated aFn14 antibody, both U87 and U251 showed expression of the Fn14 receptor via flow cytometry, with U87 expressing Fn14 on 56.4% of live cells (Figure 4A) and U251 expressing Fn14 on 98.2% of live cells (Figure 4B). However, when probed with the FITC-conjugated aFn14 antibody of a different clone, U87 showed expression of Fn14 only on 14.1% of live cells (Figure 4C) compared to U251 that demonstrated a higher expression of Fn14 on 85.3% of live cells (Figure 4D). Our results demonstrate that U251 consistently exhibit higher expression of Fn14 than U87 cells. The findings also demonstrate that although Fn14 is a suitable target for GBM, the antibody clone should be considered to facilitate success of tumor cell targeting with the nanoparticles. As our results will show later, however, even low aFn14 binding can retain enough PBNPs on cells to induce thermal death in response to PTT.

Next, we assessed whether aFn14-PBNPs can target U87 tumor cells using ICP-MS. In this study, we measured the amount of ^{57}Fe elemental isotope in solution following acid digestion of U87 cells exposed to either aFn14-PBNP or PBNP for various contact times. Increased ^{57}Fe, a main component of PBNPs, correlates with increased nanoparticle attachment to the surface of the GBM cells. U87 cells incubated with aFn14-PBNP nanoparticles showed an increase in signal over control or vehicle-treated cells. This increase was observed at 15 min (1.90 ± 0.60 ppb/1.0×10^6 cells) and progressively increased with increasing contact time reaching a maximum at 120 min (6.10 ± 0.76 ppb/1.0×10^6 cells), indicating stronger aFn14-PBNP binding to the U87 cells over time (Figure 4E). Importantly, there was a significant increase in ^{57}Fe signal in the aFn14-PBNP condition after 60 and 120 min of incubation compared to U87 cells treated with PBNP alone ($p = 0.0171$ vs. $p = 0.0177$, respectively). Our results suggest that the addition of the aFn14 antibody increases the cellular targeting capacity of the PBNPs. The targeting was enhanced even in U87 GBM tumor cells that have lower Fn14 expression levels compared to U251, suggesting that the aFn14-PBNPs can be applied for targeted PTT against U87 cells. This experiment was not repeated in the U251 cells for cost saving and logistics purposes, but we expect similarly increased targeting of U251 cells by aFn14-PBNPs compared to PBNPs alone.

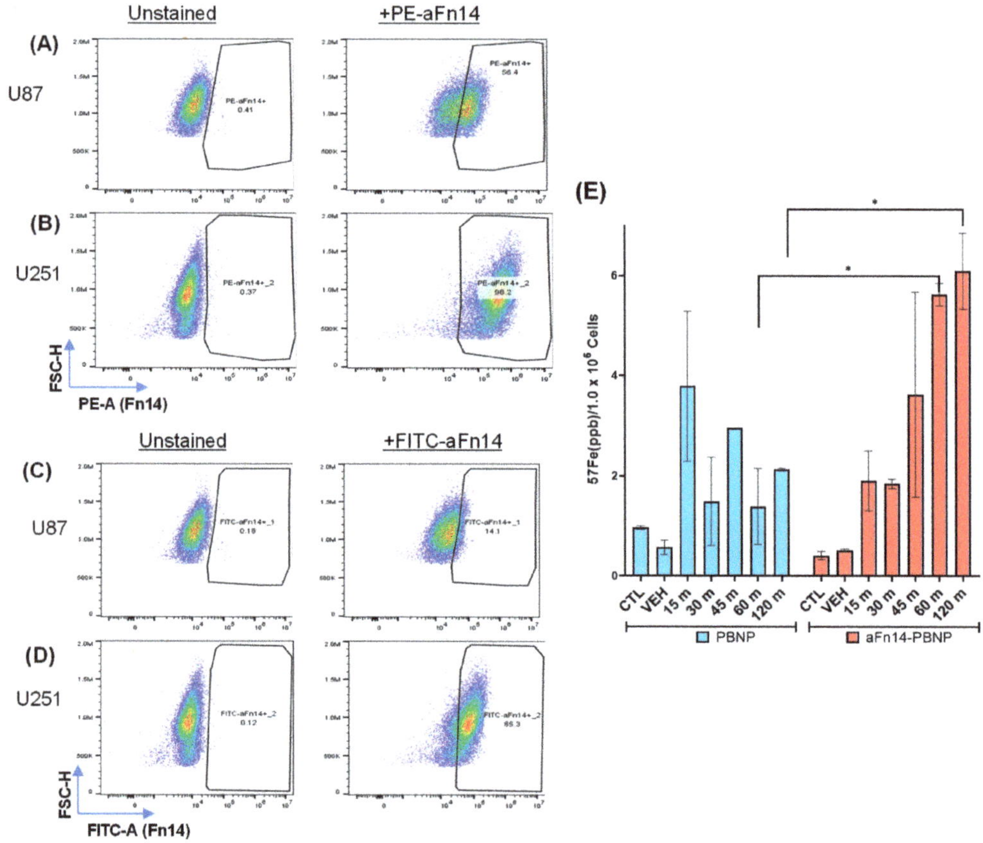

Figure 4. GBM tumor cells differentially express Fn14 that can be successfully targeted by aFn14-PBNPs. (**A**,**B**) Flow cytometry of (**A**) U87 and (**B**) U251 cells stained with PE-conjugated aFn14 antibody showing percent of cells expressing Fn14, compared to unstained. (**C**,**D**) Flow cytometry of both cell lines using the FITC-conjugated aFn-14 antibody compared to unstained. (**E**) ICP-MS of human U87 GBM cells incubated with 0.15 mg/mL PBNP (left) or aFn14-PBNP (right) for various time points, gating on the 57-Fe isotope. Each aFn14-PBNP group was compared to its respective PBNP group using a Welch's *t*-test for significance. * = $p < 0.05$. CTL = control (U87 GBM cells alone). VEH = vehicle (U87 GBM cells with addition of water).

3.5. aFn14-PBNP Is an Effective Targeted PTT Agent for GBM Tumor Cells

Thus far, we have demonstrated that conjugation to aFn14 does not negatively impact the photothermal properties of PBNPs, that aFn14-PBNPs can induce thermal and immunogenic cell death in two glioblastoma cell lines, and that aFn14-PBNPs can bind U87 cells. Next, we tested whether aFn14-PBNP-tumor cell binding provides a benefit over PBNPs when cells are washed to remove unbound particles before undergoing PTT (Figure 5). The goal of this study was to mimic, in a simple and cost-effective manner, how the blood or lymphatic system can wash PBNPs out of tumors.

In our ICP-MS study, we observed that U87 cells incubated with aFn14-PBNPs retained the largest number of particles 2 h after incubation (Figure 4E). Therefore, for this study, we incubated U87 and U251 cells with 5.8 mg/mL PBNPs, 5.8 mg/mL aFn14-PBNPs, 3 μg/mL free FITC-conjugated aFn14 (corresponding to the concentration of aFn14 used in the aFn14-PBNP condition), or a vehicle control (water) for 2 h at 37 °C. After this targeting step, the tumor cells were washed twice to remove any unbound nanoparticles and/or antibody. A sample of tumor cells were taken to measure FITC expression, which indicates aFn14-PBNP binding, just before PTT. The tumor cells were then irradiated with an 808 nm laser for 10 min and rested overnight (Figure 5A). Cell death was then assessed via flow cytometry. Because the purpose of the washing step was to remove untargeted particles after the contact time, the concentration of PBNPs and aFn14-PBNPs used had to be increased from that in Figures 2 and 3 to attain a sufficient concentration of cell bound aFn14-PBNP concentration for thermal heating above 45 °C. Various concentrations and laser powers were tested to determine that 5.8 mg/mL was the optimal concentration for use in the U251 cell line for this experiment based on heating curves and thermal dose (Figure S7).

Both cell lines yielded higher heating curves (Figure 5B,C), thermal dose (Figure 5D,E), and cell death (Figure 5F,G) in response to PTT when incubated with aFn14-PBNPs compared to PBNPs alone, indicating that the antibody provided an advantage over PBNPs alone in effecting targeted PTT in the tumor cells. However, U251 cells consistently yielded higher heating curves (e.g., 68.25 °C for U251 vs. 60.20 °C for U87 at 2.0 W), thermal doses (e.g., 8.21 log(CEM43) for U251 vs. 5.73 log(CEM43) for U87 at 2.0 W), and cell death (42.90% viability for U251 vs. 4.63% viability for U87 at 2.0 W after 24 h) when treated with aFn14-PBNPs than U87 cells (Figure 5B–G). These results are consistent with previous findings that indicated that the antibodies used for targeting on aFn14-PBNP did not bind to the U87 cells as well as the U251 cells (Figure 4A–D). These results were also reflected in the flow cytometry analysis of the cells evaluated before PTT where only a negligible increase in the percent of FITC-positive U87 cells was observed after aFn14-PBNP treatment compared to PBNP treatment (Figure 5H,J). In contrast, there was a much more noticeable increase in the percent of FITC-positive U251 cells treated with aFn14-PBNP compared to PBNP (Figure 5I,K). However, despite the negligible presence of aFn14-PBNP on U87 as measured via flow cytometry, the U87 cells treated with aFn14-PBNPs still yielded a higher thermal dose (5.74 log(CEM43) at 2.0 W) and significant decrease in viability (42.90% live) compared to those treated with PBNP (−4.59 log(CEM43) at 2.0 W and 96.05% live) post-PTT. These results indicate that even an amount of antibody binding that is too low to be detected by flow cytometry can generate heating and elicit thermally induced cell death. These trends were further accentuated in studies with U251 tumor cells where treatment with aFn14-PBNPs yielded an even higher thermal dose (8.21 log(CEM43) at 2.0 W) compared to the PBNP-treated U251 cells (−0.60 log(CEM43) at 2.0 W) and an even larger significant decrease in U251 viability following PTT (79.4% for PTT with PBNP vs. 4.63% for PTT with aFn14-PBNP). Together, these results clearly indicate that aFn14-PBNPs are capable of inducing thermal death in response to PTT as a function of Fn14-based targeting. Overall, because they are retained on the surface of aFn14-expressing cells, the aFn14-PBNPs provide an advantage over PBNPs alone in an environment where the particles are at risk of being washed out of the tumor by external forces.

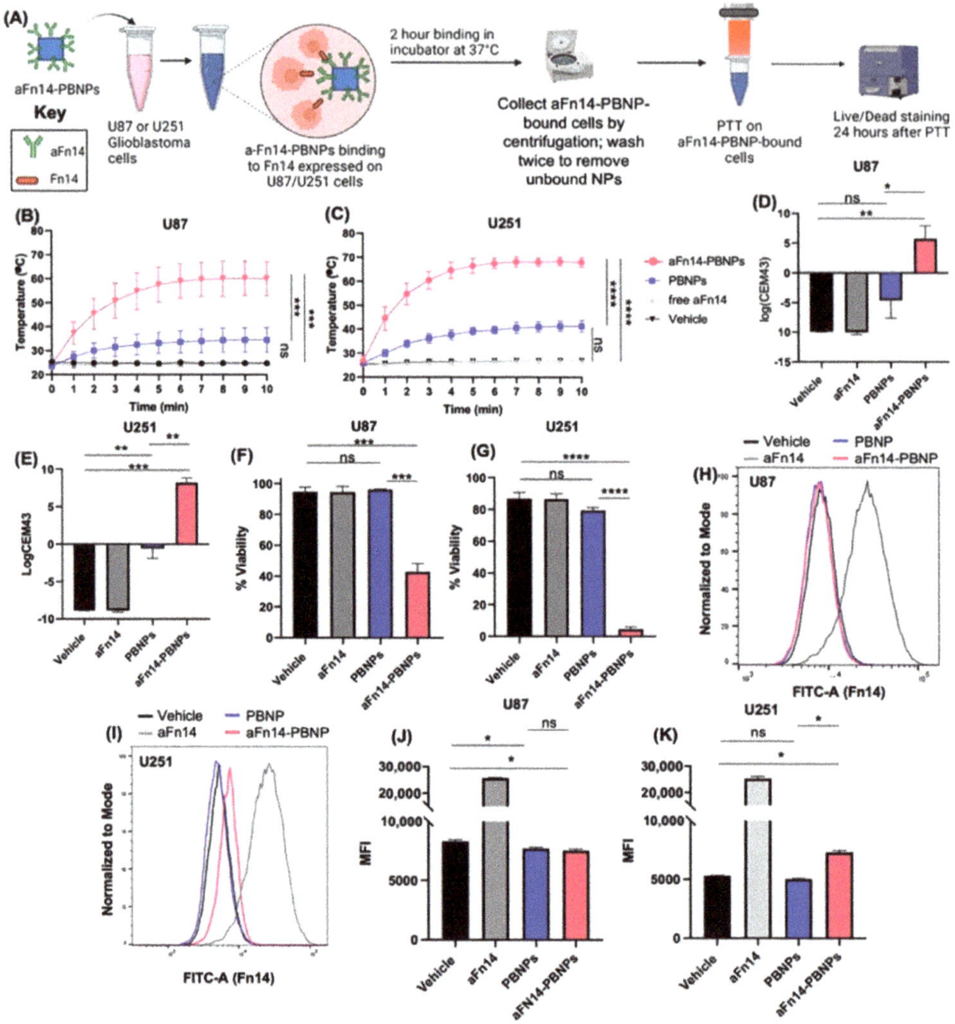

Figure 5. aFn14-PBNP is an effective targeted PTT agent for GBM tumor cells. (**A**) Experimental setup schematic, described in Methods 2.10: U87 or U251 cells were incubated with PBNPs, aFn14-PBNPs, free FITC-conjugated aFn14, or a vehicle control (water) for 2 h at 37 °C. Cells were then washed to remove unbound particles and irradiated with an 808 nm laser. Cells were plated for 24 h and analyzed via flow cytometry for viability. (**B**,**C**) Heating curves of U251 and U87 cell lines during PTT. (**D**,**E**) Thermal doses imparted on the cell lines during PTT. (**F**,**G**) Viability of tumor cells 24 h post-PTT. (**H–K**) Flow cytometric analysis done just before PTT of aFn14 binding in U87 and U251 cells. (**H**,**J**) No aFn14 binding is detectable in the aFn14-PBNP condition in U87 cells as indicated by the lack of a population shift or increase in MFI from the vehicle control to the aFn14-PBNP condition. (**I**,**K**) Some aFn14 is detectable in the aFn14-PBNP condition in U251 cells as indicated by a population shift and increase in MFI. (**B–K**) More aFn14-PBNPs remained bound to the U251 cells than to the U87 cells, which is reflected in the higher heating curves, thermal dose, and lower viability in this cell line (**B–G**,**I**,**K**) A one-way ANOVA was used to test for significance using Tukey's HSD test for multiple comparisons. * = $p < 0.05$ ** = $p < 0.01$ *** = $p < 0.001$ **** = $p < 0.0001$. ns = not significant.

4. Discussion

For the past two decades, PTT has been widely investigated as an experimental ablative cancer therapy. While PTT for GBM in vitro and in vivo has been reported, including using gold nanorods [54], graphene oxide [55], and indocyanine green nanoparticles [56], to the best of our knowledge, this is the first study that utilizes a PBNP-based platform for treating GBM tumor cells. Additionally, we also report the first successful synthesis of a PBNP therapeutic modality targeting Fn14, a cytokine receptor highly upregulated in GBM compared to surrounding tissue in the tumor microenvironment [23,24,30,57,58]. Not only is our method novel, it is also simple and cost-effective. Required materials consist exclusively of inexpensive reagents commonly found in a cell culture and nanomaterials laboratory: $FeCl_3{\bullet}6H_2O$, citric acid, $K_4[Fe(CN)_6]{\bullet}3H_2O$, MES buffer, EDC solution, sulfo-NHS solution, and the aFn14 antibody. The synthesis and conjugation protocols are straightforward and take just one day to complete. From a practical standpoint, these qualities make our method an attractive one to pursue further as it can be widely replicated and adapted for little cost (detailed protocols are provided in the Supplementary Methods).

Our facile and cost-effective synthesis scheme yielded aFn14-PBNPs that retained the targeting antibody (aFn14), exhibited stable size distributions and surface charges for up to 21 days, and maintained the optical properties (UV-Vis-NIR spectrum) of unconjugated PBNPs (Figure 1). aFn14-PBNPs also retained the photothermal properties of unconjugated PBNPs, exhibiting similar laser power-dependent heating and the ability to administer a range of thermal doses to the GBM tumor cell lines U87 and U251 in vitro (Figure 2). These effects are due to the photothermal mechanism of laser excitation of the particles followed by phonon-phonon relaxations that release heat into the surroundings. The synthesis of nanoparticles with consistent critical quality attributes (e.g., size distributions, stability, PTT properties) are imperative for use in preclinical and clinical studies.

PTT using aFn14-PBNP elicited thermally induced and immunogenic cell death in both U87 and U251 tumor cells, with the biochemical correlates of ICD being expressed at higher levels after irradiation with higher laser powers (1.5 W and 2 W) (Figure 3). Further, PTT with aFn14-PBNP resulted in a unique tumor cellular immunophenotype consisting of thermal dose-dependent enhanced expression of CD137L, CD80, CD86, and CD40. These cellular and molecular signatures are indicative of a tumor cell death phenotype that can be combined with complementary immunotherapies to generate a robust antitumor immune response and potentiate the abscopal effect. To this end, aFn14-PBNP-PTT should be investigated in conjunction with co-localized or co-administered immunotherapeutic agents including immunological adjuvants (e.g., toll-like receptor agonists), monoclonal antibodies (e.g., anti-PD-1), and/or immune cell therapies for treating GBM in syngeneic animal models of the disease. However, it is important to note that the enhanced expression of these immune markers on the GBM cells would need to be verified in in vivo GBM tumors after PTT to assess the therapeutic potential of these PTT-induced molecular modulations. The success of previous combination approaches, including our own [9,40,43,59], in syngeneic models of neuroblastoma, melanoma, and breast cancer provide us with the compelling rationale to pursue these studies in the context of GBM.

Conjugation to aFn14 also provided a benefit to PBNPs by retaining PBNPs on the surface of aFn14-expressing cells, as evidenced by flow cytometry, ICP-MS, and PTT studies (Figures 4 and 5). We observed that aFn14-PBNPs were more effective in targeting U251 cells compared to U87 cells, which was likely due to the decreased aFn14 binding in the U87 compared to U251 (Figure 4C,D). Despite this lower binding, both ICP-MS and PTT conducted after washing unbound particles from cells demonstrated that the aFn14-PBNPs bound more strongly to U87 than unconjugated PBNPs (Figures 4 and 5), resulting in approximately 12 times more killing when PTT was conducted on the U87 post-wash (5% vs. 60% killing) (Figure 5F). These studies provide evidence for the benefit of antibody-mediated targeting.

However, our studies are not without limitations. First, it is possible that aFn14 binding to Fn14 could induce TWEAK-like signaling along the NFkB pathway, leading to a more metastatic tumor phenotype. These results occurred in the aFn14 gold nanoparticle study conducted by Aido et. al. [29]. In this situation, PTT may mitigate much of the potential for metastatic conversion, as most of the cells would be ablated soon after contact. However, the exact binding location of the antibody on aFn14 would need to be considered and validated in order to confirm that a blocking, not activating, effect is achieved by antibody-Fn14 binding. Second, to ensure the success of our PBNP-based approach for tumor cells with similar or lower Fn14 expression levels as U87, we will explore other available antibody clones or targeting moieties, such as aptamers, as targeting agents on our nanoparticles [60–62]. We also cannot discount the fact that lower Fn14-expressing tumor cells (i.e., U87 in this study) may require a higher loading of the targeting antibody to ensure efficient binding. However, synthesis with increased antibody loading will have to be performed in the context of maintaining the critical quality attributes of the final nanoparticles in terms of size, stability, and other functional properties. Third, our in vitro binding studies (Figures 4 and 5) are not an exact replication of the forces occurring within the tumor. An in vivo xenograft model would provide a more accurate representation of the kinetics of how PBNPs may be washed out of tumors.

If in vivo models are to be used in the future to continue these studies, we will need to consider how the aFn14-PBNPs will be administered, such as intratumorally or intravenously. If administered intravenously, we will need to monitor where the particles accumulate in organs besides the tumor, such as in the liver or kidney. We will also need to test our theory that the particles will have the proper size distribution to cross the blood–brain barrier. For either administration route, we will also need to take into consideration tumor volume, as the accumulation of the aFn14-PBNPs at the tumor site can be highly variable depending on this parameter. Finally, to administer this targeted approach for animal models of GBM, the nanoparticles and the laser will have to be implemented using technologies and methods common to neurosurgery. One option to administer the nanoparticles to the CNS is to use convection enhanced delivery to place the aFn14-PBNP in the vicinity of the GBM tumors in the CNS [63,64]. This will avoid having to utilize systemic infusion of the nanoparticles that will then have to cross the blood–brain barrier. The second component is to excite the delivered nanoparticles with an interstitially placed laser fiber similar to those clinically used for laser interstitial thermal therapy [65–68]. Our research group has completed a thorough evaluation of this interstitial PTT approach (we term this as I-PTT) in subcutaneous models of neuroblastoma and proposes to implement I-PTT for GBM in animal models in the near future.

5. Conclusions

In conclusion, we demonstrate that aFn14 can be conjugated to PBNPs to generate aFn14-PBNPs with quality attributes (size, charge, stability, UV-Vis-NIR) that retain PBNP photothermal properties. We also demonstrate that aFn14-PBNP-PTT elicits thermal and immunogenic cell death in GBM tumor cells. We show that aFn14-PBNP can target GBM tumor cells based on Fn14 targeting capabilities of the PBNPs. Finally, we show that when unbound particles are removed from cells pre-PTT, aFn14-PBNP-PTT is more effective than PBNP-PTT in triggering GBM tumor cell death. Altogether, these findings present a novel and cost-effective aFn14-PBNP nanotherapeutic for inducing thermal and immunogenic cell death in GBM tumor cells. Future studies might consider using an in vivo model, combination therapies, or interstitially administered PTT.

Supplementary Materials: The following are available online at https://www.mdpi.com/article/10.3390/nano12152645/s1, Section S1: Supplementary Figures; Figure S1: Attachment efficiency of aFn14 to PBNPs based on (a) standard curve of fluorescence intensity compared to (b) fluorescence intensity of synthesis supernatants; Figure S2: Expression levels of calreticulin and HMGB1 in U87 tumor cells. (a) Gating strategy. (b,c) Flow cytometry histogram and dot plots of cell surface expression of (b) calreticulin (c) and intracellular expression of HMGB1 in the human U87 GBM cell line following PTT with aFn14-PBNP at various laser powers; Figure S3: Expression levels of calreticulin and HMGB1 in U251 tumor cells; Figure S4: Flow cytometry histograms and dot plots of expression of (a) HLA-A,B,C, (b) HLA-DR, (c) PD-L1, (d) B7H6, (e) GD2, (f) CD137L, (g) CD80, (h) CD86, and (i) CD40 in the human U87 GBM cell line both before and following PTT with aFn14-PBNP at various laser powers; Figure S5: Flow cytometry histograms and dot plots of expression of the above markers in the human U251 GBM cell line; Figure S6. PTT using aFn14-PBNP triggers thermal and immunogenic cell death in a manner similar to PBNPs in U87 tumor cells, (a) viability, (b,c) MHC expression, (d) tumor specific antigen expression (e,f) immune checkpoint inhibitor expression, (g–j) T cell costimulatory markers; Figure S7: Determination of the concentration of aFn14-PBNP needed for the targeted PTT assay. (a) Schematic. (b,c) Heating curves (b) and the corresponding thermal doses administered (c) to U251 after incubation with various concentrations of aFn14-PBNPs during PTT. (d,e) Heating curves (d) and corresponding thermal doses administered (e) to U251 after incubation with various concentrations of PBNPs during PTT. Section S2: Step-by-Step Protocols; S2.1. Synthesis of PBNPs; S2.2. Synthesis of bioconjugated anti-Fn14-PBNP; S2.3. Attachment efficiency of anti-Fn14 to PBNPs; S2.4. Characterization of anti-Fn14-PBNP; S2.5. Cell lines and culture; S2.6. Characterization of the PTT properties of anti-Fn14-PBNP; S2.7. Elucidation of the glioblastoma tumor cell phenotype post-PTT; S2.8. Determining anti-Fn14 binding to U87 and U251 cells via flow cytometry; S2.9. Quantifying the attachment of anti-Fn14-PBNP to U87 tumor cells; S2.10. Assessing the efficacy of using anti-Fn14-PBNP for targeted PTT of glioblastoma tumor cells.

Author Contributions: Conceptualization, M.A.T., N.F.B., D.K.L., P.B.B. and R.F.; methodology, M.A.T., N.F.B., D.K.L., P.B.B. and R.F.; validation, M.A.T., N.F.B., D.K.L., P.B.B. and R.F.; formal analysis, M.A.T., N.F.B., D.K.L., P.B.B. and R.F.; investigation, M.A.T., N.F.B., D.K.L., P.B.B. and R.F.; resources, R.F.; writing—original draft preparation, M.A.T., N.F.B., D.K.L., P.B.B. and R.F.; writing—review and editing, M.A.T., N.F.B., D.K.L., P.B.B. and R.F.; supervision, P.B.B. and R.F.; funding acquisition, R.F. All authors have read and agreed to the published version of the manuscript.

Funding: This research was funded by the George Washington University Cancer Center. Research reported in this publication was funded in part by the National Cancer Institute of the National Institutes of Health under Award Number R37CA226171. The content is solely the responsibility of the authors and does not necessarily represent the official views of the National Institutes of Health. The authors would like to acknowledge the George Washington University Flow cytometry core facility. All schematic figures were created with BioRender.com (accessed on 16 June 2022).

Acknowledgments: The authors of this study acknowledge Richard Ash, Research Scientist and Manager of the Plasma Mass Spectrometry Laboratory at the University of Maryland, College Park, MD, USA for performing the Inductively Coupled Plasma Mass Spectroscopy.

Conflicts of Interest: The authors declare no conflict of interest. The funders had no role in the design of the study; in the collection, analyses, or interpretation of data; in the writing of the manuscript, or in the decision to publish the results.

References

1. Ostrom, Q.T.; Cioffi, G.; Gittleman, H.; Patil, N.; Waite, K.; Kruchko, C.; Barnholtz-Sloan, J.S. CBTRUS Statistical Report: Primary Brain and Other Central Nervous System Tumors Diagnosed in the United States in 2012–2016. *Neuro-Oncol.* **2019**, *21*, v1–v100. [CrossRef]
2. Miller, K.D.; Ostrom, Q.T.; Kruchko, C.; Patil, N.; Tihan, T.; Cioffi, G.; Fuchs, H.E.; Waite, K.A.; Jemal, A.; Siegel, R.L.; et al. Brain and Other Central Nervous System Tumor Statistics, 2021. CA. *Cancer J. Clin.* **2021**, *71*, 381–406. [CrossRef]
3. Stupp, R.; Mason, W.P.; van den Bent, M.J.; Weller, M.; Fisher, B.; Taphoorn, M.J.B.; Belanger, K.; Brandes, A.A.; Marosi, C.; Bogdahn, U.; et al. Radiotherapy plus Concomitant and Adjuvant Temozolomide for Glioblastoma. *N. Engl. J. Med.* **2005**, *352*, 987–996. [CrossRef] [PubMed]
4. Chinot, O.L.; Wick, W.; Mason, W.; Henriksson, R.; Saran, F.; Nishikawa, R.; Carpentier, A.F.; Hoang-Xuan, K.; Kavan, P.; Cernea, D.; et al. Bevacizumab plus Radiotherapy-Temozolomide for Newly Diagnosed Glioblastoma. *N. Engl. J. Med.* **2014**, *370*, 709–722. [CrossRef]
5. Razavi, S.-M.; Lee, K.E.; Jin, B.E.; Aujla, P.S.; Gholamin, S.; Li, G. Immune Evasion Strategies of Glioblastoma. *Front. Surg.* **2016**, *3*, 11. [CrossRef]
6. Brown, N.F.; Carter, T.J.; Ottaviani, D.; Mulholland, P. Harnessing the Immune System in Glioblastoma. *Br. J. Cancer* **2018**, *119*, 1171–1181. [CrossRef]
7. O'Rourke, D.M.; Nasrallah, M.P.; Desai, A.; Melenhorst, J.J.; Mansfield, K.; Morrissette, J.J.D.; Martinez-Lage, M.; Brem, S.; Maloney, E.; Shen, A.; et al. A Single Dose of Peripherally Infused EGFRvIII-Directed CAR T Cells Mediates Antigen Loss and Induces Adaptive Resistance in Patients with Recurrent Glioblastoma. *Sci. Transl. Med.* **2017**, *9*. [CrossRef]

8. Akhavan, D.; Alizadeh, D.; Wang, D.; Weist, M.R.; Shepphird, J.K.; Brown, C.E. CAR T Cells for Brain Tumors: Lessons Learned and Road Ahead. *Immunol. Rev.* **2019**, *290*, 60–84. [CrossRef]
9. Cano-Mejia, J.; Bookstaver, M.; Sweeney, E.; Jewell, C.; Fernandes, R. Prussian Blue Nanoparticle-Based Antigenicity and Adjuvanticity Trigger Robust Antitumor Immune Responses against Neuroblastoma. *Biomater. Sci.* **2019**, *7*, 1875–1887. [CrossRef]
10. Sweeney, E.; Cano-Mejia, J.; Fernandes, R. Photothermal Therapy Generates a Thermal Window of Immunogenic Cell Death in Neuroblastoma. *Small* **2018**, *14*, 1800678. [CrossRef]
11. Loo, C.; Lowery, A.; Halas, N.; West, J.; Drezek, R. Immunotargeted Nanoshells for Integrated Cancer Imaging and Therapy. *Nano Lett.* **2005**, *5*, 709–711. [CrossRef]
12. Roper, D.K.; Ahn, W.; Hoepfner, M. Microscale Heat Transfer Transduced by Surface Plasmon Resonant Gold Nanoparticles. *J. Phys. Chem. C Nanomater. Interfaces* **2007**, *111*, 3636–3641. [CrossRef] [PubMed]
13. Huang, X.; Jain, P.K.; El-Sayed, I.H.; El-Sayed, M.A. Plasmonic Photothermal Therapy (PPTT) Using Gold Nanoparticles. *Lasers Med. Sci.* **2008**, *23*, 217–228. [CrossRef]
14. Kamath, A.A.; Friedman, D.D.; Akbari, S.H.A.; Kim, A.H.; Tao, Y.; Luo, J.; Leuthardt, E.C. Glioblastoma Treated With Magnetic Resonance Imaging-Guided Laser Interstitial Thermal Therapy: Safety, Efficacy, and Outcomes. *Neurosurgery* **2019**, *84*, 836–843. [CrossRef]
15. Bastiancich, C.; Da Silva, A.; Estève, M.-A. Photothermal Therapy for the Treatment of Glioblastoma: Potential and Preclinical Challenges. *Front. Oncol.* **2021**, *10*, 610356. [CrossRef]
16. Fernandez Cabada, T.; Sanchez Lopez de Pablo, C.; Martinez Serrano, A.; del Pozo Guerrero, F.; Serrano Olmedo, J.J.; Ramos Gomez, M. Induction of Cell Death in a Glioblastoma Line by Hyperthermic Therapy Based on Gold Nanorods. *Int. J. Nanomedicine* **2012**, *7*, 1511–1523. [CrossRef]
17. Gonçalves, D.P.N.; Rodriguez, R.D.; Kurth, T.; Bray, L.J.; Binner, M.; Jungnickel, C.; Gür, F.N.; Poser, S.W.; Schmidt, T.L.; Zahn, D.R.T.; et al. Enhanced Targeting of Invasive Glioblastoma Cells by Peptide-Functionalized Gold Nanorods in Hydrogel-Based 3D Cultures. *Acta Biomater.* **2017**, *58*, 12–25. [CrossRef]
18. Botella, P.; Ortega, I.; Quesada, M.; Madrigal, R.F.; Muniesa, C.; Fimia, A.; Fernández, E.; Corma, A. Multifunctional Hybrid Materials for Combined Photo and Chemotherapy of Cancer. *Dalton Trans. Camb. Engl. 2003* **2012**, *41*, 9286–9296. [CrossRef]
19. Robinson, J.T.; Tabakman, S.M.; Liang, Y.; Wang, H.; Casalongue, H.S.; Vinh, D.; Dai, H. Ultrasmall Reduced Graphene Oxide with High Near-Infrared Absorbance for Photothermal Therapy. *J. Am. Chem. Soc.* **2011**, *133*, 6825–6831. [CrossRef]
20. Li, Z.-J.; Li, C.; Zheng, M.-G.; Pan, J.-D.; Zhang, L.-M.; Deng, Y.-F. Functionalized Nano-Graphene Oxide Particles for Targeted Fluorescence Imaging and Photothermy of Glioma U251 Cells. *Int. J. Clin. Exp. Med.* **2015**, *8*, 1844–1852.
21. Zheng, X.; Xing, D.; Zhou, F.; Wu, B.; Chen, W.R. Indocyanine Green-Containing Nanostructure as near Infrared Dual-Functional Targeting Probes for Optical Imaging and Photothermal Therapy. *Mol. Pharm.* **2011**, *8*, 447–456. [CrossRef] [PubMed]
22. Keyvan Rad, J.; Mahdavian, A.R.; Khoei, S.; Shirvalilou, S. Enhanced Photogeneration of Reactive Oxygen Species and Targeted Photothermal Therapy of C6 Glioma Brain Cancer Cells by Folate-Conjugated Gold-Photoactive Polymer Nanoparticles. *ACS Appl. Mater. Interfaces* **2018**, *10*, 19483–19493. [CrossRef] [PubMed]
23. Tran, N.; McDonough, W.; Donohue, P.; Winkles, J.; Berens, T.; Koss, K. The Human Fn14 Receptor Gene Is Up-Regulated in Migrating Glioma Cells in Vitro and Overexpressed in Advanced Glial Tumors. *Am. J. Pathol.* **2003**, *162*, 1313–1321. [CrossRef]
24. Tran, N.L.; McDonough, W.S.; Savitch, B.A.; Fortin, S.P.; Winkles, J.A.; Symons, M.; Nakada, M.; Cunliffe, H.E.; Hostetter, G.; Hoelzinger, D.B.; et al. Increased Fibroblast Growth Factor-Inducible 14 Expression Levels Promote Glioma Cell Invasion via Rac1 and Nuclear Factor-KappaB and Correlate with Poor Patient Outcome. *Cancer Res.* **2006**, *66*, 9535–9542. [CrossRef]
25. Fortin, S.P.; Ennis, M.J.; Savitch, B.A.; Carpentieri, D.; McDonough, W.S.; Winkles, J.A.; Loftus, J.C.; Kingsley, C.; Hostetter, G.; Tran, N.L. Tumor Necrosis Factor-Like Weak Inducer of Apoptosis (TWEAK) Stimulation of Glioma Cell Survival Is Dependent Upon Akt2 Function. *Mol. Cancer Res. MCR* **2009**, *7*, 1871–1881. [CrossRef] [PubMed]
26. Perez, J.G.; Tran, N.L.; Rosenblum, M.G.; Schneider, C.S.; Connolly, N.P.; Kim, A.J.; Woodworth, G.F.; Winkles, J.A. The TWEAK Receptor Fn14 Is a Potential Cell Surface Portal for Targeted Delivery of Glioblastoma Therapeutics. *Oncogene* **2016**, *35*, 2145–2155. [CrossRef]
27. Fortin Ensign, S.P.; Mathews, I.T.; Eschbacher, J.M.; Loftus, J.C.; Symons, M.H.; Tran, N.L. The Src Homology 3 Domain-Containing Guanine Nucleotide Exchange Factor Is Overexpressed in High-Grade Gliomas and Promotes Tumor Necrosis Factor-like Weak Inducer of Apoptosis-Fibroblast Growth Factor-Inducible 14-Induced Cell Migration and Invasion via Tumor Necrosis Factor Receptor-Associated Factor 2. *J. Biol. Chem.* **2013**, *288*, 21887–21897. [CrossRef]
28. Winkles, J.A.; Tran, N.L.; Berens, M.E. TWEAK and Fn14: New Molecular Targets for Cancer Therapy? *Cancer Lett.* **2006**, *235*, 11–17. [CrossRef]
29. Aido, A.; Zaitseva, O.; Wajant, H.; Buzgo, M.; Simaite, A. Anti-Fn14 Antibody-Conjugated Nanoparticles Display Membrane TWEAK-Like Agonism†. *Pharmaceutics* **2021**, *13*, 1072. [CrossRef]
30. Schneider, C.; Perez, J.; Cheng, E.; Zhang, C.; Panagiotis, M.; Hanes, J.; Winkles, J.; Woodworth, G.; Kim, A. Minimizing the Non-Specific Binding of Nanoparticles to the Brain Enables Active Targeting of Fn14-Positive Glioblastoma Cells. *Biomaterials* **2015**, *42*, 42–51. [CrossRef]
31. Kik, K.; Bukowska, B.; Sicińska, P. Polystyrene Nanoparticles: Sources, Occurrence in the Environment, Distribution in Tissues, Accumulation and Toxicity to Various Organisms. *Environ. Pollut.* **2020**, *262*, 114297. [CrossRef]

32. Arvizo, R.; Bhattacharya, R.; Mukherjee, P. Gold Nanoparticles: Opportunities and Challenges in Nanomedicine. *Expert Opin. Drug Deliv.* **2010**, *7*, 753–763. [CrossRef]
33. Singh, V.K.; Romaine, P.L.P.; Seed, T.M. Medical Countermeasures for Radiation Exposure and Related Injuries: Characterization of Medicines, FDA-Approval Status and Inclusion into the Strategic National Stockpile. *Health Phys.* **2015**, *108*, 607–630. [CrossRef] [PubMed]
34. Hussar, D.A. New Drugs 05, Part I. *Nursing* **2005**, *35*, 54–61. [CrossRef] [PubMed]
35. Hoffman, H.A.; Chakrabarti, L.; Dumont, M.F.; Sandler, A.D.; Fernandes, R. Prussian Blue Nanoparticles for Laser-Induced Photothermal Therapy of Tumors. *RSC Adv.* **2014**, *4*, 29729–29734. [CrossRef]
36. Wu, X.; Luo, T. Effect of Electron-Phonon Coupling on Thermal Transport across Metal-Nonmetal Interface —A Second Look. *EPL Europhys. Lett.* **2015**, *110*, 67004. [CrossRef]
37. Dacarro, G.; Taglietti, A.; Pallavicini, P. Prussian Blue Nanoparticles as a Versatile Photothermal Tool. *Mol. Basel Switz.* **2018**, *23*, 1414. [CrossRef] [PubMed]
38. Galluzzi, L.; Vitale, I.; Warren, S.; Adjemian, S.; Agostinis, P.; Martinez, A.B.; Chan, T.A.; Coukos, G.; Demaria, S.; Deutsch, E.; et al. Consensus Guidelines for the Definition, Detection and Interpretation of Immunogenic Cell Death. *J. Immunother. Cancer* **2020**, *8*, e000337. [CrossRef]
39. Galluzzi, L.; Buqué, A.; Kepp, O.; Zitvogel, L.; Kroemer, G. Immunogenic Cell Death in Cancer and Infectious Disease. *Nat. Rev. Immunol.* **2017**, *17*, 97–111. [CrossRef]
40. Ruiz-Justiz, A.J.; Cano-Mejia, J.; Fernandes, R. DAMPs-Coated Prussian Blue Nanoparticles as Photothermal-Nanoimmunotherapy Agents for Cancer. *FASEB J.* **2019**, *33*, 510.2. [CrossRef]
41. Sweeney, E.E.; Burga, R.A.; Li, C.; Zhu, Y.; Fernandes, R. Photothermal Therapy Improves the Efficacy of a MEK Inhibitor in Neurofibromatosis Type 1-Associated Malignant Peripheral Nerve Sheath Tumors. *Sci. Rep.* **2016**, *6*, 37035. [CrossRef] [PubMed]
42. Shukla, A.; Cano-Mejia, J.; Andricovich, J.; Burga, R.A.; Sweeney, E.E.; Fernandes, R. An Engineered Prussian Blue Nanoparticles-Based Nanoimmunotherapy Elicits Robust and Persistent Immunological Memory in a TH-MYCN Neuroblastoma Model. *Adv. NanoBiomed Res.* **2021**, *1*, 2100021. [CrossRef] [PubMed]
43. Balakrishnan, P.B.; Ledezma, D.K.; Cano-Mejia, J.; Andricovich, J.; Palmer, E.; Patel, V.A.; Latham, P.S.; Yvon, E.S.; Villagra, A.; Fernandes, R.; et al. CD137 Agonist Potentiates the Abscopal Efficacy of Nanoparticle-Based Photothermal Therapy for Melanoma. *Nano Res.* **2021**. [CrossRef]
44. Vojtech, J.M.; Cano-Mejia, J.; Dumont, M.F.; Sze, R.W.; Fernandes, R. Biofunctionalized Prussian Blue Nanoparticles for Multimodal Molecular Imaging Applications. *JoVE J. Vis. Exp.* **2015**, e52621. [CrossRef] [PubMed]
45. Sapareto, S.A.; Dewey, W.C. Thermal Dose Determination in Cancer Therapy. *Int. J. Radiat. Oncol. Biol. Phys.* **1984**, *10*, 787–800. [CrossRef]
46. Yarmolenko, P.S.; Moon, E.J.; Landon, C.; Manzoor, A.; Hochman, D.W.; Viglianti, B.L.; Dewhirst, M.W. Thresholds for Thermal Damage to Normal Tissues: An Update. *Int. J. Hyperth. Off. J. Eur. Soc. Hyperthermic Oncol. N. Am. Hyperth. Group* **2011**, *27*, 320–343. [CrossRef]
47. Van Rhoon, G.C.; Samaras, T.; Yarmolenko, P.S.; Dewhirst, M.W.; Neufeld, E.; Kuster, N. CEM43°C Thermal Dose Thresholds: A Potential Guide for Magnetic Resonance Radiofrequency Exposure Levels? *Eur. Radiol.* **2013**, *23*, 2215–2227. [CrossRef]
48. Betzer, O.; Shilo, M.; Opochinsky, R.; Barnoy, E.; Motiei, M.; Okun, E.; Yadid, G.; Popovtzer, R. The Effect of Nanoparticle Size on the Ability to Cross the Blood-Brain Barrier: An in Vivo Study. *Nanomedicine* **2017**, *12*, 1533–1546. [CrossRef]
49. Ceña, V.; Játiva, P. Nanoparticle Crossing of Blood–Brain Barrier: A Road to New Therapeutic Approaches to Central Nervous System Diseases. *Nanomedicine* **2018**, *13*, 1513–1516. [CrossRef]
50. Inouye, H.; Kirschner, D.A. Membrane Interactions in Nerve Myelin. I. Determination of Surface Charge from Effects of PH and Ionic Strength on Period. *Biophys. J.* **1988**, *53*, 235–245. [CrossRef]
51. Sekhri, P.; Ledezma, D.K.; Shukla, A.; Sweeney, E.E.; Fernandes, R. The Thermal Dose of Photothermal Therapy Generates Differential Immunogenicity in Human Neuroblastoma Cells. *Cancers* **2022**, *14*, 1447. [CrossRef]
52. Mateus, A.; Kurzawa, N.; Becher, I.; Sridharan, S.; Helm, D.; Stein, F.; Typas, A.; Savitski, M.M. Thermal Proteome Profiling for Interrogating Protein Interactions. *Mol. Syst. Biol.* **2020**, *16*, e9232. [CrossRef] [PubMed]
53. Moussa, M.; Goldberg, S.N.; Kumar, G.; Levchenko, T.; Torchilin, V.; Ahmed, M. Effect of Thermal Dose on Heat Shock Protein Expression after Radio-Frequency Ablation with and without Adjuvant Nanoparticle Chemotherapies. *Int. J. Hyperth. Off. J. Eur. Soc. Hyperthermic Oncol. N. Am. Hyperth. Group* **2016**, *32*, 829–841. [CrossRef] [PubMed]
54. Youssef, Z.; Yesmurzayeva, N.; Larue, L.; Jouan-Hureaux, V.; Colombeau, L.; Arnoux, P.; Acherar, S.; Vanderesse, R.; Frochot, C. New Targeted Gold Nanorods for the Treatment of Glioblastoma by Photodynamic Therapy. *J. Clin. Med.* **2019**, *8*, 2205. [CrossRef] [PubMed]
55. Akhavan, O.; Ghaderi, E. Graphene Nanomesh Promises Extremely Efficient in Vivo Photothermal Therapy. *Small Weinh. Bergstr. Ger.* **2013**, *9*, 3593–3601. [CrossRef]
56. Zhu, M.; Sheng, Z.; Jia, Y.; Hu, D.; Liu, X.; Xia, X.; Liu, C.; Wang, P.; Wang, X.; Zheng, H. Indocyanine Green-Holo-Transferrin Nanoassemblies for Tumor-Targeted Dual-Modal Imaging and Photothermal Therapy of Glioma. *ACS Appl. Mater. Interfaces* **2017**, *9*, 39249–39258. [CrossRef] [PubMed]

57. Tran, N.L.; McDonough, W.S.; Savitch, B.A.; Sawyer, T.F.; Winkles, J.A.; Berens, M.E. The Tumor Necrosis Factor-like Weak Inducer of Apoptosis (TWEAK)-Fibroblast Growth Factor-Inducible 14 (Fn14) Signaling System Regulates Glioma Cell Survival via NFkappaB Pathway Activation and BCL-XL/BCL-W Expression. *J. Biol. Chem.* **2005**, *280*, 3483–3492. [CrossRef]
58. Winkles, J.A. The TWEAK-Fn14 Cytokine-Receptor Axis: Discovery, Biology and Therapeutic Targeting. *Nat. Rev. Drug Discov.* **2008**, *7*, 411–425. [CrossRef]
59. Ledezma, D.K.; Balakrishnan, P.B.; Cano-Mejia, J.; Sweeney, E.E.; Hadley, M.; Bollard, C.M.; Villagra, A.; Fernandes, R. Indocyanine Green-Nexturastat A-PLGA Nanoparticles Combine Photothermal and Epigenetic Therapy for Melanoma. *Nanomaterials* **2020**, *10*, 161. [CrossRef]
60. Wadajkar, A.S.; Dancy, J.G.; Roberts, N.B.; Connolly, N.P.; Strickland, D.K.; Winkles, J.A.; Woodworth, G.F.; Kim, A.J. Decreased Non-Specific Adhesivity, Receptor Targeted (DART) Nanoparticles Exhibit Improved Dispersion, Cellular Uptake, and Tumor Retention in Invasive Gliomas. *J. Control. Release Off. J. Control. Release Soc.* **2017**, *267*, 144–153. [CrossRef] [PubMed]
61. Michaelson, J.S.; Amatucci, A.; Kelly, R.; Su, L.; Garber, E.; Day, E.S.; Berquist, L.; Cho, S.; Li, Y.; Parr, M.; et al. Development of an Fn14 Agonistic Antibody as an Anti-Tumor Agent. *mAbs* **2011**, *3*, 362–375. [CrossRef]
62. Li, G.; Zhang, Z.; Cai, L.; Tang, X.; Huang, J.; Yu, L.; Wang, G.; Zhong, K.; Cao, Y.; Liu, C.; et al. Fn14-Targeted BiTE and CAR-T Cells Demonstrate Potent Preclinical Activity against Glioblastoma. *OncoImmunology* **2021**, *10*, 1983306. [CrossRef]
63. Young, J.S.; Bernal, G.; Polster, S.P.; Nunez, L.; Larsen, G.F.; Mansour, N.; Podell, M.; Yamini, B. Convection Enhanced Delivery of Polymeric Nanoparticles Encapsulating Chemotherapy in Canines with Spontaneous Supratentorial Tumors. *World Neurosurg.* **2018**, *117*, e698–e704. [CrossRef] [PubMed]
64. Wang, Y.; Jiang, Y.; Wei, D.; Singh, P.; Yu, Y.; Lee, T.; Zhang, L.; Mandl, H.K.; Piotrowski-Daspit, A.S.; Chen, X.; et al. Nanoparticle-Mediated Convection-Enhanced Delivery of a DNA Intercalator to Gliomas Circumvents Temozolomide Resistance. *Nat. Biomed. Eng.* **2021**, *5*, 1048–1058. [CrossRef] [PubMed]
65. Holste, K.G.; Orringer, D.A. Laser Interstitial Thermal Therapy. *Neuro-Oncol. Adv.* **2020**, *2*. [CrossRef]
66. Bozinov, O.; Yang, Y.; Oertel, M.F.; Neidert, M.C.; Nakaji, P. Laser Interstitial Thermal Therapy in Gliomas. *Cancer Lett.* **2020**, *474*, 151–157. [CrossRef]
67. Traylor, J.I.; Patel, R.; Muir, M.; de Almeida Bastos, D.C.; Ravikumar, V.; Kamiya-Matsuoka, C.; Rao, G.; Thomas, J.G.; Kew, Y.; Prabhu, S.S. Laser Interstitial Thermal Therapy for Glioblastoma: A Single-Center Experience. *World Neurosurg.* **2021**, *149*, e244–e252. [CrossRef]
68. Barnett, G.H.; Voigt, J.D.; Alhuwalia, M.S. A Systematic Review and Meta-Analysis of Studies Examining the Use of Brain Laser Interstitial Thermal Therapy versus Craniotomy for the Treatment of High-Grade Tumors in or near Areas of Eloquence: An Examination of the Extent of Resection and Major Complication Rates Associated with Each Type of Surgery. *Stereotact. Funct. Neurosurg.* **2016**, *94*, 164–173. [CrossRef]

nanomaterials

MDPI

Article

Modulation of Macrophage Polarization by Carbon Nanodots and Elucidation of Carbon Nanodot Uptake Routes in Macrophages

Andrew Dunphy [1], Kamal Patel [1], Sarah Belperain [1], Aubrey Pennington [1], Norman H. L. Chiu [2,3], Ziyu Yin [3], Xuewei Zhu [4], Brandon Priebe [1], Shaomin Tian [5], Jianjun Wei [3], Xianwen Yi [6,7] and Zhenquan Jia [1,*]

1 Department of Biology, The University of North Carolina at Greensboro 312 Eberhart Building, 321 McIver Street, Greensboro, NC 27402-617, USA; amdunphy@uncg.edu (A.D.); kkpatel3@uncg.edu (K.P.); srbelper@uncg.edu (S.B.); a_pennin@uncg.edu (A.P.); bmpriebe@uncg.edu (B.P.)
2 Department of Chemistry and Biochemistry, University of North Carolina at Greensboro, Greensboro, NC 27412, USA; nhchiu@uncg.edu
3 Department of Nanoscience, Joint School of Nanoscience and Nanoengineering, University of North Carolina at Greensboro, Greensboro, NC 27401, USA; z_yin@uncg.edu (Z.Y.); j_wei@uncg.edu (J.W.)
4 Department of Internal Medicine, Section on Molecular Medicine, Wake Forest School of Medicine, Winston-Salem, NC 27101, USA; xwzhu@wakehealth.edu
5 Department of Microbiology & Immunology, University of North Carolina, Chapel Hill, NC 27599, USA; shaomin_tian@med.unc.edu
6 Lineberger Comprehensive Cancer Center, University of North Carolina, Chapel Hill, NC 27599, USA; xianwen_yi@med.unc.edu
7 McAllister Heart Institute, University of North Carolina, Chapel Hill, NC 27599, USA
* Correspondence: z_jia@uncg.edu; Tel.: +1-336-334-5391; Fax: +1-336-334-5839

Citation: Dunphy, A.; Patel, K.; Belperain, S.; Pennington, A.; Chiu, N.H.L.; Yin, Z.; Zhu, X.; Priebe, B.; Tian, S.; Wei, J.; et al. Modulation of Macrophage Polarization by Carbon Nanodots and Elucidation of Carbon Nanodot Uptake Routes in Macrophages. *Nanomaterials* **2021**, *11*, 1116. https://doi.org/10.3390/nano11051116

Academic Editor: Daniela Guarnieri

Received: 8 April 2021
Accepted: 23 April 2021
Published: 26 April 2021

Abstract: Atherosclerosis represents an ever-present global concern, as it is a leading cause of cardiovascular disease and an immense public welfare issue. Macrophages play a key role in the onset of the disease state and are popular targets in vascular research and therapeutic treatment. Carbon nanodots (CNDs) represent a type of carbon-based nanomaterial and have garnered attention in recent years for potential in biomedical applications. This investigation serves as a foremost attempt at characterizing the interplay between macrophages and CNDs. We have employed THP-1 monocyte-derived macrophages as our target cell line representing primary macrophages in the human body. Our results showcase that CNDs are non-toxic at a variety of doses. THP-1 monocytes were differentiated into macrophages by treatment with 12-*O*-tetradecanoylphorbol-13-acetate (TPA) and co-treatment with 0.1 mg/mL CNDs. This co-treatment significantly increased the expression of CD 206 and CD 68 (key receptors involved in phagocytosis) and increased the expression of CCL2 (a monocyte chemoattractant and pro-inflammatory cytokine). The phagocytic activity of THP-1 monocyte-derived macrophages co-treated with 0.1 mg/mL CNDs also showed a significant increase. Furthermore, this study also examined potential entrance routes of CNDs into macrophages. We have demonstrated an inhibition in the uptake of CNDs in macrophages treated with nocodazole (microtubule disruptor), N-phenylanthranilic acid (chloride channel blocker), and mercury chloride (aquaporin channel inhibitor). Collectively, this research provides evidence that CNDs cause functional changes in macrophages and indicates a variety of potential entrance routes.

Keywords: carbon nanodots; macrophages; polarization; phagocytosis; uptake routes

1. Introduction

Cardiovascular disease (CVD) has more clinical implications than any other condition worldwide. Globally, CVD accounts for one-third of all deaths [1]. In the United States, over 600,000 humans die of CVD per year, representing a quarter of all American deaths [2]. For these reasons, devoting resources and research into ameliorating the mortality caused by CVD is of principal priority. CVD can be expressed in several types such as ischemic heart disease, cerebrovascular disease, or coronary artery diseases, which is usually caused by atherosclerosis [3]. Atherosclerosis is the build-up of plaque in artery walls. This condition leads to a decrease in blood flow to tissues. With the ever-increasing mortality rate due to CVD, it is crucial to develop new methods of treatment. The biggest challenge still is understanding the ramifications of the development of atherosclerosis.

Macrophages play a key role in the onset of the disease state. Free oxygen radicals modify low-densitiy lipoprotieins (LDL) into ox-LDL. Upon injury to the endothelium of blood vessels by ox-LDL, circulating monocytes differentiate into pro-inflammatory (M1) or anti-inflammatory (M2) macrophages. In progressing lesions, M1 macrophages engulf excess ox-LDL. In the process, these macrophages become lipid-laden and lose mobility, finally proceeding to settle en masse on the bed of arteries as plaque. Dysregulated plaque build-up in arteries results in a several fatal long-term health issues. Due to their crucial role as mediators in atherogenesis, as well as their involvement in several aspects of the immune response, macrophages are popular targets in vascular research and therapeutic treatment.

In recent years, interest in the development of nanoparticles for biological application has risen. A key area of intrigue revolves around the interaction of nanoparticles with some aspects of the immune response, resulting in their induction or repression [4]. Their use also extends to imaging macrophages and disease states such as atherosclerotic lesions [5,6]. Carbon nanodots (CNDs) are particles of particular interest for a variety of reasons. These particles tend to be smaller than 10 nm in size, have an sp2 hybridization, and are quasi-spherical [7]. An essential characteristic of these nanodots is their high hydrophilicity, which is made apparent by the presence of several functional groups in their surface such as ether, carbonyl, hydroxyl, etc. This hydrophilicity allows for a very biocompatible particle that is ready to interact with various organic or inorganic species [7]. CNDs also have photoluminescent properties. This, combined with their hydrophilicity, makes CNDs useful in sensing other particles. Their luminescent characteristics are defined by their individual size, shape, functional groups, and other factors. Upon excitation by UV to visible light, CNDs emission wavelengths range from UV to near-infrared [7].

CNDs have been synthesized with scavenging properties and proved themselves capable ex vivo scavengers of free radicals, one of which is 2,2-diphenyl-1-picrylhydrazyl radicals (DPPH) [8]. In this assay, the successful conversion of DPPH into a stable DPPH-H complex is due to antioxidant activity. This leads to a change in color from violet to light yellow, which can be quantified by ultraviolet-visible spectroscopy. Zhang et al. demonstrated a dose-dependent increase in DPPH scavenging using N,S-codoped CNDs [8]. These nanoparticles were synthesized using citric acid, α-lipoic acid, and urea precursors through a hydrothermal method [9]. In vitro, CNDs have also demonstrated radical scavenging ability. By way of Di-Chloro Di-Hydrofuran Fluorescein Di-Acetate (DCFH-DA) assay and NBT (Nitro Blue Tetrazolium) reduction assay, Das et al. showcased CNDs (synthesized by microwave irradiation of date molasses) scavenging ability of hydroxyl and superoxide free radicals [10]. Altogether, these results denote the antioxidant propensity of CNDs and evidence their potential for biological utilization.

Atherosclerosis is a long-standing inflammatory disease characterized by the narrowing of arteries due to a build-up of plaque. The overproduction of ROS and its subsequent oxidative stress play a key role in its initiation. Macrophages play an essential role as intermediators of the disease state by differentiating into a pro-inflammatory state, secreting cytokines and eventually becoming foam cells. As the concrete source for plaque build-up, macrophages signify an area of interest. CNDs are a prospective choice for biomedical implementation, having shown usefulness in ROS scavenging, biosensing, and drug delivery. However, currently, no account exists that indicates whether or not CNDs have any ability to affect the M1/M2 polarization of macrophages. In this study, we examined the effects of CNDs on the expression of M1/M2 biomarkers and phagocytic activity of macrophages, as well as potential entrance routes.

2. Materials and Methods

2.1. Cell Culture

The THP-1 cell line (ATCC® TIB-202™, Manassas, VA, USA) cells were cultured in Roswell Park Memorial Institute 1640 Medium (RPMI 1640) fortified with 10% Fetal Bovine Serum (FBS) and 1% Penicillin–Streptomycin. This cell line was obtained from the ATCC (Manassas, VA, USA) and grown in Cellstar® Filter Cap 75 cm^2 cell-culture treated, filter screw cap flasks in humidified incubators programmed to 37 °C and 5% CO_2. Corresponding media was renewed every 2 days and cells were split into a new passage upon 85–90% confluence.

2.2. CNDs Synthesis and Characterization

The CNDs preparation using citric acid and ethylenediamine (EDA) as precursors was synthesized based on an adaption of a previously published microwave-assisted method [11]. In a 100 mL beaker, 0.96 citric acid 99.5% (Sigma-Aldrich, St. Louis, MO, USA) was dissolved with 10.0 mL DDI water and then mixed with 1 mL (0.8980 g/mL) EDA 99% (ACROS Organics, Fair Lawn, NJ, USA) under vigorous stirring for 30 s. The beaker containing the clear and colorless solution was covered with a watch glass and heated using a domestic microwave oven (1200 W) for 5.0 min. After the elapsed time, the beaker and contents were allowed to cool to room temperature. The brownish-orange crystalline product was diluted with 10.0 mL DDI water and then dialyzed using a 500–1000 Da MWCO Spectra Por Float-A-Lyzer G2 (Repligen, Waltham, MA, USA). The resultant clear, brownish-orange aqueous solution was lyophilized using Labconco FreeZone Plus 12 (Labconco, Kansas City, MO, USA) to obtain the dried carbon nanodot product. UV-Vis spectroscopy of CNDs was performed by Cary® Eclipse TM Fluorescence Spectrophotometer (Agilent, Santa Clara, CA, USA). Upon dilution to 2 mg/mL in DI-H_2O, CNDs were measured for fluorescence in a quartz cuvette to determine excitation and emission wavelengths. The surface chemistry of CNDS was characterized by carbon 1s X-ray photoelectron spectroscopy (XPS, ESCALAB 250 Xi, Thermo Fisher, West Sussex, UK).

2.3. Monocyte Differentiation into Macrophages

THP-1 cells were cultured in cell plates containing RPMI 1640 Medium (Sigma-Aldrich, St. Louis, MO, USA) or Hank's Balanced Salt Solution (HBSS) with calcium, magnesium, and glucose (Sigma-Aldrich, St. Louis, MO, USA). Monocyte differentiation into macrophages was induced with 3 ng/µL of 12-O-tetra-decanoylphorbol-13-acetate (TPA) in an incubation period of 72 h period. Cells were lifted by cell scraper and the supernatant placed in 50 mL Falcon tubes. Cell plates were rinsed with 2 mL PBS and also added to the supernatant. Then, cells were pelleted by centrifuge.

2.4. CNDs Treatment

For cell viability measured by trypan blue staining and flow cytometry ViaCount assays, THP-1 cells were treated with 3 ng/mL TPA in the presence or absence of 0.01, 0.1, 0.3, or 0.6 mg/mL CNDs in RPMI media for 72 h. The media was refreshed, and then, the cells were incubated for another 72 h. For other experiments, THP-1 monocytes were cultured in cell plates and co-treated with 0.1 mg/mL CNDs and 3 ng/µL TPA for 72 h in RPMI media. Incubation occurred in 37 °C/5% CO_2 incubators. Surrounding media was decanted and replaced with new media, followed by another incubation period of 72 h, after which plates were rinsed with 2 mL PBS and also added to the supernatant. Then, cells were pelleted by centrifuge.

2.5. Cell Count (Trypan Blue)

Before and after differentiation and CNDs treatments, cells were counted. Monocytes and macrophages were centrifuged at 300 *g* for 5 min at 4 °C and resuspended in either PBS or respective media. After resuspension, cells were counted using a hemocytometer.

Trypan blue was used to count viable, unstained cells, and the resulting concentration was also calculated.

2.6. RNA Extraction

THP-1 cells were cultured in appropriate media in Corning® cell culture treated plates (Corning Life Sciences, Durham, NC, USA). Upon treatment and incubation, adhered cells were lifted sing a cell scraper. The media was extracted into 50 mL tubes. The cell plates were rinsed twice with 1 × PBS to ensure no treatment media remained and also to extract any remaining cells. Then, cells were then centrifuged at 300 *g* for 5 min at 4 °C. The supernatant media was decanted, and the resulting cell pellets were treated with 1 mL of ambion TRIzol® (Thermo Fisher Scientific, Waltham, MA, USA). The resulting solution was pipetted into 1 mL Eppendorf tubes. Then, 200 µL of chloroform were added, which was followed by agitation, and then the solution was centrifuged at 12,000 rcf for 15 min. The top aqueous phase was transferred to another set of 1 mL Eppendorf tubes and then combined with 500 µL isopropanol and agitated before centrifuging again at 12,000 rcf for 10 min. The resulting pellet (RNA) was washed with 1 mL 75% ethanol and centrifuged at 7400 rcf for 5 min twice. Then, the pellet was resuspended in 10–15 µL of DEPC H_2O.

2.7. CDNA Synthesis

After RNA extraction, the resulting RNA was quantified by way of a Thermo Scientific™ Nanodrop 2000 (Thermo Fisher Scientific, Waltham, MA, USA). Then, RNA was diluted to a concentration of 500 ng/µL. Then, 2 µL of diluted RNA were mixed with 5 µL of 5× Buffer, 1.25 µL of ddNTP, 1.25 µL of Random Primer, 14.875 µL of DEPC H_2O, and 0.625 µL of MMLV-Reverse Transcriptase. Using Applied Biosystems™ Veriti™ 96-Well Thermal Cycler (Thermo Fisher Scientific, Waltham, MA, USA), the 25 µL solution was converted to cDNA.

2.8. Quantitative Real-Time Polymerase Chain Reaction

Once cDNA was synthesized using the methods above, the resulting cDNA was probed for a selection of M1/M2 biomarkers as mentioned previously, using H_GAPDH as a housekeeping gene. This was performed by mixing 1 µL of cDNA with 10 µL of Power SYBR® Green PCR Master Mix, 2 µL of 5 µM Forward Primer (Table 1), 2 µL of 5 µM Reverse Primer (Table 1), 2 µL of 1:10 diluted cDNA, and 5 µL of DEPC H_2O. The Applied Biosystems™ StepOnePlus™ Real-Time PCR system (Thermo Fisher Scientific, Waltham, MA, USA) was employed and ran for 40 cycles. Each individual cycle constituted a 95 °C phase for 15 s, a 58 °C phase for 60 s, and a 60 °C phase for 15 s. Comparative threshold values were evaluated in order to quantify gene expression.

Table 1. Primer sequences for qrt-PCR reactions.

Target	Forward Primer	Reverse Primer
GAPDH	5′-AGA ACG GGA AGC TTG TCA TC-3′	5′-GGA GGC ATT GCT GAT GAT CT-3′
IL-8	5′-CTC TGT GTG AAG GTG CAG TT-3′	5′ –AAA CTT CTC CAC AAC CCT CTG-3′
CCL-2	5′-GCT CAG CCA GAT GCA ATC AA-3′	5-GGT TGT GGA GTG AGT GGT CAA G-3′
CD68	5′-TCAGCTTTGGATTCATGCAG-3′	5′-AGGTGGACAGCTGGTGAAAG-3′
IL-10	5′-CTAACCTCATTCCCCAACCA-3′	5′-GTAGAGACGGGGTTTCACCA-3′
TNF-α	5′-CTATCTGGGAGGGGTCTTCC-3′	5′-GGTTGAGGGTGTCTGAAGGA-3′

2.9. Vybrant™ Phagocytosis Assay Kit (V-6694)

First, 4 × 10^6 THP-1 cells were grown in cell culture plates in corresponding media. Differentiation into macrophages was induced by administering 3 ng/µL TPA with an incubation period of 72 h (with or without co-treatment with a CNDs concentration of 0.1 mg/mL) at 37 °C, 5% CO2. After harvesting and pelleting cells, concentration was re-suspended in HBSS to 2 × 10^6 cells/mL. Then, 1 mL of control cells were treated with

1000 ng/mL TPA in order to activate cells to serve as a positive control. Next, 100 µL of cell suspension was added to 5 wells per sample, plus 50 µL of HBSS (negative control wells contained 200 µL of HBSS). Cells were left to incubate for 18 h in 35 °C/5% CO_2 incubators. This incubation period allows macrophages to settle. Then, HBSS was removed, and 200 µL of fluorescently labeled *E. coli* suspension was administered for 2 h. Upon the removal of suspension, cells were treated for 60 s with 100 µL of Trypan Blue suspension. Immediate removal of suspension followed. The phagocytic activities of cells were quantified using a BioTek™ Synergy 2.0 plater reader (BioTek Instruments, Winooski, VT, USA).

2.10. ViaCount Flow Cytometry

Cells were cultured with the necessary incubation times and treatments. Next, cells were harvested by cell scraping and placed into tubes. The cell concentration was adjusted to 5×10^6/mL. Then, 80 µL of cells were treated with 20 µL of ViaCount Reagent for 10 min at room temperature, upon which 500 µL of cold PBS was added. The samples were analyzed for viability using a Guava® easyCyte™ Flow Cytometer (Luminex Corporation, Austin, TX, USA).

2.11. CNDs Uptake

THP-1 human monocyte-derived macrophages were grown to 85–90% confluence with corresponding media in clear cell plates and pre-treated with or without the following inhibitors for 30 min: Cytochalasin A or D (5 µg/mL), chlorpromazine (10 µg/mL), genistein (200 µM), nocodazole (20 µM), phenylglyoxal (100 µg/mL), amiloride hydrochloride (50 uM), n-phenylanthranilic acid (0.1mM), niflumic acid (10 mM), ebselen (15 µM), amiodarone hydrochloride (10 µM), chlorpromazine HCl (0.1 mg/mL), mercury chloride (0.075 mM), and copper sulfate (100 µM). Then, cells were treated with a concentration of 0.1 mg/mL CNDs for 24 h. Cells were harvested and resuspended in PBS. Fluorescence was read at 360/460 top 400 nm in a well plate reader (Synergy 2.0).

3. Results

3.1. Characterization of CNDs

The UV-visible CNDs spectrum shows a shoulder peak at 240 nm, which is consistent with π–$\pi*$ transitions of C–C and C = C bonds in sp^2 hybrid regions. The main peak at 350 nm comes from the n–$\pi*$ transitions of C=O moieties (Figure 1A). The fluorescence emission spectra at different excitation wavelengths starting from 220 to 400 nm with 20 nm intervals were conducted. The strongest emission peak is centered at 450 nm with an excitation wavelength of 350 nm. XPS was used to examine the surface functional groups of CNDs. The XPS survey spectra demonstrates characteristic peaks corresponding to C1s (284.5 eV), O1s (531.6 eV), and N 1s (399.5 eV), which confirms the presence of C, O, and N elements. Based on the high-resolution C1s XPS spectra, four components were detected at 284.5 eV (C=C/C-C, 54.69%), 285.8 eV (C-O-C/C-OH, 19.68%), 286.8 eV (C-N, 9.83%), and 288.3 eV (C=O, 15.8%). The high-resolution N1s XPS spectrum shows three peaks at 399.5, 400.3, and 401.3 eV, which can be attributed to pyridinic, pyrrolic, and graphitic nitrogen atoms. The results of deconvolution treatment for the high-resolution O1s spectrum of the sample shows two peaks, located at 531.6 eV and 532.8 eV, respectively, which were attributed to C–OH/C–O–C and C = O.

3.2. Differentiation of THP-1 Monocytes

THP-1 monocytes were treated with 0, 1, 3, and 10 ng/mL TPA for 72 h in RPMI media so as to stimulate the cells into differentiation. Then, the media was replaced, and cells were allowed to incubate for an additional 72 h. By way of qrt-PCR, expression of CD 206 (a macrophage differentiation marker) was assessed. As demonstrated by Figure 2, an increase in the expression of CD 206 ($p < 0.05$) was observed in these TPA-treated cells.

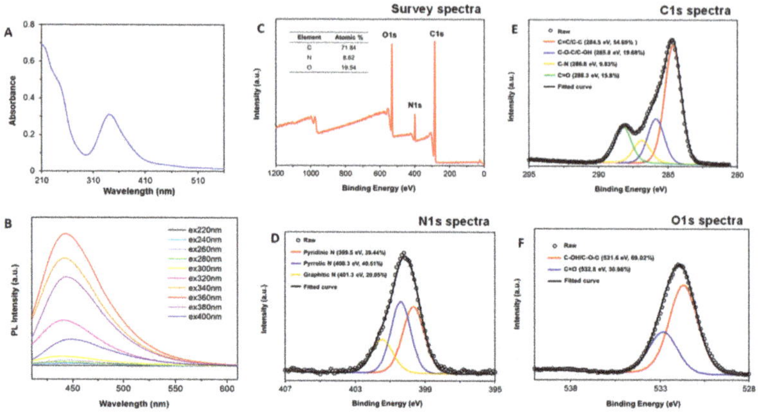

Figure 1. Characterization of CNDs. (**A**) Absorption spectrum. (**B**) CNDs show emission peak of ≈450 nm with excitation wavelengths from 220 to 400 nm. (**C–F**), X-ray photoelectron spectrum signals.

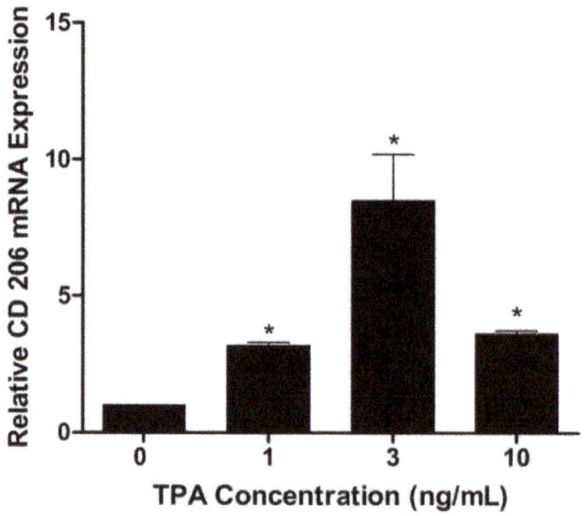

Figure 2. Increase in CD206 expression in THP-1 human monocyte-derived macrophages. THP-1 cells (3.3×10^6) were treated with 0, 1, 3, and 10 ng/mL of 12-O-tetradecanoylphorbol 13-acetate (TPA) in RPMI media for 72 h, upon which media was refreshed. Then, cells were left to incubate and mature for another 72 h. RNA was isolated, converted to cDNA, and probed for CD206 using SYBR green qRT-PCR reagents via Biosystems™ StepOnePlus™ Software v2.3. GAPDH was the housekeeping gene. All data represent mean ± SEM. ($n = 3$, *, $p < 0.05$ vs. control).

3.3. Cell Viability Determined by Trypan Blue Cell Counts and ViaCount Flow Cytometry

In order to analyze the effect of CNDs on the viability of THP-1 monocyte-derived macrophages, cell counts in a hemocytometer were performed using Trypan Blue and ViaCount Flow Cytometry. The general concept behind the cell counts is that Trypan Blue can enter cells with a compromised membrane [12]. ViaCount reagent demonstrates effects on cell viability by using two DNA-binding dyes. One stains DNA in all cells, the other specifically binds to DNA in dead cells. THP-1 cells were treated with CNDs concentrations ranging from 0.01 to 0.6 mg/mL for 72 h. After refreshing media, and an additional incubation period of 72 h, cells were analyzed with both methods. The Trypan Blue cell counts demonstrate a significant decrease ($p < 0.05$) in cell viability only at a concentration of 0.6 mg/mL (Figure 3). Flow cytometric analysis demonstrates a significant reduction ($p < 0.05$) in cell viability at 0.6 mg/mL CNDs (Figure 4), and also that the percentage of live cells in the upper left quadrant only differ significantly ($p < 0.05$) between untreated cells and cells treated with the same CNDs concentration (Figure 4).

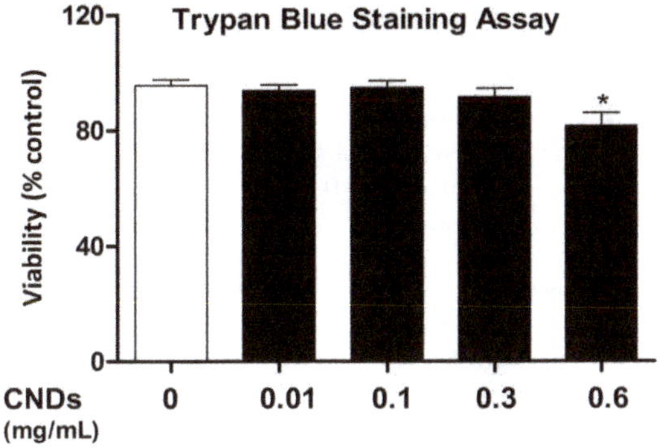

Figure 3. Effect of CNDs on cell viability (Trypan Blue). THP-1 cells were treated with 3 ng/mL TPA in the presence or absence of 0.01, 0.1, 0.3, or 0.6 mg/mL CNDs in RPMI media for 72 h, upon which media was refreshed. Then, cells were left to incubate for another 72 h. Cells were harvested, and a cell count performed using a hemocytometer and Trypan Blue. All data represent mean ± SEM ($n = 3$, *, $p < 0.05$ vs. control).

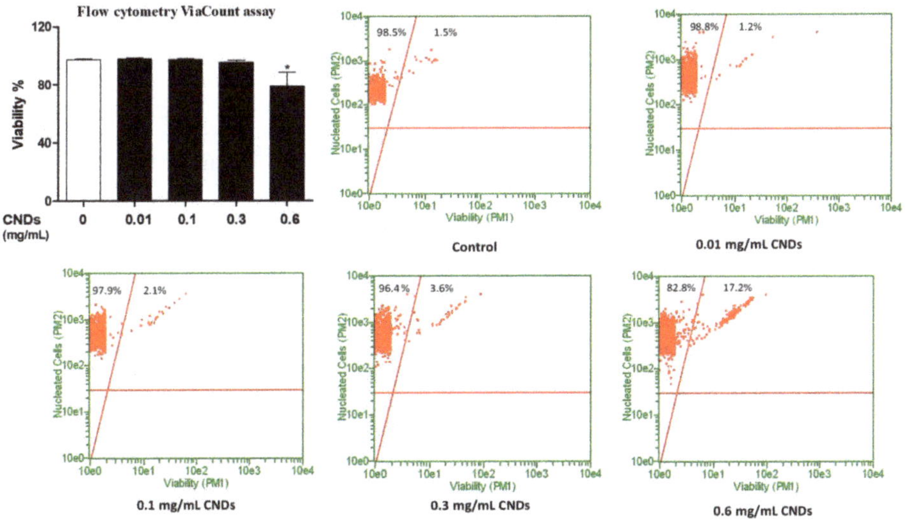

Figure 4. Effect of CNDs on cell viability (ViaCount). THP-1 cells were treated with 3 ng/mL TPA in the presence or absence of 0.01, 0.1, 0.3, or 0.6 mg/mL CNDs in RPMI media for 72 h, upon which media was refreshed. Then, cells were left to incubate for another 72 h, after which cells were harvested and treated with ViaCount reagent. Then, a viability analysis was performed using a Guava® easyCyte™ Flow Cytometer (Single Sample System). All data represent mean ± SEM. ($n = 3$, *, $p < 0.05$ vs. control).

3.4. Expression of M1/M2 Biomarkers in Macrophages as Affected by CNDs

M1 (pro-inflammatory) macrophages play a crucial intermediary role in the atherosclerosis disease state. The effect of CNDs on the expression of M1 or M2 biomarkers was analyzed by PCR. THP-1 monocytes were co-treated with 3 ng/mL TPA and 0.1 mg/mL CNDs for 72 h. Then, these cells had their media refreshed, followed by an additional incubation period of 72 h. Afterwards, cells were harvested, RNA isolated, cDNA synthesized, and analyzed for expression of genes by qrt-PCR.

As previously mentioned, CD206 is a recognized M2 biomarker. IL-10 cytokine, which suppresses the immune response, was also analyzed as an M2 biomarker [13]. A selection of M1 biomarkers was included in the analysis. IL-8 and TNF-α are all well-established pro-inflammatory cytokines, and they are also regarded as M1 biomarkers. CCL2 serves as a macrophage chemoattractant [14,15]. CD68 is a surface receptor classified as an M1 biomarker. Our results indicate a significant increase ($p < 0.05$) in CD 206, CD 68, and CCL2 expression in cells treated with 0.1 mg/mL CNDs. No significant effect was observed in the expression of TNF-alpha, IL-8, and IL-10 (Figure 5).

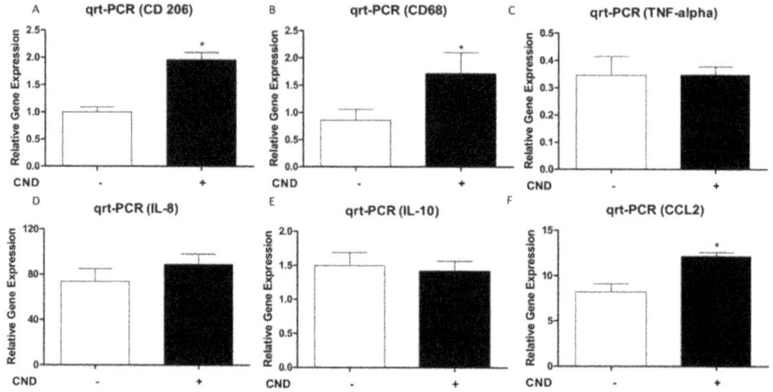

Figure 5. Effect of CNDs on expression of M1/M2 biomarkers in macrophages. THP-1 cells (1×10^6) were treated with 3 ng/mL TPA in the presence or absence of 0.1 mg/mL CNDs in RPMI media for 72 h, upon which media was refreshed. Then, cells were left to incubate and mature for another 72 h. RNA was isolated, converted to cDNA, and probed for CD206 (**panel A**), CD68 (**panel B**), TNF-alpha (**panel C**), IL-8 (**panel D**), IL-10 (**panel E**), and CCL2 (**panel F**) using SYBR green qRT-PCR reagents via an Applied Biosystems™ StepOne™ Real-Time PCR System. GAPDH was the housekeeping gene. All data represent mean ± SEM. (n = 4–6, *, $p < 0.05$ vs. control).

3.5. CNDs Effect on the Phagocytic Activity of Macrophages

Phagocytic activity is an essential function of macrophages. Macrophages that absorb an excess of ox-LDL turn into foam cells, which are the main component of the necrotic plaque that settles on an artery bed [16]. In order to analyze the effect of CNDs on the phagocytic function of THP-1 monocyte-derived macrophages, THP-1 cells were co-treated with 3 ng/mL TPA and with or without CNDs at 0.1 mg/mL with similar incubation periods as denoted previously. Cells were harvested and incubated in a 96-well plate for 18 h. Before this incubation period, a sample of control cells was treated with 1000 ng/mL TPA so as to activate macrophages (positive control). Next, cells were treated for 2 h with a suspension of fluorescent-labeled *Escherichia coli*. Lastly, cells were treated with a Trypan Blue suspension for 1 min before analysis in a plate reader. Our results indicate that THP-1 monocytes treated with 0.1 mg/mL CNDs during the differentiation process exhibit a significant increase ($p < 0.05$) in phagocytic activity (Figure 6).

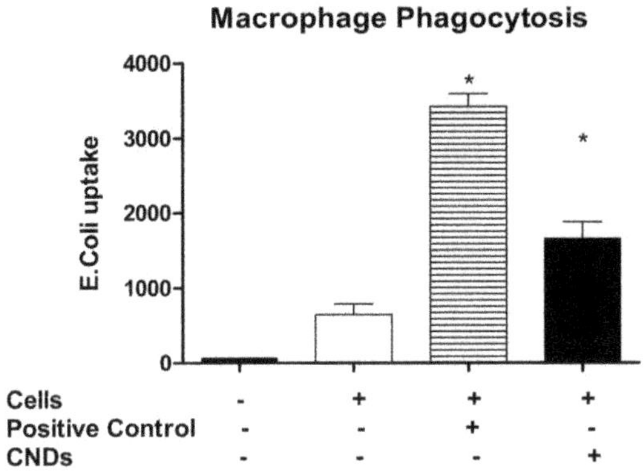

Figure 6. Increase in phagocytic activity in CND co-treated cells. THP-1 cells (2×10^6) were treated with 3 ng/mL TPA in the presence or absence of 0.1 mg/mL CNDs. Cells were incubated for a period of 72 h, upon which media was refreshed, and followed by another incubation period of 72 h. Cells were harvested, and a sample of control cells were treated with 1000 ng/mL TPA to serve as a positive control. Cells were distributed in a 96-well plate and left to incubate for 18 h. Treatment of cells with fluorescent *E. coli* suspension followed for 2 h, upon which the suspension was removed. Cells were finally treated with Trypan Blue. The removal of Trypan Blue preceded the reading in a BioTek™ Synergy 2.0 plate reader. All data represent mean ± SEM. ($n = 5$, *, $p < 0.05$ vs. control).

3.6. Potential Uptake Routes of CNDs into Macrophages

In order to exert intracellular effects, xenobiotics often need to cross the plasma membrane. Nanoparticles are an example of xenobiotics, and recently, uptake routes have become characterized. Hara et al. demonstrated that pre-treatment of THP-1 monocyte-derived macrophages with cytochalasin D, a potent inhibitor of actin polymerization, led to a decrease in the uptake of nano-silica particles [17]. This supports the notion that nanoparticles may mainly enter the cell through phagocytosis.

In order to characterize potential uptake routes of CNDs into macrophages, THP-1 monocytes with 3 ng/µL TPA with incubation periods were differentiated as described previously. Treatment with or without a variety of chemical inhibitors (Table 2) for 30 min ensued (with the exception of mercury chloride for 15 min) before treating cells with 0.1 mg/mL CNDs. Then, cells were harvested, placed in a 96-well plate, and analyzed for fluorescence in a plate reader. Our results indicate significant inhibition ($p < 0.05$) of CNDs uptake with the use of Nocodazole, mercury chloride, and N-phenylanthranilic acid (Figure 7a–c). All other inhibitors used did not show a significant inhibition in CNDs uptake (Figure 8a–k).

Table 2. Inhibitors used for potential uptake routes of CNDs.

Inhibitor Name	Abbrev	Function
4-Aminopyridine ~98%	$C_5H_6N_2$	Ion channel blocker (K^+) [18]
Amiodarone Hydrochloride	Amiodarone HCL	Non-selective ion channel blocker [19]
Barium Chloride Anhydrous	$BaCL_2$	Ion channel blocker (K^+) [18]
Chlorpromazine HCL	Chlorpromazine HCL	Suppresses clathrin disassembly [20,21]
Cobalt (II) Chloride	$CoCL_2$	Ion channel blocker (Ca^+) [22]
Copper Sulfate	Cu	hAQP3 Aquaporins [23]
Cytochalasin A	Cyt	Actin disruptor [21]
Ebselen	Ebselen	Inhibits mammalian H^+, K^+–ATPase [24]
Genstein	Genstein	Inhibits tyrosine kinase receptors [21]
Mercury Chloride	Mercury Chloride	hAQPI Aquaporins [23]
N-Phenlanthranilic Acid	N-Phen	Ion channel blocker (Cl^-) [25]
Niflumic Acid	Niflumic Acid	Ion channel blocker (Cl^-)
Nocodazole	Phenylglyoxal	Actin and microtubule disruptor [21]
Phenylglyoxal	Phenylglyoxal	Selective inhibitor of phagocytosis [26]

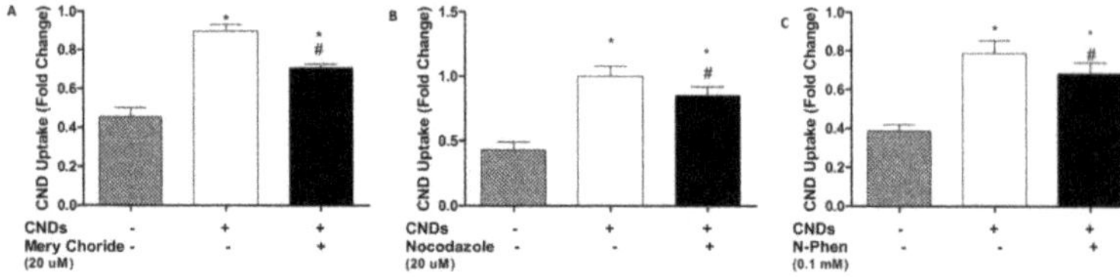

Figure 7. Effect of mercury chloride (**A**), nocodazole (**B**) and N-phen (**C**) on the uptake of CNDs into macrophages. THP-1 human monocyte-derived macrophages were treated for 30 min with or without inhibitors (for 15 min). Next, cells were treated with 0.1 mg/mL CNDs for 24 h. Lastly, cells were harvested and placed in a 96-well plate. Fluorescence analysis ensued in a BioTek™ Synergy 2.0 plate reader. All data represent mean ± SEM. (n = 4, *, $p < 0.05$ vs. control, #, $p < 0.05$ vs. CND treatment only).

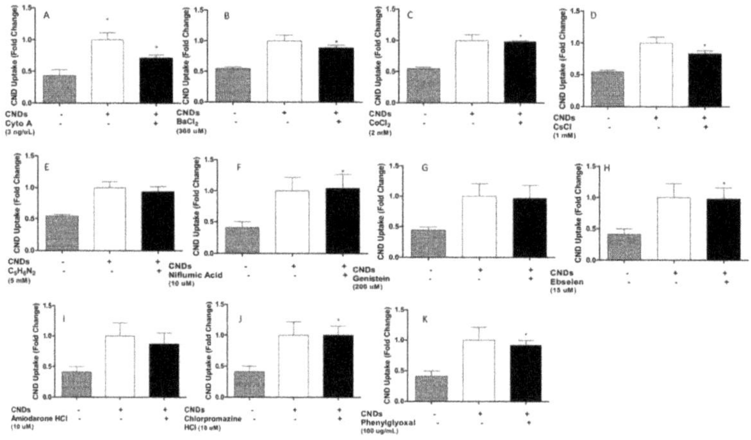

Figure 8. Chemical inhibitors' effect on uptake of CNDs into macrophages (**Panel A**: cyto A; **Panel B**: $BaCl_2$; **Panel C**: $CoCl_2$; **Panel D**: CsCl; **Panel E**: $C_5H_6N_2$; **Panel F**: Niflumic Acid; **Panel G**: Genistein; **Panel H**: Ebselen; **Panel I**: Amiodarone HCl; **Panel J**: Chlorpromazine HCL; **Panel K**: Phenylglyoxal). THP-1 monocyte-derived macrophages (1×10^6) were treated for 30 min with or without inhibitors. Next, cells were treated with 0.1 mg/mL CNDs for 24 h. Lastly, cells were harvested and placed in a 96-well plate. Fluorescence analysis ensued in a BioTek™ Synergy 2.0 plate reader. All data represent mean ± SEM. (n = 3, *, $p < 0.005$ vs control).

With data indicating inhibition in CNDs uptake of cells treated with nocodazole, n-phenylanthranilic acid, and mercury chloride, the effects of these inhibitors on cell viability were determined using flow cytometry. Cells were differentiated as previously described. Treatment with the previously mentioned inhibitor concentrations followed. After refreshing culture media (so as to remove the presence of the inhibitors), cells were left to incubate for 24 h. Cell viability was tested using previously mentioned Trypan Blue cell count and ViaCount protocols. Our results indicate no significant decrease in the viability of macrophages at any designated concentration of each inhibitor in both the Trypan Blue (Figure 9a) and ViaCount analyses (Figure 9b,c). Representative flow cytometric analysis demonstrates no change in the percentage of live cells present in the upper left quadrant for any cells treated with inhibitors (Figure 9c).

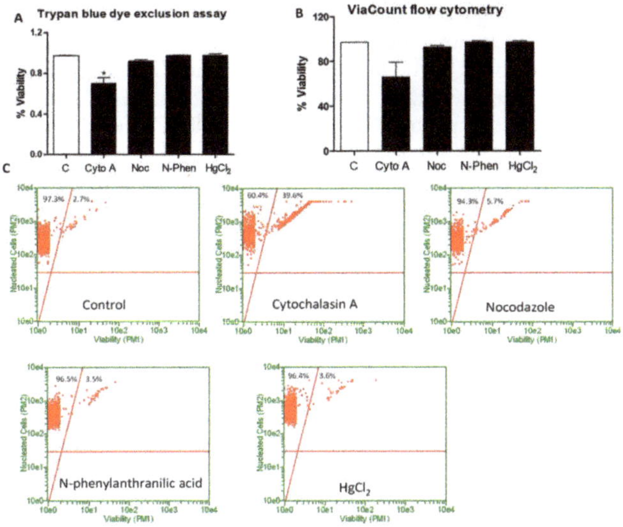

Figure 9. Effect of chemical inhibitors on the viability of cells (ViaCount). THP-1 monocyte-derived macrophages (1×10^6) cultured as mentioned previously, then harvested, resuspended in HBSS, and treated for 30 min with cytochalasin A (3 µg/mL), nocodazole (20 mM), N-phenylanthranilic acid (0.1 mM), or mercury chloride (0.075 mM) for 15 min. With media refreshed. **Panel A**: Then, cells were incubated for a period of 24 h, after which a viability analysis was performed with a hemocytometer cell count using Trypan Blue. **Panel B,C**, cells were incubated for a period of 24 h before treatment with ViaCount reagent. Then, a viability analysis was performed using a Guava® easyCyte™ Flow Cytometer (Single Sample System). All data represent mean ± SEM. (n = 3, *, $p < 0.05$ vs. control).

4. Discussion

Macrophages play an important role as mediators of atherosclerosis. For this reason, they are highly sought targets when studying the disease state. CNDs are recently discovered carbon-based nanomaterials reported to have sizes of 10 nm or less, and they also exhibit favorable qualities for use in biomedical application [27]. Collectively, our study represents an initiatory attempt at understanding the interactions of CNDs and macrophages involved in atherosclerosis. Our analysis included studying changes in macrophage biomarker expression. In addition, we studied the effect of CNDs on the phagocytic activity of macrophages. Lastly, we investigated possible uptake routes of this nanoparticle.

Macrophages play a crucial intermediary role in the atherosclerosis disease state. The overabundance of settling macrophages and foam cells, due to an excess of lipoprotein ingestion, leads to the emergence of plaque. These macrophages exacerbate the inflammatory microenvironment by secreting pro-inflammatory cytokines to different cell types [28]. A side effect of this process is an excessive dysregulation of macrophage polarization, causing circulating monocytes to differentiate into pro-inflammatory macrophages (M1) in abundance.

As a model, THP-1 human monocyte-derived macrophages were utilized. These monocytes exhibit a homogenous genetic background and differentiate into adheren macrophages upon exposure to 12-O-tetradecanoylphorbol-13-acetate (TPA). The cell line is resembling of primary monocytes/macrophages, which made it an ideal model for our purposes [29]. These macrophages are characterized by an increase in the expression of scavenger receptors while simultaneously reducing LDL receptor expression [30]. Due to their ability to absorb modified lipoproteins and convert to foam cells, THP-1 monocyte-derived macrophages act as a representative model to study macrophage involvement in atherogenesis. In fact, this model has seen extensive use in recent years, appearing in several in vitro studies regarding monocyte/macrophage drug transport, signaling, and function [31]. The favorable increase in CD 206 (a macrophage biomarker) expression observed at 3 ng/μL TPA confirmed monocyte differentiation (Figure 2). With this result, and previously mentioned properties, THP-1 monocyte-derived macrophages became a useful model to analyze the effects of CNDs on the phagocytic activity of macrophages and their expression of biomarkers.

Phagocytic activity is among the most important functions of macrophages. As a form of endocytosis, phagocytosis is defined by the use of a cell membrane to engulf extracellular particles, allowing them entrance into the cell's cytoplasm. As key players of the immune system, macrophages ingest a variety of particles including microbes, modified lipids, and even dead cells entirely [32]. The phagocytic function of macrophages, as well as other roles in immunological responses, makes macrophages a popular target for therapeutic testing. Despite this popularity, no research has been committed to studying the effects of CNDs on the phagocytic activity of primary macrophages. Our study provides a novel insight into this matter. As shown in Figure 6, THP-1 monocyte-derived macrophages that were treated with 0.1 mg/mL CNDs during the differentiation process exhibit an increase in phagocytic activity.

The expressions of macrophage biomarkers, from treatment with CNDs during the differentiation process of THP-1 monocyte-derived macrophages, were analyzed to further understand if this boost in phagocytic function favors M1 or M2 polarization. Circulating monocytes that are activated through receptor-ligand binding differentiate into M1 (pro-inflammatory) or M2 (anti-inflammatory) macrophages. This is typically dependent on the immunological response in need. M1 macrophages typically eliminate xenobiotics through phagocytosis and promote the local inflammatory environment. In an atherogenic state, M1 macrophages aim to phagocytose modified lipoproteins in an effort to clear cholesterol. They also extend the inflammatory response by secreting several pro-inflammatory cytokines such as TNF-alpha, IL-1, IL-6, and IL-12 [33]. These cytokines signal additional circulating monocytes to differentiate into M1 macrophages, as well as a host of other cell types involved in immunity. In addition to the previously mentioned cytokines, M1 macrophages exhibit a variety of biomarkers. In vitro studies identify M1 macrophages by the up-regulation of certain receptors such as CD 64, 68, and 80 [34,35]. Though crucial for host defense, the functions of M1 macrophages can be expropriated during disease states, resulting in dysregulated inflammation [36]. In contrast, M2 macrophages promote tissue repair, clear cellular debris, and secrete anti-inflammatory cytokines [37]. The presence of M2 macrophages is associated with regressing plaques. Biomarkers of M2 macrophages include anti-inflammatory cytokines such as IL-4, IL-10, and IL-13, as well as a variety of surface receptors that include CD 206, CD 23, and CD 163. As a commonality, both types exhibit phagocytic function.

Cells treated with 0.1 mg/mL CNDs demonstrated a significant increase in CCL-2 and CD 68, which are both considered M1 biomarkers (Figure 5b,f). Several studies have demonstrated that M1 macrophages accumulate cholesterol via modified, atherogenic LDL (e.g., ox-LDL) as opposed to native LDL [28]. Modified LDL is internalized through phagocytosis. In the case of atherosclerosis, M1 macrophages recognize ox-LDL by means of scavenger receptors including scavenger receptor A, CD 36, and CXCL16 [28,38]. CD 68 and its mouse ortholog macrosialin have also been recognized as receptors for ox-LDL [39]. The excessive uptake of cholesterol from modified lipoproteins leads to a dysregulation of lipid metabolism within M1 macrophages. This dysregulation results in a build-up of free cholesterol, which is toxic unlike other forms such as cholesteryl ester [38]. Among the effects of free cholesterol is the activation of stress responses in the endoplasmic reticulum, which prevents the re-esterification of cholesterol. Normally, macrophages submit cholesterol through a process of esterification that permits a series of transporters to expel them from the cell [28]. Thus, ER stress promotes the build-up of free cholesterol in macrophages, which in turn furthers the creation of foam cells. This knowledge, combined with the increase in both M1 biomarkers observed, would seem to suggest CNDs tilt the polarization of macrophages toward M1.

Our results also demonstrated a significant increase in the expression of CD 206, which is a prominent M2 biomarker (Figure 5a). This receptor has functionality in the phagocytosis of different bacteria [13]. The increase observed in expression of this receptor may very well explain the increase observed in phagocytic activity, considering that CD 206 recognizes *E. coli* [40]. Additionally, CD 206 serves as a regulator of adipocyte progenitors [41]. This result seemingly counters the increase in M1 biomarkers mentioned previously and suggests polarization toward M2 phenotypes.

In recent years, interest in the development of various carbon nanoparticles in the biological application has risen. For example, diamond-like carbon (DLC) nanofilm is a promising material for application in medical implants, with high mechanical and chemical inertness and biocompatibility [42,43]. Both cell culture and animal experiments have shown that DLC coating does not cause toxicity and inflammation [44]. The colloidal solution of nanocarbon has recently been shown to inhibit bacteria's growth without affecting the viability of eukaryotic animal cells [44]. Studies by Jelinek et al. have shown that Ge-doped DLC layers with low doping levels are not cytotoxic, while for higher doping levels, Ge has been proven to be cytotoxic, which is related to the production of reactive oxygen species [42,43]. Carbon nanotubes (CNTs) are promising candidates in nanomedicine in treating various diseases [45]. However, these nanotubes have been shown to exert various toxic effects on various cells. Multi-walled carbon nanotubes (MWCNT) cause dose-dependent cytotoxicity when its concentration in THP-1 monocyte-derived macrophages is higher than 25 μg/mL, and the concentration in lung epithelial cell-derived A549 cells is higher than 100 μg/mL [46]. In our studies, trypan blue cell count and ViaCount flow cytometry assays showed that CNDs have no effect on cell viability of THP-1 monocyte-derived macrophages at concentrations that do not exceed 0.3 mg/mL (300 μg/mL). These results suggested that CNDs showed relatively low toxicity and better biocompatibility compared to carbon nanotubes.

To further deepen our understanding of the interaction of CNDs and macrophages, the final aims of this study examined potential uptake routes of CNDs into THP-1 monocyte-derived macrophages. Previous studies have denoted the involvement of actin, microtubules, and endocytic pathways in the uptake of nanoparticles: (i) Dos Santos et al. showed that the use of chlorpromazine, genistein, nocodazole, and cytochalasin A inhibited the uptake of carboxylated polystyrene nanoparticles via clathrin-mediated endocytosis in various cell lines [21,47]. Chlorpromazine suppresses clathrin disassembly and receptor recycling in the cell membrane. Genistein specifically inhibits tyrosine kinase receptors involved in calveolae-mediated endocytosis. Nocodazole and cytochalasin A disrupt microtubule and actin filaments. (ii) Park et al. demonstrated that amiloride successfully inhibited the uptake of hydrophobically modified glycol chitosan nanoparticles (HGC-NPs). Amiloride inhibits macropinocytosis by suppressing Na^+/H^+ exchange [20]. These studies suggest that nanoparticles may enter cells primarily through endocytic pathways. Nonetheless, no research has been committed to utilizing these inhibitors to characterize the potential uptake routes of CNDs into macrophages.

In addition to the previously mentioned inhibitors, our study employed a variety of chemical inhibitors designed to cover multiple cellular entrance routes. Among the list were mercury chloride ($HgCl_2$), which is known to inhibit aquaporin channels [21]. Barium chloride and 4-aminopyridine also served to block potassium channels [18]. The extensive list of inhibitors also included niflumic acid, ebselen, and phenylglyoxal. Uptake analysis demonstrated significant inhibition in the uptake of CNDs when macrophages were treated with nocodazole, N-phenylanthranilic acid, and mercury chloride ($HgCl_2$) (Figure 7). Changes were also observed with other inhibitors; however, no significant trend could be established (Figure 8). Treatment with cytochalasin A demonstrated inhibition of CNDs uptake. However, upon performing cell viability tests, it was discovered that cytochalasin A had adverse effects on macrophages (Figure 9). This likely represents the observed inhibition of CNDs uptake as a causation of cell death, which would reduce the fluorescent signal of CNDs, giving the appearance of uptake inhibition.

The observed inhibition of CNDs uptake as a result of treatment with nocodazole suggests that CNDs can gain entrance into cells through endocytic pathways. N-phenylanthranilic acid acts as a chloride channel blocker in cell membranes. The CNDs utilized in this study exhibit negatively charged surface functional groups. Given that chloride is a negatively charged molecule, the passage of CNDs through this channel has merit. This result also gives rise to an interesting notion. Though small even in the nanoparticle scale ($\approx$10 nm), CNDs are still relatively large in comparison to chloride ions ($\approx$0.2 nm). Our results suggest that depending on the surface groups tailored to CNDs, size may not be an issue in gaining entrance into cells through ion channels. Nanoparticles have demonstrated the capability of binding to carrier proteins in order to enter plant cells through aquaporins, ion channels, or endocytosis [48]. These findings explain the inhibition of uptake observed with treatment of macrophages with $HgCl_2$ and suggest that CNDs may pass through aquaporins in similar fashion.

Our results showed that CNDs are taken up by THP-1 monocyte-derived macrophages. However, in vivo, it remains unclear whether CNDs can enter macrophages and accumulate in the aorta or atherosclerotic plaque. In addition, the efficacy of the interaction of nanoparticles and their target cell is judged not just by their ability to enter a cell but also by the time it takes to be metabolized or released. It is not yet clear whether CNDs are released from macrophages or other cells or tissues and whether this nanoparticle can be metabolized into different molecules, which remains to be further examined in the future.

In summary, our results provide novel evidence of the interaction of CNDs and macrophages involved in atherosclerosis. CNDs were confirmed to be non-toxic in concentrations that do not exceed 0.3 mg/mL by performing Trypan Blue cell counts and ViaCount flow cytometry. Our PCR results indicate a significant increase in the expression of at least one M2 biomarker (CD 206) and increases in M1 biomarkers CCL2 and CD 68. Two of these biomarkers are involved in the phagocytic function of macrophages. Although no fixed conclusions can yet be assumed regarding how CNDs affect macrophage polarization, our phagocytosis assay results indicate that CNDs treatment during the differentiation process boosts phagocytic activity, which is possibly due to the scavenging of ROS. Lastly, we also determined potential cellular uptake routes of CNDs. Results showcased inhibitions of CNDs uptake in cells treated with nocodazole, n-phenylanthranilic acid, and mercury chloride, providing evidence for entrance routes in the form of endocytosis, chloride, and water channels. Due to their crucial role as mediators in atherogenesis and their involvement in several aspects of the immune response, macrophages are popular targets in vascular research and therapeutic treatment [49]. Upon injury to the endothelium of blood vessels by ox-LDL, circulating monocytes differentiate into pro-inflammatory (M1) or anti-inflammatory (M2) macrophages [49]. In progressing lesions, M1 macrophages engulf excess ox-LDL [49]. The effect of CNDs on the polarization of macrophages has not been examined. Our results would provide new information on the potential applications of novel CNDs to modulate macrophages' polarization, which is a promising treatment strategy for atherosclerosis, a chronic progressive inflammatory disease. Collectively, these results yield a deeper understanding in the interaction between macrophages involved in atherosclerosis and CNDs.

Author Contributions: Z.J. designed the experiments. A.D., K.P., S.B. and A.P. completed the experiments and processed the experimental data. A.D. and Z.J. drafted the manuscript. X.Z., S.T. and X.Y. contributed to analysis and interpretation of results. N.H.L.C., Z.Y., B.P., and J.W. contributed nanoparticle synthesis and characterization. All authors have read and agreed to the published version of the manuscript.

Funding: The work was supported in part by grants from the U.S. National Institutes of Health (1R15HL129212-01A1 and 1R15HL150664-01A1). The US National Science Foundation (NSF) (Grant #: 1832134) also provided the financial support to JW and ZY for CNDs synthesis and characterization.

Institutional Review Board Statement: "Not applicable" for studies not involving humans or animals.

Informed Consent Statement: "Not applicable" for studies not involving humans.

Data Availability Statement: The data used to support the findings of this study are available from the corresponding author upon request.

Conflicts of Interest: The authors declare that they have no conflicts of interest.

References

1. Wahab, A.; Dey, A.K.; Bandyopadhyay, D.; Katikineni, V.; Chopra, R.; Vedantam, K.S.; Devraj, M.; Chowdary, A.K.; Navarengom, K.; Lavie, C.J.; et al. Obesity, Systemic Hypertension, and Pulmonary Hypertension: A Tale of Three Diseases. *Curr. Probl. Cardiol.* **2021**, *46*, 100599. [CrossRef] [PubMed]
2. Virani, S.S.; Alonso, A.; Benjamin, E.J.; Bittencourt, M.S.; Callaway, C.W.; Carson, A.P.; Chamberlain, A.M.; Chang, A.R.; Cheng, S.; Delling, F.N.; et al. Heart Disease and Stroke Statistics-2020 Update: A Report from the American Heart Association. *Circulation* **2020**, *141*, e139–e596. [CrossRef] [PubMed]
3. Joseph, P.; Leong, D.; McKee, M.; Anand, S.S.; Schwalm, J.D.; Teo, K.; Mente, A.; Yusuf, S. Reducing the Global Burden of Cardiovascular Disease, Part 1: The Epidemiology and Risk Factors. *Circ. Res.* **2017**, *121*, 677–694. [CrossRef] [PubMed]
4. Rezaei, R.; Safaei, M.; Mozaffari, H.R.; Moradpoor, H.; Karami, S.; Golshah, A.; Salimi, B.; Karami, H. The Role of Nanomaterials in the Treatment of Diseases and Their Effects on the Immune System. *Open Access Maced. J. Med. Sci.* **2019**, *7*, 1884–1890. [CrossRef]
5. Ovais, M.; Guo, M.; Chen, C. Tailoring Nanomaterials for Targeting Tumor-Associated Macrophages. *Adv. Mater.* **2019**, *31*, e1808303. [CrossRef] [PubMed]
6. Weissleder, R.; Nahrendorf, M.; Pittet, M.J. Imaging macrophages with nanoparticles. *Nat. Mater.* **2014**, *13*, 125–138. [CrossRef] [PubMed]
7. Roy, P.; Chen, P.-C.; Periasamy, A.P.; Chen, Y.-N.; Chang, H.-T. Photoluminescent carbon nanodots: Synthesis, physicochemical properties and analytical applications. *Mater. Today* **2015**, *18*, 447–458. [CrossRef]
8. Zhang, W.; Zeng, Z.; Wei, J. Electrochemical Study of DPPH Radical Scavenging for Evaluating the Antioxidant Capacity of Carbon Nanodots. *J. Phys. Chem. C* **2017**, *121*, 18635–18642. [CrossRef]
9. Zhang, W.; Chavez, J.; Zeng, Z.; Bloom, B.; Sheardy, A.; Ji, Z.; Yin, Z.; Waldeck, D.H.; Jia, Z.; Wei, J. Antioxidant Capacity of Nitrogen and Sulfur Codoped Carbon Nanodots. *ACS Appl. Nano Mater.* **2018**, *1*, 2699–2708. [CrossRef]
10. Das, B.; Dadhich, P.; Pal, P.; Srivas, P.K.; Bankoti, K.; Dhara, S. Carbon nanodots from date molasses: New nanolights for the in vitro scavenging of reactive oxygen species. *J. Mater. Chem. B* **2014**, *2*, 6839–6847. [CrossRef]
11. Hu, Q.; Meng, X.; Choi, M.M.; Gong, X.; Chan, W. Elucidating the structure of carbon nanoparticles by ultra-performance liquid chromatography coupled with electrospray ionisation quadrupole time-of-flight tandem mass spectrometry. *Anal. Chim. Acta* **2016**, *911*, 100–107. [CrossRef]
12. Riss, T.; Niles, A.; Moravec, R.; Karassina, N.; Vidugiriene, J. Cytotoxicity Assays: In Vitro Methods to Measure Dead Cells. In *Assay Guidance Manual*; Markossian, S., Sittampalam, G.S., Grossman, A., Brimacombe, K., Arkin, M., Auld, D., Austin, C., Baell, J., Bejcek, B., Caaveiro, J.M.M., et al., Eds.; Eli Lilly and Company: Bethesda, MD, USA; Indianapolis, IN, USA, 2004.
13. Shrivastava, R.; Shukla, N. Attributes of alternatively activated (M2) macrophages. *Life Sci.* **2019**, *224*, 222–231. [CrossRef] [PubMed]
14. Cooper, A.M.; Khader, S.A. IL-12p40: An inherently agonistic cytokine. *Trends Immunol.* **2007**, *28*, 33–38. [CrossRef]
15. Ruytinx, P.; Proost, P.; Van Damme, J.; Struyf, S. Chemokine-Induced Macrophage Polarization in Inflammatory Conditions. *Front. Immunol.* **2018**, *9*, 1930. [CrossRef]
16. Yu, X.H.; Fu, Y.C.; Zhang, D.W.; Yin, K.; Tang, C.K. Foam cells in atherosclerosis. *Clin. Chim. Acta* **2013**, *424*, 245–252. [CrossRef] [PubMed]
17. Hara, K.; Shirasuna, K.; Usui, F.; Karasawa, T.; Mizushina, Y.; Kimura, H.; Kawashima, A.; Ohkuchi, A.; Matsuyama, S.; Kimura, K.; et al. Interferon-tau attenuates uptake of nanoparticles and secretion of interleukin-1beta in macrophages. *PLoS ONE* **2014**, *9*, e113974. [CrossRef]
18. Romero, F.; Palacios, J.; Jofre, I.; Paz, C.; Nwokocha, C.R.; Paredes, A.; Cifuentes, F. Aristoteline, an Indole-Alkaloid, Induces Relaxation by Activating Potassium Channels and Blocking Calcium Channels in Isolated Rat Aorta. *Molecules* **2019**, *24*. [CrossRef] [PubMed]
19. Roden, D.M. Pharmacogenetics of Potassium Channel Blockers. *Card. Electrophysiol. Clin.* **2016**, *8*, 385–393. [CrossRef]
20. Park, S.; Lee, S.J.; Chung, H.; Her, S.; Choi, Y.; Kim, K.; Choi, K.; Kwon, I.C. Cellular uptake pathway and drug release characteristics of drug-encapsulated glycol chitosan nanoparticles in live cells. *Microsc. Res. Tech.* **2010**, *73*, 857–865. [CrossRef]
21. Dos Santos, T.; Varela, J.; Lynch, I.; Salvati, A.; Dawson, K.A.; Schnur, J.M.E. Effects of Transport Inhibitors on the Cellular Uptake of Carboxylated Polystyrene Nanoparticles in Different Cell Lines. *PLoS ONE* **2011**, *6*, e24438. [CrossRef]
22. Wu, D.; Yotnda, P. Induction and testing of hypoxia in cell culture. *J. Vis. Exp.* **2011**, *54*, 2899. [CrossRef]
23. Alejandra, R.; Natalia, S.; Alicia, E.D. The blocking of aquaporin-3 (AQP3) impairs extravillous trophoblast cell migration. *Biochem. Biophys. Res. Commun.* **2018**, *499*, 227–232. [CrossRef]

24. Kjellerup, L.; Gordon, S.; Cohrt, K.O.; Brown, W.D.; Fuglsang, A.T.; Winther, A.L. Identification of Antifungal H(+)-ATPase Inhibitors with Effect on Plasma Membrane Potential. *Antimicrob. Agents Chemother.* **2017**, *61*. [CrossRef]
25. Martin, D.K.; Boneham, G.C.; Pirie, B.L.; Collin, H.B.; Campbell, T.J. Chloride ion channels are associated with adherence of lymphatic endothelial cells. *Microvasc. Res.* **1996**, *52*, 200–209. [CrossRef]
26. Wieth, J.O.; Bjerrum, P.J.; Borders, C.L., Jr. Irreversible inactivation of red cell chloride exchange with phenylglyoxal, and arginine-specific reagent. *J. Gen. Physiol.* **1982**, *79*, 283–312. [CrossRef]
27. Anwar, S.; Ding, H.; Xu, M.; Hu, X.; Li, Z.; Wang, J.; Liu, L.; Jiang, L.; Wang, D.; Dong, C.; et al. Recent Advances in Synthesis, Optical Properties, and Biomedical Applications of Carbon Dots. *ACS Appl. Bio Mater.* **2019**, *2*, 2317–2338. [CrossRef]
28. Bobryshev, Y.V.; Ivanova, E.A.; Chistiakov, D.A.; Nikiforov, N.G.; Orekhov, A.N. Macrophages and Their Role in Atherosclerosis: Pathophysiology and Transcriptome Analysis. *Biomed. Res. Int.* **2016**, *2016*, 9582430. [CrossRef] [PubMed]
29. Chanput, W.; Peters, V.; Wichers, H. THP-1 and U937 Cells. In *The Impact of Food Bioactives on Health: In vitro and ex vivo Models*; Verhoeckx, K., Cotter, P., Lopez-Exposito, I., Kleiveland, C., Lea, T., Mackie, A., Requena, T., Swiatecka, D., Wichers, H., Eds.; Springer: New York, NY, USA, 2015; pp. 147–159. [CrossRef]
30. Qin, Z. The use of THP-1 cells as a model for mimicking the function and regulation of monocytes and macrophages in the vasculature. *Atherosclerosis* **2012**, *221*, 2–11. [CrossRef] [PubMed]
31. Chanput, W.; Mes, J.J.; Wichers, H.J. THP-1 cell line: An in vitro cell model for immune modulation approach. *Int. Immunopharmacol.* **2014**, *23*, 37–45. [CrossRef] [PubMed]
32. Gordon, S.; Pluddemann, A. Tissue macrophages: Heterogeneity and functions. *BMC Biol.* **2017**, *15*, 53. [CrossRef]
33. Chistiakov, D.A.; Myasoedova, V.A.; Revin, V.V.; Orekhov, A.N.; Bobryshev, Y.V. The impact of interferon-regulatory factors to macrophage differentiation and polarization into M1 and M2. *Immunobiology* **2018**, *223*, 101–111. [CrossRef]
34. Ren, S.; Fan, X.; Peng, L.; Pan, L.; Yu, C.; Tong, J.; Zhang, W.; Liu, P. Expression of NF-kappaB, CD68 and CD105 in carotid atherosclerotic plaque. *J. Thorac. Dis.* **2013**, *5*, 771–776. [CrossRef]
35. Tarique, A.A.; Logan, J.; Thomas, E.; Holt, P.G.; Sly, P.D.; Fantino, E. Phenotypic, functional, and plasticity features of classical and alternatively activated human macrophages. *Am. J. Respir. Cell Mol. Biol.* **2015**, *53*, 676–688. [CrossRef]
36. Tugal, D.; Liao, X.; Jain, M.K. Transcriptional control of macrophage polarization. *Arter. Thromb. Vasc. Biol.* **2013**, *33*, 1135–1144. [CrossRef] [PubMed]
37. Bi, Y.; Chen, J.; Hu, F.; Liu, J.; Li, M.; Zhao, L. M2 Macrophages as a Potential Target for Antiatherosclerosis Treatment. *Neural Plast.* **2019**. [CrossRef] [PubMed]
38. Moore, K.J.; Koplev, S.; Fisher, E.A.; Tabas, I.; Bjorkegren, J.L.M.; Doran, A.C.; Kovacic, J.C. Macrophage Trafficking, Inflammatory Resolution, and Genomics in Atherosclerosis: JACC Macrophage in CVD Series (Part 2). *J. Am. Coll. Cardiol.* **2018**, *72*, 2181–2197. [CrossRef]
39. Chistiakov, D.A.; Killingsworth, M.C.; Myasoedova, V.A.; Orekhov, A.N.; Bobryshev, Y.V. CD68/Macrosialin: Not just a histochemical marker. *Lab. Investig.* **2017**, *97*, 4–13. [CrossRef]
40. Schulz, D.; Severin, Y.; Zanotelli, V.R.T.; Bodenmiller, B. In-Depth Characterization of Monocyte-Derived Macrophages using a Mass Cytometry-Based Phagocytosis Assay. *Sci. Rep.* **2019**, *9*, 1925. [CrossRef] [PubMed]
41. Nawaz, A.; Aminuddin, A.; Kado, T.; Takikawa, A.; Yamamoto, S.; Tsuneyama, K.; Igarashi, Y.; Ikutani, M.; Nishida, Y.; Nagai, Y.; et al. CD206(+) M2-like macrophages regulate systemic glucose metabolism by inhibiting proliferation of adipocyte progenitors. *Nat. Commun.* **2017**, *8*, 286. [CrossRef]
42. Jelinek, M.; Kocourek, T.; Jurek, K.; Jelinek, M.; Smolková, B.; Uzhytchak, M.; Lunov, O. Preliminary Study of Ge-DLC Nanocomposite Biomaterials Prepared by Laser Codeposition. *Nanomaterials* **2019**, *9*. [CrossRef] [PubMed]
43. Zemek, J.; Jiricek, P.; Houdkova, J.; Ledinsky, M.; Jelinek, M.; Kocourek, T. On the Origin of Reduced Cytotoxicity of Germanium-Doped Diamond-Like Carbon: Role of Top Surface Composition and Bonding. *Nanomaterials* **2021**, *11*. [CrossRef]
44. Barkhudarov, E.M.; Kossyi, I.A.; Anpilov, A.M.; Ivashkin, P.I.; Artem'ev, K.V.; Moryakov, I.V.; Misakyan, M.A.; Christofi, N.; Burmistrov, D.E.; Smirnova, V.V.; et al. New Nanostructured Carbon Coating Inhibits Bacterial Growth, but Does Not Influence on Animal Cells. *Nanomaterials* **2020**, *10*. [CrossRef] [PubMed]
45. Negri, V.; Pacheco-Torres, J.; Calle, D.; López-Larrubia, P. Carbon Nanotubes in Biomedicine. *Top. Curr. Chem.* **2020**, *378*, 15. [CrossRef] [PubMed]
46. Keshavan, S.; Andón, F.T.; Gallud, A.; Chen, W.; Reinert, K.; Tran, L.; Fadeel, B. Profiling of Sub-Lethal in Vitro Effects of Multi-Walled Carbon Nanotubes Reveals Changes in Chemokines and Chemokine Receptors. *Nanomaterials* **2021**, *11*. [CrossRef] [PubMed]
47. Ismail, M.; Bokaee, S.; Morgan, R.; Davies, J.; Harrington, K.J.; Pandha, H. Inhibition of the aquaporin 3 water channel increases the sensitivity of prostate cancer cells to cryotherapy. *Br. J. Cancer* **2009**, *100*, 1889–1895. [CrossRef] [PubMed]
48. Rico, C.M.; Majumdar, S.; Duarte-Gardea, M.; Peralta-Videa, J.R.; Gardea-Torresdey, J.L. Interaction of nanoparticles with edible plants and their possible implications in the food chain. *J. Agric. Food Chem.* **2011**, *59*, 3485–3498. [CrossRef]
49. Zhang, Z.; Tang, J.; Cui, X.; Qin, B.; Zhang, J.; Zhang, L.; Zhang, H.; Liu, G.; Wang, W.; Zhang, J. New Insights and Novel Therapeutic Potentials for Macrophages in Myocardial Infarction. *Inflammation* **2021**. [CrossRef]

nanomaterials

Article

Peptide-Functionalized Nanoparticles-Encapsulated Cyclin-Dependent Kinases Inhibitor Seliciclib in Transferrin Receptor Overexpressed Cancer Cells

Guan Zhen He [1] and Wen Jen Lin [1,2,*]

1 School of Pharmacy, College of Medicine, National Taiwan University, Taipei 10050, Taiwan; r07423010@ntu.edu.tw
2 Drug Research Center, College of Medicine, National Taiwan University, Taipei 10050, Taiwan
* Correspondence: wjlin@ntu.edu.tw; Tel.: +886-2-33668765; Fax: +886-2-23919098

Abstract: Seliciclib, a broad cyclin-dependent kinases (CDKs) inhibitor, exerts its potential role in cancer therapy. For taking advantage of overexpressive transferrin receptor (TfR) on most cancer cells, T7 peptide, a TfR targeting ligand, was selected as a targeting ligand to facilitate nanoparticles (NPs) internalization in cancer cells. In this study, poly(D,L-lactide-co-glycolide) (PLGA) was conjugated with maleimide poly(ethylene glycol) amine (Mal-PEG-NH_2) to form PLGA-PEG-maleimide copolymer. The synthesized copolymer was used to prepare NPs for encapsulation of seliciclib which was further decorated by T7 peptide. The result shows that the better cellular uptake was achieved by T7 peptide-modified NPs particularly in TfR-high expressed cancer cells in order of MDA-MB-231 breast cancer cells > SKOV-3 ovarian cancer cells > U87-MG glioma cells. Both SKOV-3 and U87-MG cells are more sensitive to encapsulated seliciclib in T7-decorated NPs than to free seliciclib, and that IC_{50} values were lowered for encapsulated seliciclib.

Keywords: seliciclib; T7 peptide; nanoparticles; TfR-overexpressed cancer cells

Citation: He, G.Z.; Lin, W.J. Peptide-Functionalized Nanoparticles-Encapsulated Cyclin-Dependent Kinases Inhibitor Seliciclib in Transferrin Receptor Overexpressed Cancer Cells. *Nanomaterials* **2021**, *11*, 772. https://doi.org/10.3390/nano11030772

Academic Editors: Fiore P. Nicoletta and Francesca Iemma

Received: 20 February 2021
Accepted: 16 March 2021
Published: 18 March 2021

1. Introduction

Cyclin-dependent kinases (CDKs) are a group of threonine/kinases and play regulatory roles in cell cycle or transcription, which requires binding of subunits known as cyclins. Nevertheless, among 21 CDKs, only a certain subset of the CDK/cyclin complex is directly involved in driving the cell cycle, namely, interphase CDKs (CDK2, CDK4 and CDK6) and a mitotic CKD (as known as CDK1) [1]. The development of cancer is characterized by mysregulated CDKs, which contribute to cell cycle defects including unscheduled proliferation, genomic instability and chromosomal instability. In addition, many cancers are uniquely dependent on specific CDKs and, hence, selectively sensitive to inhibition of them. CDK2 is a core cell-cycle component that is essentially active from the late G1-phase and throughout the S-phase; otherwise, CDK2 is expressed at a lower level in most normal tissues [2]. The rationale to target CDK2 for treatment of malignancies includes the indispensable role of CDK2 in proliferation and overexpression of its binding partners, cyclin A or E in several cancers, such as ovary cancer, breast cancer and glioma etc. [3–6]. Moreover, overexpression of cyclin E has been reported to correlate with the tumor formation in mice and poor prognosis in patients with different cancer types [7]. Owing to the exclusivity of cyclin E for CDK2 and its deregulation in some cancers, CDK2 is an attractive target in cancer therapy.

Seliciclib, also known as (R)-roscovitine or CYC202, is a broad range purine analog inhibitor, which mainly inhibits CDK1, CDK2, CDK5, CDK7 and CDK9 instead of CDK4 and CDK6. Preclinical studies have shown great anti-cancer potential of seliciclib [8–12]. However, clinical trials showed limited benefits in a series of cancer therapies [13,14]. This might be attributed to dependence of CDKs in different stages of tumor development and its rapid metabolism, which limited the maintenance of drug concentration within therapeutic window [15].

In addition to active pharmaceutical ingredients, nanomedicine provides supportive components which improve drug bioavailability as well as aid in drug protection, site-specific activation and cellular uptake [16]. The modification of nanoparticles' (NPs) surfaces to actively target the overexpressed biomolecules on the surface of a tumor provides specific binding and more efficient internalization of a drug through receptor-mediated endocytosis [17]. Cell-penetrating peptides (CPPs) and short-chain peptides as targeting ligands have been applied to facilitate nanocarriers crossing the blood brain barrier (BBB) and target to glioblastoma as well [18]. CPPs are defined as short chain peptides (no more than 30 amino acids) possessing ability not only to translocate themselves into cells but facilitate the cargo complex entering the targeting sites. The applications of CPPs involve fields of inflammation, central nervous system disorders, ocular disorders and cancer treatment [19]. In the strategy against tumor development, CPPs play an important role in circumvention of the barrier constructed by the tumor and its microenvironment. The application of CPPs in cancer therapy has drawn lots of attention in areas including triple-negative breast cancer, ovarian cancer, colorectal cancer etc. [20–22]. Recently, the cationic Tat-peptide modified nanoformulations have been demonstrated to deliver antiviral drugs as well as vaccines for treatment of SARS-CoV-2 infections [23].

Transferrin receptor (TfR) is a 180 kDa membrane glycoprotein, which can import iron by binding transferrin. TfR is classified into two subtypes, TfR1 (known as CD71) and TfR2. TfR1 is a homodimeric type II transmembrane glycoprotein expressed ubiquitously on the surface of most cells while TfR2 is mainly expressed in the liver. TfR1 is expressed on malignant cells at levels about 100-fold higher than those on normal cells, and its expression can be correlated with either tumor stage or cancer progression [24]. Targeting TfR of cancer cells promotes the delivery of therapeutic agents and blocks the natural function of the receptor leading to cancer cell death [25]. T7 peptide (HAIYPRH), composed of seven peptides, has been reported to specifically bind to TfR with high affinity (Kd of 10 nM). Due to different binding site from Tf, endogenous Tf will not compete with the uptake of T7 peptide-modified nanocarriers. Meanwhile, Tf facilitating T7 uptake in vivo has been confirmed which attracts T7 peptide being applied in a cancer-targeting drug delivery system [26–28] In this study, we took advantage of NPs and targeting ability of T7 peptide to facilitate seliciclib uptake by cancer cells and achieve a better cytotoxicity effect in TfR-overexpressed cancer cells.

2. Materials and Methods

2.1. Materials

1-(3-Dimethylaminopropyl)-3-ethylcarbodiimide hydrochloride (EDC), *N*-ethyldiisopropylamine (*N,N*-diisopropylethlamine) (DIEA, 99%), and thiazolyl blue tetrazolium bromide (MTT, 98%) were from Alfa Aesar (Echo Chemical Co., Ltd., Heysham, UK). *N*-hydroxysuccinimide (NHS, 98%) and poly(vinyl alcohol) (PVA, 88% hydrolyzed, 20,000–30,000 g/mol) were from Acros Organics Co., Inc. (Fair Lawn, NJ, USA). Poly(D,L-lactide-co-glycolide) 50:50 (PLGA, ~52,000 g/mol) was from Evonik Industries (Birmingham, AL, USA). Maleimide poly(ethylene glycol) amine (Mal-PEG-amine, 5000 g/mol) was provided by Hunan Hua Teng Pharmaceutical Co., Ltd. (Merelbeke, Belgium). FITC-NHS (MW 473.4 g/mol) was from Thermo Fisher Scientific Inc. (Hudson, NH, USA). FITC-Cys-T7 peptide (1498.71 g/mol, FITC-Cys-His-Ala-Ile-Tyr-Pro-Arg-His-OH) was from Kelowna International Scientific Inc. (Taipei, Taiwan). Anti-human CD71 (transferrin receptor) monoclonal antibody, allophycocyanin (APC) and mouse IgG1 kappa Isotype Control APC were from eBioscience, Inc. (Vienna, Austria). Seliciclib (purity > 99%) was from LC Laboratories (Woburn, MA, USA). A549, MDA-MB-231, SKOV-3 and U87-MG cell lines were from Bioresource Collection and Research Center (Hsinchu, Taiwan).

2.2. Synthesis and Characterization of Poly(D,L-Lactide-Co-Glycolide)-Poly(Ethylene Glycol) (PLGA-PEG)-Maleimide Copolymer

PLGA-PEG-maleimide was synthesized by two steps. In the first step, activation of PLGA to PLGA-NHS was performed based on previous method with modification which was subsequently conjugated with NH_2-PEG-maleimdie [29–31]. Briefly, PLGA-NHS was reacted with NH_2-PEG-maleimide (molar ratio 1:2) in chloroform at room temperature for 24 h in dark. The synthesized PLGA-PEG-maleimide was precipitated with cold methanol/ether co-solvent (1:4 v/v) and centrifuged under 4000 rpm at 4 °C for 10 min. The precipitate was re-dissolved in chloroform and precipitated by co-solvent three times. Finally, the product was dried in vacuum desiccator for 24 h. The yield and the molar mass were determined. The molar mass of PLGA-PEG-maleimide was determined by size exclusion chromatography (SEC) equipped with a refractive index detector (RI 2031 Plus, Jasco, Tokyo, Japan) and a Styragel® HR 4E column (7.8 mm × 300 mm, Waters, Milford, MA, USA). The mobile phase was high-performance liquid chromatography (HPLC)-graded chloroform and the flow rate was set at 1 mL/min at 35 °C. The copolymer was dissolved in chloroform and filtered through a 0.22 μm polytetrafluoroethylene (PTFE) syringe filter prior to injection. The polystyrene standards were used to construct the calibration curve by plotting the logarithm of the nominal molar mass versus the retention time. The molar mass of PLGA-PEG-maleimide was then calculated based on the calibration curve. The PEGylation efficiency was determined by Equation (1) based on the indicating peaks of PLGA-PEG-maleimide shown in ^{1}H nuclear magnetic resonance (NMR) spectrum.

$$\text{PEGylation efficiency (mol\%)} = \frac{\dfrac{\text{Area (3.62 ppm)}}{4 \times \frac{\text{MW of PEG}}{\text{MW of EG monomer}}}}{\dfrac{\text{Area (1.55 ppm)} + \text{Area (4.80 ppm)} + \text{Area (5.20 ppm)}}{6 \times \frac{\text{MW of PLGA}}{\text{MW of (LA monomer + GA monomer)}}}} \times 100\% \quad (1)$$

2.3. Preparation and Characterization of Seliciclib-Loaded Nanoparticles (NPs)

PLGA-PEG-maleimide copolymer was used to prepare NPs for encapsulation of seliciclib. The single emulsion solvent evaporation method was applied to prepare seliciclib-loaded NPs (seliciclib@NPs) [32]. Briefly, seliciclib and copolymer (1:5 w/w) were dissolved in dichloromethane, and seliciclib-copolymer mixture was added into pH 7.4 phosphate-buffered saline (PBS) containing 0.5% polyvinyl alcohol solution (O/W 1:10 v/v) under sonication in an ice bath followed by magnetic stir for 4 h. The residual organic solvent was eliminated by rotary evaporator under reduced pressure at 30 °C. The seliciclib@PLGA-PEG-maleimide NPs (seliciclib@PPM NPs) were collected after centrifugation and washed with water twice. To prepare seliciclib loaded T7 peptide-conjugated NPs (seliciclib@PPM NPs-Cys-T7), seliciclib@PPM NPs and Cys-T7 peptide (molar ratio 1:2) were incubated in pH 7.4 PBS for 2 h and then collected after centrifugation followed by lyopilization [33]. The yields of seliciclib-loaded NPs were calculated. The peptide conjugation efficiency was calculated by using Equation (2) where the fluorescence of fluorescein isothiocyanate (FITC)-labeled T7 peptide was determined.

$$\text{Cys} - \text{T7 peptide conjugation ratio (mol\%)} = \frac{\dfrac{\text{Conc. of FITC labeled Cys}-\text{T7 peptide}}{\text{MW of FITC labeled Cys}-\text{T7 peptide (1498 .71 g/mol)}}}{\dfrac{\text{Conc. of FITC labeled PLGA}-\text{PEG}-\text{maleimide}-\text{Cys}-\text{T7}}{\text{MW of FITC labeled PLGA}-\text{PEG}-\text{maleimide}-\text{Cys}-\text{T7 (59,700 g/mol)}}} \quad (2)$$

The particle size and zeta potential were determined by zetasizer (Nano-ZS90, Malvern Instruments, Worcestershire, UK). The amount of seliciclib encapsulated by NPs was determined by HPLC detected at 290 nm. The drug loading (DL) and encapsulation efficiency (EE) were calculated. The morphology of NPs was observed by transmission electron microscope (TEM) (Hitachi H-7650, Hitachi High-Technologies Corporation, Tokyo, Japan). The NPs suspension was dripped on copper grids (300 mesh Formvar/carbon coated) and pended for 30 s. The excess solution was removed by filter paper. The phosphotungstic acid solution (2% w/v) was added and placed for another 30 s to stain the NPs. After removing the excess fluid, the sample was proceeded for observation with

50 K and 400 K magnifications. For the stability study, the freshly prepared seliciclib@NPs was mixed with sucrose solution followed by lyophilization overnight. The lyophilized seliciclib@NPs was stored at −20 °C, and the samples were collected at day 0, 7, 14, 21 and 28 after lyophilization. Each sample was resuspended in deionized water, and the particle size, polydispersity index (PdI), and zeta potential were measured.

In vitro release of seliciclib from seliciclib@PPM NPs and seliciclib@PPM NPs-Cys-T7 was conducted in pH 7.4 PBS release medium at 37 °C. The lyophilized seliciclb@NPs were loaded in the dialysis bag (cut off MW 12,000–14,000 g/mol) and immersed in release medium under shaking at 100 rpm. The medium was withdrawn at each specific time point, and same volume of fresh release medium was added. The collected samples were subjected to centrifugation and the concentration of seliciclib in the medium was quantified by the HPLC method.

2.4. Determination of Transferrin Receptor Expression Level

In order to confirm the expression level of transferrin receptor (TfR) on the surface of cancer cells, the APC-conjugated anti-human CD71 monoclonal antibody and mouse IgG1 kappa isotype control were applied. Briefly, the cells were trypsinized from a 10 cm dish and collected, and the cell pellet was resuspended in staining buffer. The anti-human CD71 antibody or mouse IgG1 kappa isotype was added into the tube and reacted for 1 h followed by centrifugation and washing with staining buffer for three times. The cell pellet was then resuspended in staining buffer and analyzed by FACSCalibur (Becton Dickinson, Franklin Lakes, NJ, USA). A total of 10,000 events were analyzed, and the upper limit of the IgG isotype control was set no more than 1% of the events with non-specific binding. The M1 gated (%) and relative mean fluorescence intensity (relative MFI) calculated by Equation (3) were obtained.

$$\text{Relative MFI} = \frac{\left(\text{MFI}_{\text{Anti-CD71}} - \text{MFI}_{\text{IgG1 isotype}}\right)}{\text{MFI}_{\text{no treatment (unstained)}}} \tag{3}$$

2.5. Cellular Uptake Study

The cellular uptake of peptide-conjugated PPM NPs-Cys-T7 was investigated, and the peptide-free PLGA-PEG NPs (PP NPs) were served as the control group. The amount of NPs uptake by cells was determined by mean fluorescence intensity (MFI) derived from fluorescence probe FITC. MDA-MB-231 cells, SKOV-3 cells, U87-MG cells or A549 cells were uniformly seeded in 24-well plates at a density of 2 × 10^5 cells/well in medium (McCoy's 5A medium for SKOV-3 cells; Dulbecco's Modified Eagle medium (DMEM) for others) containing 10% bovine growing serum and 1% PSA, and incubated for 24 h. The PP NPs and PPM NPs-Cys-T7 were added and incubated in 5% CO_2 at 37 °C for 2 h. The cells were washed with PBS for three times and trypsinized followed by centrifugation. Finally, the cells were collected and the fluorescence intensity was measured by flow cytometer (BD FACSCalibur Becton Dickinson, Franklin Lakes, NJ, USA). A total of 10,000 events were analyzed for each sample, and the cellular uptake efficiency in terms of relative mean fluorescence intensity was calculated as follows:

$$\text{Cellular uptake efficiency} = \frac{\text{MFI}_{\text{PPM NPs-Cys-T7}} - \text{MFI}_{\text{PP NPs}}}{\text{MFI}_{\text{non-treatment}}} \tag{4}$$

2.6. Fluorescence Microscopy

MDA-MB-231, SKOV-3, U87-MG and A549 cells were seeded at a density of 2 × 10^5 cells/well on coverslip in 6-well plates. After 24-h incubation, the medium was removed, and PBS was added to wash the cells, further replaced by a final concentration of 1 mg/mL of NPs in free DMEM medium except McCoy's 5A medium for SKOV-3 cells. After incubation at 37 °C in 5% CO_2 for 2 h, cells were washed with cold PBS and fixed with cold methanol. After washing with PBS, the nucleus was subsequently stained with

4′,6-diamidino-2-phenylindole (DAPI), followed PBS washing cycle. Finally, coverslip was covered on the slide with Fluoromount gel. The sample slides were imaged by a fluorescence microscope (Zeiss AxioImager. A1, Jena, Germany).

2.7. Cytotoxicity of Seliciclib@NPs (Nanoparticles)

MDA-MB-231 cells, SKOV-3 cells, U87-MG cells or A549 cells were uniformly seeded in 96-well plates at a density of 1×10^4 cells/well in medium (McCoy's 5A medium for SKOV-3 cells; DMEM medium for others) containing 10% bovine growing serum and 1% PSA. After 24 h incubation, the medium was replaced with various concentrations of free seliciclib or seliciclib@NPs (1–50 μg/mL). The cells were further incubated in 5% CO_2 at 37 °C for 48 h. Subsequently, the MTT solution was added into each well for another 4-h incubation. Finally, the supernatant was removed, and dimethyl sulfoxide (DMSO) was added to dissolve the formazan crystal. The absorbance was measured at 570 nm and 690 nm by a microplate reader (SpectraMax Paradigm, Molecular Devices, San Jose, CA, USA) and cytotoxicity was interpreted as follows:

$$\text{Cytotoxicity } (\%) = \left[1 - \frac{[OD_{570\ nm} - OD_{690\ nm}]_{sample}}{[OD_{570\ nm} - OD_{690\ nm}]_{control}}\right] \times 100\% \quad (5)$$

2.8. Statistical Analysis

All statistical analysis was conducted by SigmaPlot 12.5 (Softhome International, Inc., Taipei, Taiwan). One-way analysis of variance (ANOVA) and unpaired Student's *t*-test were applied, and the statistical significance was defined as $p < 0.05$.

3. Results and Discussion

3.1. Characterization of PLGA-PEG-Maleimide Copolymer

Figure 1A illustrates the ^1H NMR spectrum of PLGA-PEG-maleimide. The signals at δ 1.55 ppm, δ 4.80 ppm, and δ 5.20 ppm derived from PLGA represent -CH3 protons of lactide (a), -CH2 protons of glycolide (b), and -CH proton of lactide (c), respectively. On the other hand, the signal at δ 3.62 ppm represents -CH2- protons of PEG-maleimide (d). All of these indicating peaks in the ^1H NMR spectrum imply the successful conjugation of PEG-maleimide onto PLGA. The yield of synthesized PLGA-PEG-maleimide was 53.4 ± 3.4%. The molar mass of PLGA-PEG-maleimide was determined by size exclusion chromatography (SEC) (Figure 1B). The weight-average molar mass (M_w), number-average molar mass (M_n), and dispersity of synthesized PLGA-PEG-maleimide were 59,700 ± 2600 g/mol, 32,000 ± 1900 g/mol and 1.87 ± 0.03, respectively. The M_w increased approximately 8000 g/mol after conjugation with PEG compared to commercial PLGA, and the PEGylation efficiency was 60.4 ± 4.0 mol%.

3.2. Characterization of Seliciclib-Loaded NPs

PLGA-PEG-maleimide NPs (PPM NPs) were prepared followed by functionalized with T7 peptide via maleimide-thiol linkage, and the peptide conjugation efficiency was 26.9 ± 4.8 mol%. The anticancer drug, seliciclib, was encapsulated by peptide-free and peptide-conjugated NPs, respectively, and the characteristics of seliciclib@PPM NPs and seliciclib@PPM NPs-Cys-T7 are summarized in Table 1. The particle sizes of seliciclib@PPM NPs and seliciclib@PPM NPs-Cys-T7 were 115.7 ± 5.5 nm and 127.3 ± 0.7 nm with narrow size distribution (PdI 0.11 ± 0.03 and 0.19 ± 0.03, respectively). The increasing zeta potential from −30.8 ± 9.2 mV to −20.0 ± 4.2 mV was observed after conjugation of T7 peptide. The encapsulation efficiency (EE) of seliciclib was 64.8 ± 3.7% for seliciclib@PPM NPs and 60.0 ± 1.2% for seliciclib@PPM NPs-Cys-T7, and the corresponding drug loading (DL) was 14.9 ± 1.0% and 12.3 ± 0.5%, respectively. The TEM images illustrate these particles were separated with spherical shape (Figure 2). The size of seliciclib@NPs observed in TEM images was approximately 90 nm which was slightly smaller than the ones measured by dynamic light scattering (DLS) method (115.7 nm to 127.3 nm). It might be attributed to the difference in sample preparation process. For DLS method, the NPs suspension was diluted with deionized water followed by measurement directly. In contrast, the NPs sample for TEM imaging was subjected to the drying process to retain the particles on the mesh for observation. This dehydration process induced the shrinkage of outer PEG hydrophilic layer and smaller size of particles was observed.

(A)

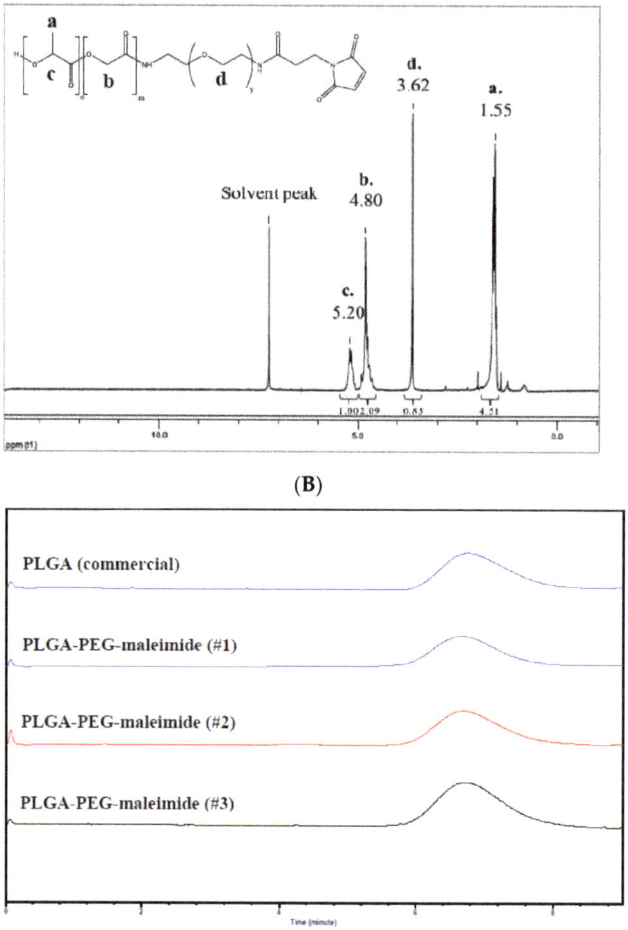

Figure 1. (**A**) ^{1}H nuclear magnetic resonance (NMR) spectrum and (**B**) size exclusion chromatograms of poly(D,L-lactide-co-glycolide)-poly(ethylene glycol)-maleimide (PLGA-PEG-mal, n = 3).

Table 1. The characteristics of seliciclib@PLGA-PEG-maleimide nanoparticles (seliciclib@PPM NPs) and seliciclib@PPM NPs-Cys-T7.

	Seliciclib@PPM NPs	Seliciclib@PPM NPs-Cys-T7
Particle size (nm)	115.7 ± 5.5	127.3 ± 0.7
PdI	0.11 ± 0.03	0.19 ± 0.03
Zeta potential (mV)	−30.8 ± 9.2	−20.0 ± 4.2
Yield (%)	72.5 ± 3.6	81.3 ± 1.7
EE (%)	64.8 ± 3.7	60.0 ± 1.2
DL (%)	14.9 ± 1.0	12.3 ± 0.5
Peptide conjugation (mol%)	-	26.9 ± 4.8

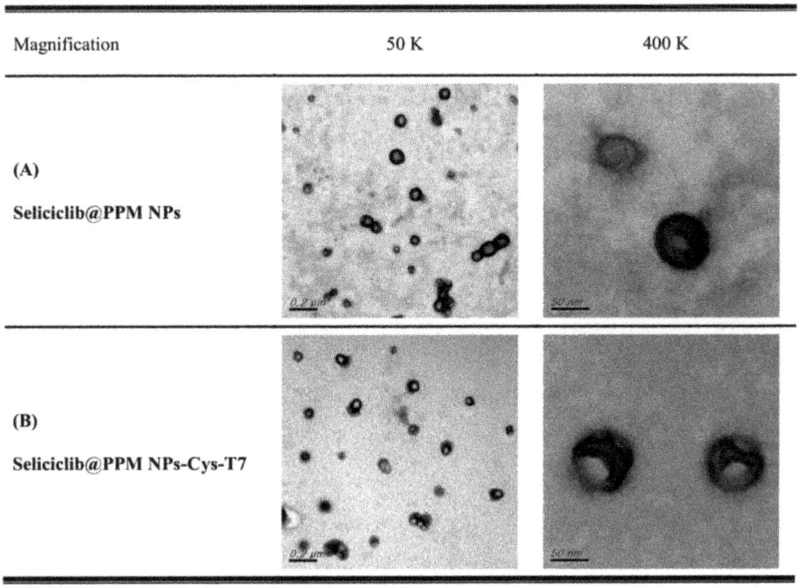

Figure 2. Transmission electron microscope (TEM) images of (**A**) seliciclib@PPM NPs and (**B**) seliciclib@PPM NPs-Cys-T7 with magnification 50 K (scale bar: 200 nm) and 400 K (scale bar: 50 nm).

3.3. *Stability and Release of Seliciclib@NPs*

The lyophilized NPs were stored at −20 °C for 28 days. The samples were collected at specific time points and re-dispersed in deionized H_2O for particle size, PdI, and zeta potential measurement. All seliciclib@PPM NPs and seliciclib@PPM NPs-Cys-T7 maintained their particle size during storage (Figure 3A) and the final to initial size ratios (S_f/S_i) were within 5% (ranging from 0.96 to 0.99) with PdIs below 0.25 (Figure 3B) indicating no aggregation occurred. In addition, there was no obvious change in zeta potentials of NPs after lyophilization and reconstitution as well (Figure 3A). All of these results implied the great stability of lyophilized seliciclib@PPM NPs and seliciclib@PPM-Cys-T7 NPs in these storage conditions following reconstitution with deionized water.

The release of seliciclib from NPs under simulated physiological conditions was demonstrated in pH 7.4 PBS. Figure 3C illustrates the in vitro release of seliciclib from seliciclib@PPM NPs and seliciclib@PPM NPs-Cys-T7 in pH 7.4 PBS release medium at 37 °C. There were 81.9 ± 10.7% and 79.9 ± 3.0% of drug released from seliciclib@PPM NPs and seliciclib@PPM NPs-Cys-T7, respectively, within 96 h, and the corresponding t_{50} values (the time for 50% of seliciclib released) were 12.8 ± 4.6 and 13.2 ± 0.8 h. These results indicate that the conjugation of peptide onto NPs did not affect seliciclib release and the similar t_{50} values were observed in seliciclib@PPM NPs and seliciclib@PPM NPs-Cys-T7.

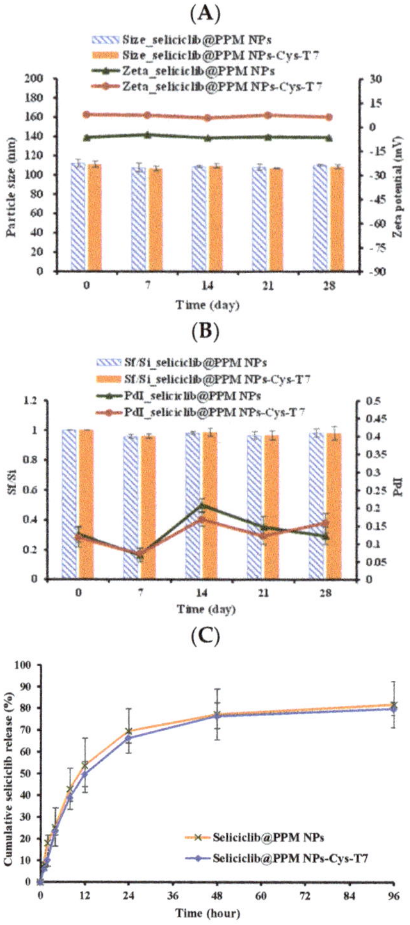

Figure 3. Stability of lyophilized seliciclib@NPs for 28 days. (**A**) Particle sizes (bar) and zeta potentials (line). (**B**) Final to initial size ratio (S_f/S_i) (bar) and PdIs (line). (**C**) Cumulative release of seliciclib from seliciclib@NPs in pH 7.4 phosphate-buffered saline (PBS) release medium at 37 °C. (n = 3, mean ± SD).

3.4. Cellular Uptake of NPs

Since TfR plays an important role in dominating cellular uptake of T7 peptide-modified NPs, the TfR expression levels in MDA-MB-231 breast cancer cells, SKOV-3 ovarian cancer cells, U87-MG glioma cells, and A549 non-small cell lung cancer cells were determined first. The degree of transferrin receptor expression was defined as low expression with 0–30% of M1 gated, medium expression with 30–60% of M1 gated and high expression with 61–100% of M1 gated [34]. Figure 4A shows MDA-MB-231, U87-MG, and SKOV-3 cancer cells had a high expression of TfR with 99.83%, 99.91% and 99.72% of M1 gated, respectively. On the other hand, A549 cancer cells had a low expression of TfR with 2.42% of M1 gated. Among three highly TfR-expressive cancer cells, MDA-MB-231 cells exhibited the highest relative MFI of 301.8 compared to 114.1 for SKOV-3 cells and 116.7 for U87-MG cells (Figure 4B).

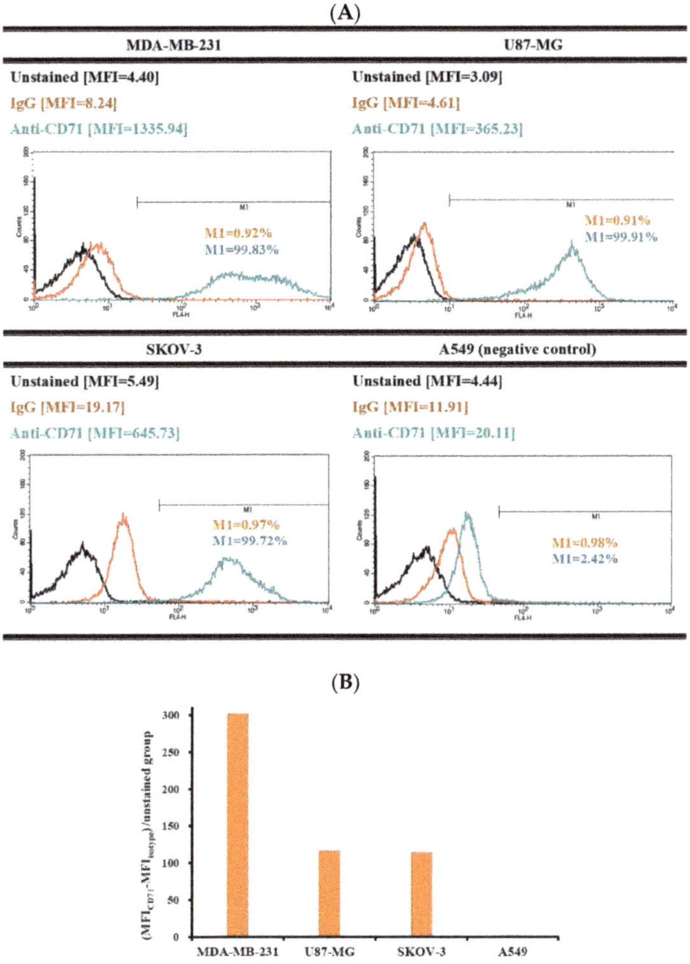

Figure 4. (**A**) The cytometric histograms of various cancer cells stained with anti-human CD71 antibody (green line) and isotype IgG control (orange line), the black line indicated the group without any treatment. (**B**) Transferrin receptor (TfR) expression levels expressed in relative mean fluorescence intensity (MFI) calculated by Equation (3).

The uptake of PP NPs and PPM NPs-Cys-T7 in various cancer cell lines for 2 h was analyzed by flow cytometer and the mean fluorescence intensity (MFI) was determined. Figure 5A shows that, generally, the uptake of T7-modified NPs (PPM NPs-Cys-T7) was elevated as compared to peptide-free NPs (PP NPs) in all cancer cells, especially in highly TfR-expressive MDA-MB-231 cells. In contrast, A549 cells were used as a negative control of TfR-expression and displayed the lowest MFI among these cancer cell lines. These results indicated the improvement of NPs uptake in TfR-overexpressed cancer cells was through T7 peptide. The cellular uptake efficiency, in terms of relative MFI calculated by Equation (4), of PPM NPs-Cys-T7 at 1.5 mg/mL in four cancer cell lines was further displayed in Figure 5B. Herein MDA-MB-231 cells exerted the highest cellular uptake efficiency followed by SKOV-3, U87-MG and A549 cells.

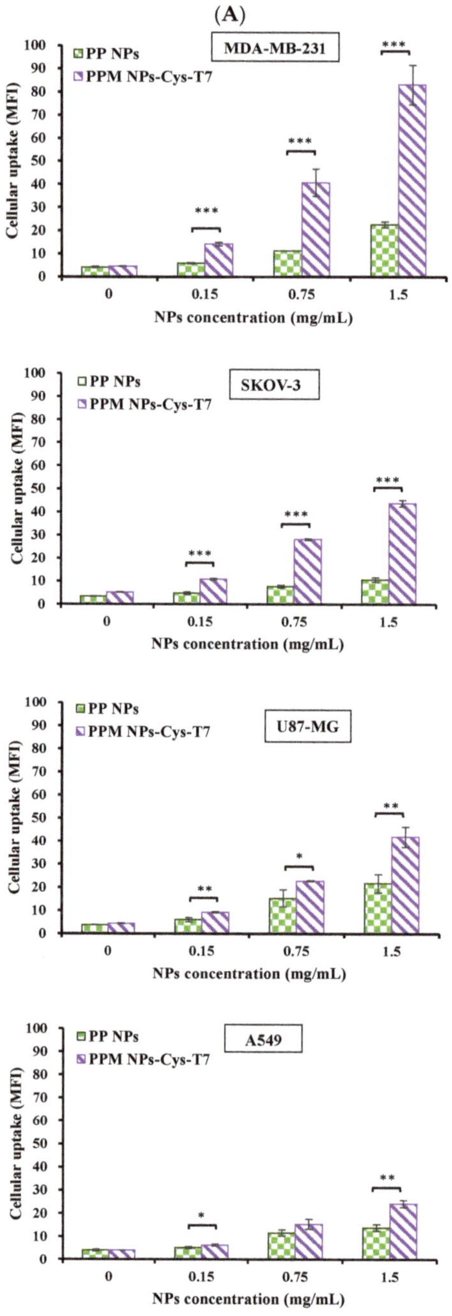

Figure 5. *Cont.*

(B)

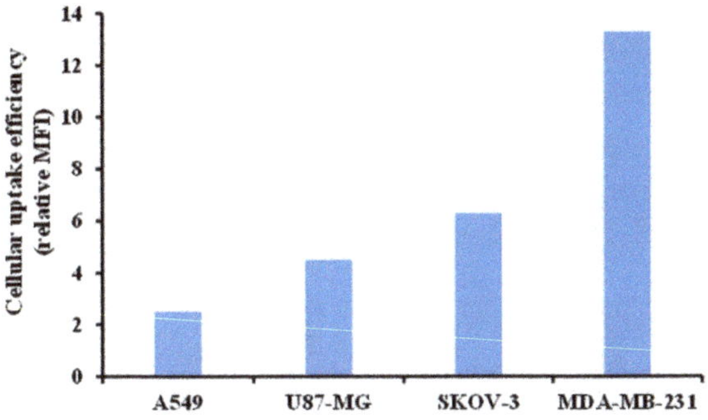

Figure 5. (**A**) Cellular uptake of PLGA-PEG (PP NPs) and PLGA-PEG-maleimide (PPM) NPs-Cys-T7 in four cancer cell lines under 5% CO_2 at 37 °C for 2 h analyzed by flow cytometer (n = 3, mean ± SD, * $p < 0.05$, ** $p < 0.01$, *** $p < 0.001$, compared to PP NPs). (**B**) Cellular uptake efficiency of PPM NPs-Cys-T7 (relative MFI calculated by Equation (3)) at 1.5 mg/mL in four cancer cell lines.

Figure 6 shows the fluorescence microscope images for cellular uptake of FITC-labeled NPs in MDA-MB-231 (Figure 6A), SKOV-3 (Figure 6B), U87-MG (Figure 6C), and A549 cells (Figure 6D), respectively. The cell nuclei, presented as blue, were stained with DAPI, and the green signals were derived from the FITC labeled NPs. The fluorescence intensity of PP NPs was very weak no matter in TfR-high expressed cancer cells (e.g., MDA-MB-231, SKOV-3, and U87-MG) or TfR-low expressed A549 cells. The images show stronger fluorescence intensity for peptide-conjugated PPM NPs-Cys-T7 than for peptide-free PP NPs, especially in highly TfR-expressed cancer cell lines (e.g., MDA-MB-231, SKOV-3, and U87-MG cells), indicating more NPs internalized into TfR-overexpressed cells via functional T7 peptide. However, there was slight difference in fluorescence intensity between PP NPs and PPM NPs-Cys-T7 in negative control A549 cells.

Since transferrin receptor plays an important role in delivery of T7 peptide-functional ized NPs into tumor cells, the correlation of TfR expression level and internalization of T7 peptide (Figure 7A) as well as cellular uptake of PPM NPs-Cys-T7 (Figure 7B) are further evaluated. It is found that the cellular uptake of PPM NPs-Cys-T7 (y) exerts high correlation to TfR expression level (×) with $R^2 \sim 0.991$ in these four cancer cells (Figure 5B). The role of T7 peptide in the presence of NPs or not is worth elucidating. The slope of the correlation line shown in Figure 5 is served as the indicator for utilization of TfR on either internalization of peptide or uptake of peptide-conjugated PPM NPs-Cys-T7. Figure 5A demonstrates the utilization of TfR with the treatment of T7 peptide alone in terms of slope 0.0182, while for the conjugation of T7 peptide onto NPs, the slope of correlation equation increased to 0.1996 (Figure 7B). This implied that PPM NPs-Cys-T7 exerts ten-fold efficient utilization of TfR as compared to T7 peptide alone. This result elucidates the important role of T7 peptide particularly in the presence of NPs. A combination of targeting peptide and nanocarriers allows PPM NPs-Cys-T7 possessing synergistic effect on delivery of seliciclib@NPs via receptor-mediated pathway to induce cancer cell apoptosis. In other words, the correct orientation of T7 peptide on NPs is feasible for recognizing TfR on receptor overexpressed tumor cells resulting in higher cellular uptake efficiency of PPM NPs-Cys-T7.

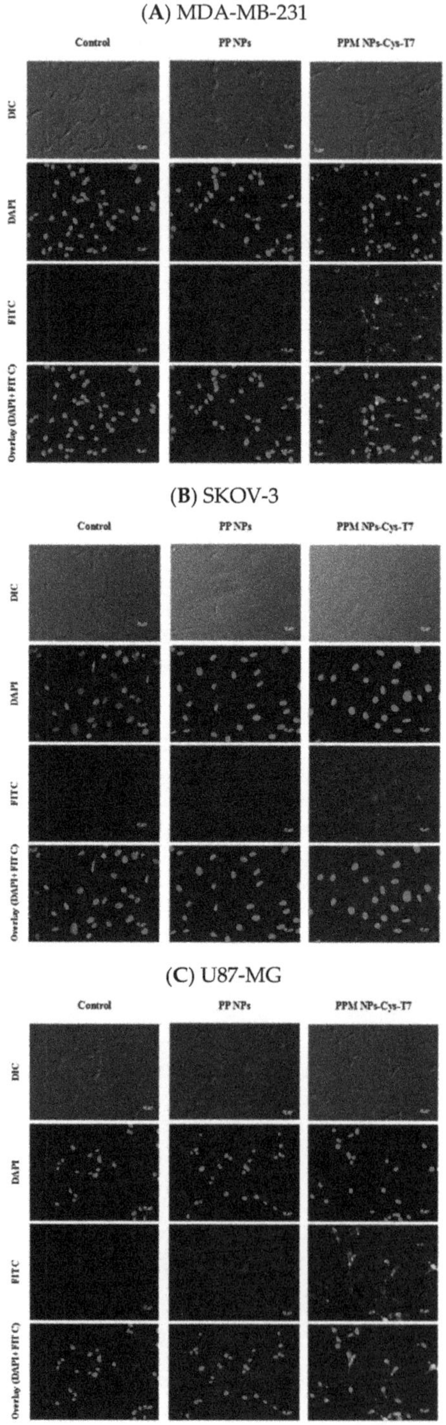

Figure 6. *Cont.*

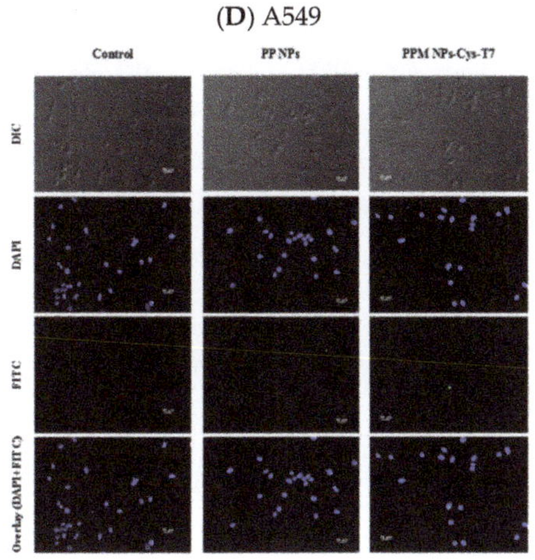

Figure 6. Fluorescence microscopic images of (**A**) MDA-MB-231, (**B**) SKOV-3, (**C**) U87-MG and (**D**) A549 cells treated with free medium only (left) and FITC-labeled PP NPs (middle) and PPM NPs-Cys-T7 (right). The blue spots indicate the nuclei stained with DAPI, and the green signals present the FITC-labeled NPs. (400×, scale bar 20 μm).

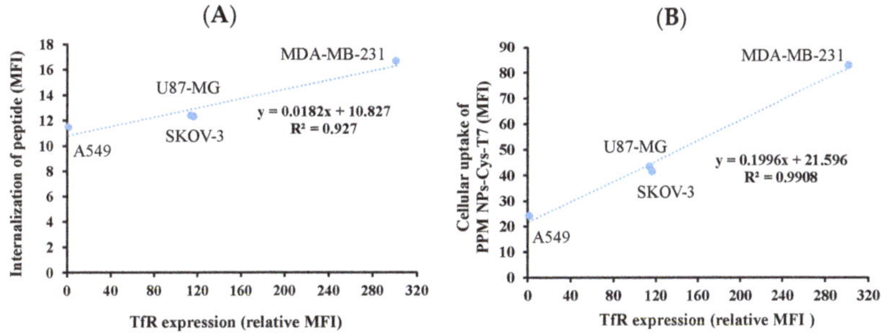

Figure 7. Correlation of TfR expression level and (**A**) internalization of T7 peptide as well as (**B**) cellular uptake of PPM NPs-Cys-T7 in four cancer cell lines.

3.5. Cytotoxicity of Seliciclib@NPs

Figure 8 shows the cytotoxicity of seliciclib, seliciclib@PPM NPs, and seliciclib@PPM NPs-Cys-T7 in MDA-MB-231, SKOV-3, U87-MG, and A549 cells for 48 h, and the corresponding IC_{50} values are summarized in Table 2. MDA-MB-231 cells exhibited high susceptibility to free seliciclib with IC_{50} 3.58 ± 0.91 μg/mL. The IC_{50} values of seliciclib@PPM NPs and seliciclib@PPM NPs-Cys-T7 in MDA-MB-231 cells were 2.49 ± 1.13 μg/mL and 2.03 ± 0.24 μg/mL, respectively, showing a decrease tendency without significant difference from that of free seliciclib. Both SKOV-3 and U87-MG cells were relatively insensitive to seliciclib with IC_{50} > 50 μg/mL. However, the IC_{50} of seliciclib@PPM NPs significantly reduced to 7.09 ± 0.25 μg/mL in SKOV-3 and 4.39 ± 0.27 μg/mL in U87-MG as compared to free seliciclib (### $p < 0.001$), which further lowered to 4.92 ± 0.19 μg/mL and 1.35 ± 0.28 μg/mL respectively by seliciclib@PPM NPs-Cy-T7 (*** $p < 0.001$). This means more efficient delivery of seliciclib by NPs, particularly the peptide-modified NPs into SKOV-3 and U87-MG cancer cells. A549 cells were not sensitive to seliciclib either with IC_{50} > 50 μg/mL. Nevertheless, the IC_{50} of seliciclib@PPM NPs significantly reduced to 3.02 ± 0.50 μg/mL as compared to free seliciclib (### $p < 0.001$) due to enhanced permeability and retention (EPR) effect of NPs. There was no significant difference in IC_{50} between seliciclib@PPM NPs and seliciclib@PPM NPs-Cys-T7 in A549, implying the conjugation of T7 peptide did not affect the cytotoxicity in cancer cells with low TfR expression.

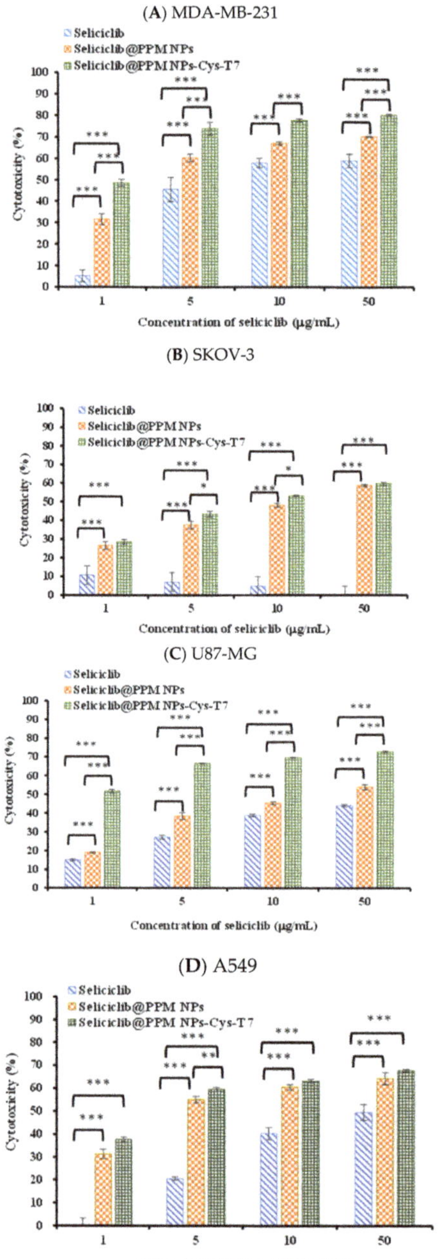

Figure 8. Cytotoxicity of seliciclib, seliciclib@PPM NPs and seliciclib@PPM NPs-Cys-T7 in (**A**) MDA-MB-231, (**B**) SKOV-3, (**C**) U87-MG and (**D**) A549 cells for 48 h. (n = 3, mean ± SD, * $p < 0.05$, ** $p < 0.01$, *** $p < 0.001$).

Table 2. The IC_{50} of seliciclib, seliciclib@PPM NPs and seliciclib@PPM NPs-Cys-T7 in four cancer cell lines for 48-h treatment. (n = 3, mean ± SD, $^{\#\#\#}$ $p < 0.001$, compared to IC_{50} of seliciclib; *** $p < 0.001$, compared to IC_{50} of seliciclib@PPM NPs).

IC_{50} (μg/mL) \ Cell Line	MDA-MB-231	SKOV-3	U87-MG	A549
Seliciclib	3.58 ± 0.91	>50	>50	>50
Seliciclib@PPM NPs	2.49 ± 1.13	7.09 ± 0.25 ###	4.39 ± 0.27 ###	3.02 ± 0.50 ###
Seliciclib@PPM NPs-Cys-T7	2.03 ± 0.24	4.92 ± 0.19 ###,***	1.35 ± 0.28 ###,***	3.09 ± 0.16 ###

4. Conclusions

To the best of our knowledge, this study was the first to encapsulate seliciclib in T7 peptide conjugated PLGA-PEG NPs (seliciclib@PPM NPs-Cys-T7) for cancer therapy. The advantage of NPs and the targeting ability of T7 peptide concurrently enhanced cellular uptake of peptide-conjugated PPM NPs-Cys-T7 in TfR-high expressed cancer cells in order of MDA-MB-231 > SKOV-3 > U87-MG. For seliciclib-loaded NPs, since MDA-MB-231 cells exhibited high susceptibility to free seliciclib, the further effects of NPs and T7 peptide on its cytotoxicity were limited. Instead of low susceptibility to free seliciclib by SKOV-3 and U87-MG cancer cells, the antitumor cytotoxicity in terms of IC_{50} was prominently lowered by seliciclib@NPs particularly the T7 peptide-conjugated seliciclib@PPM NPs-Cys-T7.

Author Contributions: W.J.L. and G.Z.H. conceived and designed the experiments; G.Z.H. performed the experiments, analyzed the data, wrote the original draft; W.J.L. supervised the project and revised and edited the original draft. All authors have read and agreed to the published version of the manuscript.

Funding: This research was funded by Ministry of Science and Technology in Taiwan [MOST 106-2320-B-002-008-MY3].

Acknowledgments: We thank the staff of the imaging core at the First Core Labs, National Taiwan University College of Medicine, for technical assistance.

Conflicts of Interest: The authors declare no conflict of interest.

References

1. Malumbres, M.; Barbacid, M. Cell cycle, CDKs and cancer: A changing paradigm. *Nat. Rev. Cancer* **2009**, *9*, 153–163. [CrossRef] [PubMed]
2. Tadesse, S.; Caldon, E.C.; Tilley, W.; Wang, S. Cyclin-dependent kinase 2 inhibitors in cancer therapy: An update. *J. Med. Chem.* **2019**, *62*, 4233–4451. [CrossRef]
3. Yang, L.; Fang, D.; Chen, H.; Lu, Y.; Dong, Z.; Ding, H.; Jing, Q.; Su, S.; Huang, S. Cyclin-dependent kinase 2 is an ideal target for ovary tumors with elevated cyclin E1 expression. *Oncotarget* **2015**, *6*, 20801–20812. [CrossRef] [PubMed]
4. Akli, S.; Van Pelt, C.S.; Bui, T.; Meijer, L.; Keyomarsi, K. CDK2 is required for breast cancer mediated by the low-molecular weight isoform of cyclin E. *Cancer Res.* **2011**, *71*, 3377–3386. [CrossRef] [PubMed]
5. Chakravarti, A.; Delaney, M.A.; Noll, E.; Black, P.M.; Loeffler, J.S.; Muzikansky, A.; Dyson, N.J. Prognostic and pathologic significance of quantitative protein expression profiling in human gliomas. *Clin. Cancer Res.* **2001**, *7*, 2387–2395.
6. Wang, J.; Yang, T.; Xu, G.; Liu, H.; Ren, C.; Xie, W.; Wang, M. Cyclin-dependent kinase 2 promotes tumor proliferation and induces radio resistance in glioblastoma. *Tansl. Oncol.* **2016**, *9*, 548–556. [CrossRef]
7. Keyomarsi, K.; Tucker, S.L.; Buchholz, T.A.; Callister, M.; Ding, Y.E.; Hortobagyi, G.N.; Bedrosian, I.; Knickerbocker, C.; Toyofuku, W.; Lowe, M.; et al. Cyclin E and survival in patients with breast cancer. *N. Engl. J. Med.* **2002**, *347*, 1566–1575. [CrossRef]
8. Maggiorella, L.; Deutsch, E.; Frascogna, V.; Chavaudra, N.; Jeanson, L.; Milliat, F.; Eschwege, F.; Bourhis, J. Enhancement of radiation response by roscovitine in human breast carcinoma in vitro and in vivo. *Cancer Res.* **2003**, *63*, 2513–2517.
9. Appleyard, M.V.; O'Neill, M.A.; Murray, K.E.; Paulin, F.E.M.; Bray, S.E.; Kernohan, N.M.; Levison, D.A.; Lane, D.P.; Thompson, A.M. Seliciclib (CYC202, R-roscovitine) enhances the antitumor effect of doxorubicin in vivo in a breast cancer xenograft model. *Int. J. Cancer* **2009**, *124*, 465–472. [CrossRef]
10. Molinsky, J.; Klanova, M.; Koc, M.; Beranova, L.; Andera, L.; Ludvikova, Z.; Bohmova, M.; Gasova, Z.; Strnad, M.; Ivanek, R.; et al. Roscovitine sensitizes leukemia and lymphoma cells to tumor necrosis factor-related apoptosis-inducing ligand-induced apoptosis. *Leuk. Lymphoma.* **2013**, *54*, 372–380. [CrossRef]
11. Nair, B.C.; Vallabhaneni, S.; Tekmal, R.R.; Vadlamudi, R.K. Roscovitine confers tumor suppressive effect on therapy-resistant breast tumor cells. *Breast Cancer Res.* **2011**, *13*, R80. [CrossRef]

12. Yakisich, J.S.; Vita, M.F.; Siden, A.; Tasat, D.R.; Cruz, M. Strong inhibition of replicative DNA synthesis in the developing rat cerebral cortex and glioma cells by roscovitine. *Inv. New Drugs* **2010**, *28*, 299–305. [CrossRef]
13. Benson, C.; White, J.; De Bono, J.D.; O'Donnell, A.; Raynaud, F.; Cruickshank, C.; McGrath, H.; Walton, M.; Workman, P.; Kaye, S.; et al. A phase I trial of the selective oral cyclin-dependent kinase inhibitor seliciclib (CYC202; R-roscovitine), administered twice daily for 7 days every 21 days. *Br. J. Cancer* **2007**, *96*, 29–37. [CrossRef]
14. Le Tourneau, C.; Faivre, S.; Laurence, V.; Delbaldo, C.; Vera, K.; Girre, V.; Chiao, J.; Armour, S.; Frame, S.; Green, S.R.; et al. Phase I evaluation of seliciclib (R-roscovitine), a novel oral cyclin-dependent kinases inhibitor, in patients with advanced malignancies. *Eur. J. Cancer* **2010**, *46*, 3243–3250. [CrossRef]
15. Aldosss, I.T.; Tashi, T.; Ganti, A.K. Seliciclib in malignancies. *Expert. Opin. Investig. Drugs.* **2009**, *18*, 1957–1965. [CrossRef] [PubMed]
16. Zhao, Y.; Xiong, S.; Liu, P. Polymeric nanoparticles-based brain delivery with improved therapeutic effcacy of ginkgolide B in Parkinson's disease. *Int. J. Nanomed.* **2020**, *24*, 10453–10467. [CrossRef] [PubMed]
17. Wolfram, J.; Ferrari, M. Clinical cancer nanomedicine. *Nano Today* **2019**, *25*, 85–98. [CrossRef] [PubMed]
18. Gallego, L.; Ceña, V. Nanoparticle-mediated therapeutic compounds delivery to glioblastoma. *Exp. Opin. Drug Deliv.* **2020**, *17*, 1541–1554. [CrossRef]
19. Xie, J.; Bi, Y.; Zhang, H.; Dong, S.; Teng, L.; Lee, R.J.; Yang, Z. Cell-penetrating peptides in diagnosis and treatment of human diseases: From preclinical research to clinical application. *Front. Pharmacol.* **2020**, *11*, 697. [CrossRef]
20. Nam, S.H.; Jang, J.; Cheon, D.H.; Chong, S.; Ahn, J.H.; Hyun, S.; Yu, J.; Lee, Y. pH-activated cell penetrating peptide dimers for potent delivery of anticancer drug to triple-negative breast cancer. *J. Control. Release* **2020**, *330*, 898–906. [CrossRef] [PubMed]
21. Massodi, I.; Moktan, S.; Rawat, A.; Bidwell, G.L., III; Raucher, D. Inhibition of ovarian cancer cell proliferation by a cell cycle inhibitory peptide fused to a thermal responsive polypeptide carrier. *Int. J. Cancer* **2010**, *126*, 533–544. [CrossRef] [PubMed]
22. Al-Husaini, K.; Elkamel, E.; Han, X.; Chen, P. Therapeutic potential of a cell penetrating peptide (CPP, NP1) mediated siRNA delivery: Evidence in 3D spheroids of colon cancer cells. *Can. J. Chem. Eng.* **2020**, *98*, 1240–1254. [CrossRef]
23. Mohammad, A.A.; Ahmad, A.; Mohammad, A.; AlYahya, S.; Alomary, M.N.; Al-Dossary, H.A.; Alghamdi, S. Lipid-based nano delivery of Tat-peptide conjugated drug or vaccine–promising therapeutic strategy for SARS-CoV-2 treatment. *Expert Opin. Drug Deliv.* **2020**, *17*, 1671–1674.
24. Han, L.; Huang, R.; Liu, S.; Huang, S.; Jiang, C. Peptide-conjugated PAMAM for targeted doxorubicin delivery to transferrin receptor overexpressed tumors. *Mol. Pharm.* **2010**, *7*, 2156–2165. [CrossRef] [PubMed]
25. Daniels, T.R.; Bernabeu, E.; Rodríguez, J.A.; Patel, S.; Kozman, M.; Chiappetta, D.A.; Holler, E.; Ljubimova, J.Y.; Helguera, G.; Penichet, M.L. The transferrin receptor and the targeted delivery of therapeutic agents against cancer. *Biochim. Biophys. Acta Gen. Subj.* **2012**, *1820*, 291–317. [CrossRef]
26. Wang, S.; Sun, H. Transferrin receptors targeting peptide (T7 peptide) surface-modified sorafenib nanoliposomes enchance the anti-tumor effect in colorectal cancer. *Pharm. Dev. Technol.* **2020**, *25*, 1063–1070. [CrossRef] [PubMed]
27. Bi, Y.; Liu, L.; Lu, Y.; Sun, T.; Shen, C.; Chen, X.; Chen, Q.; An, S.; He, X.; Ruan, C.; et al. T7 peptide-functionalized PEG-PLGA micelles loaded with carmustine for targeting therapy of glioma. *Appl. Mater. Interf.* **2016**, *8*, 27465–27473. [CrossRef] [PubMed]
28. Yu, M.Z.; Pang, W.H.; Yang, T.; Wang, J.; Wei, L.; Qiu, C.; Wu, Y.; Liu, W.; Wei, W.; Guo, X.; et al. Systemic delivery of siRNA by T7 peptide modified core-shell nanoparticles for targeted therapy of breast cancer. *Eur. J. Pharm. Sci.* **2016**, *92*, 39–48. [CrossRef]
29. Milane, L.; Duan, Z.; Amiji, M. Development of EGFR-targeted polymer blend nanocarriers for combination paclitaxel/lonidamine delivery to treat multi-drug resistance in human breast and ovarian tumor cells. *Mol. Pharm.* **2011**, *8*, 185–203. [CrossRef]
30. Lin, W.J.; Kao, L.T. Cytotoxic enhancement of hexapeptide-conjugated micelles in EGFR high-expressed cancer cells. *Expert Opin. Drug Deliv.* **2014**, *11*, 1537–1550. [CrossRef]
31. Vasconcelos, A.; Vega, E.; Perez, Y.; Gomara, M.J.; Garcia, M.L.; Haro, I. Conjugation of cell-penetrating peptides with poly(lactic-co-glycolic acid)-polyethylene glycol nanoparticles improves ocular drug delivery. *Int. J. Nanomed.* **2015**, *10*, 609–631.
32. Halevas, E.; Kokotidou, C.; Zaimai, E.; Moschona, A.; Lialiaris, E.; Mitraki, A.; Lialiaris, T.; Pantazaki, A. Evaluation of the hemocompatibility and anticancer potential of poly(ε-caprolactone) and poly(3-hydroxybutyrate) microcarriers with encapsulated chrysin. *Pharmaceutics* **2021**, *13*, 109. [CrossRef] [PubMed]
33. Martinez-Jothar, L.; Doulkeridou, S.; Schiffelers, R.M.; Torano, J.S.; Oliveira, S.; van Nostrum, C.F.; Hennink, W.E. Insights into maleimide-thiol conjugation chemistry: Conditions for efficient surface functionalization of nanoparticles for receptor targeting. *J. Control. Release* **2018**, *282*, 101–109. [CrossRef] [PubMed]
34. Tront, J.S.; Willis, A.; Huang, Y.; Hoffman, B.; Liebermann, D.A. Gadd45a levels in human breast cancer are hormone receptor dependent. *J. Transl. Med.* **2013**, *11*, 131–137. [CrossRef] [PubMed]

nanomaterials
Nanomaterials **2020**, *10*, 1919

MDPI

Article

Hadron Therapy, Magnetic Nanoparticles and Hyperthermia: A Promising Combined Tool for Pancreatic Cancer Treatment

Francesca Brero [1,*], Martin Albino [2], Antonio Antoccia [3], Paolo Arosio [4], Matteo Avolio [1], Francesco Berardinelli [3], Daniela Bettega [4], Paola Calzolari [4], Mario Ciocca [5], Maurizio Corti [1], Angelica Facoetti [5], Salvatore Gallo [4], Flavia Groppi [6], Andrea Guerrini [2], Claudia Innocenti [2,7], Cristina Lenardi [4,8], Silvia Locarno [4], Simone Manenti [6], Renato Marchesini [4], Manuel Mariani [1], Francesco Orsini [4], Emanuele Pignoli [9], Claudio Sangregorio [2,7,10], Ivan Veronese [4] and Alessandro Lascialfari [1,*]

1 Dipartimento di Fisica and INFN, Università degli Studi di Pavia, 27100 Pavia, Italy; matteo.avolio01@universitadipavia.it (M.A.); maurizio.corti@unipv.it (M.C.); manuel.mariani@unipv.it (M.M.)
2 Dipartimento di Chimica, Università di Firenze and INSTM, 50019 Sesto Fiorentino (FI), Italy; martin.albino@unifi.it (M.A.); andrea.guerrini@sns.it (A.G.); claudia.innocenti@unifi.it (C.I.); claudio.sangregorio@iccom.cnr.it (C.S.)
3 Dipartimento di Scienze and INFN, Università Roma Tre, 00146 Roma, Italy; antonio.antoccia@uniroma3.it (A.A.); francesco.berardinelli@uniroma3.it (F.B.)
4 Dipartimento di Fisica and INFN, Università degli Studi di Milano, 20133 Milano, Italy; paolo.arosio@unimi.it (P.A.); daniela.bettega@mi.infn.it (D.B.); paola.calzolari@unimi.it (P.C.); salvatore.gallo@unimi.it (S.G.); cristina.lenardi@mi.infn.it (C.L.); silvia.locarno@unimi.it (S.L.); renato.marchesini@alice.it (R.M.); francesco.orsini@unimi.it (F.O.); ivan.veronese@unimi.it (I.V.)
5 Fondazione CNAO, 27100 Pavia, Italy; mario.ciocca@cnao.it (M.C.); angelica.facoetti@cnao.it (A.F.)
6 Dipartimento di Fisica, Università degli Studi di Milano and INFN, Lab. LASA, 20090 Segrate (MI), Italy; flavia.groppi@mi.infn.it (F.G.); simone.manenti@mi.infn.it (S.M.)
7 ICCOM-CNR, 50019 Sesto Fiorentino (FI), Italy
8 C.I.Ma.I.Na., Centro Interdisciplinare Materiali e Interfacce Nanostrutturati, 20133 Milano, Italy
9 Fondazione IRCSS Istituto Nazionale dei tumori, 20133 Milano, Italy; emanuele.pignoli@istitutotumori.mi.it
10 INFN, Sezione di Firenze, 50019 Sesto Fiorentino (FI), Italy
* Correspondence: francesca.brero01@universitadipavia.it (F.B.); alessandro.lascialfari@unipv.it (A.L.); Tel.: +39-0382-987-483 (F.B. & A.L.)

Received: 31 August 2020; Accepted: 18 September 2020; Published: 25 September 2020

Abstract: A combination of carbon ions/photons irradiation and hyperthermia as a novel therapeutic approach for the in-vitro treatment of pancreatic cancer BxPC3 cells is presented. The radiation doses used are 0–2 Gy for carbon ions and 0–7 Gy for 6 MV photons. Hyperthermia is realized via a standard heating bath, assisted by magnetic fluid hyperthermia (MFH) that utilizes magnetic nanoparticles (MNPs) exposed to an alternating magnetic field of amplitude 19.5 mTesla and frequency 109.8 kHz. Starting from 37 °C, the temperature is gradually increased and the sample is kept at 42 °C for 30 min. For MFH, MNPs with a mean diameter of 19 nm and specific absorption rate of 110 ± 30 W/$g_{Fe_3O_4}$ coated with a biocompatible ligand to ensure stability in physiological media are used. Irradiation diminishes the clonogenic survival at an extent that depends on the radiation type, and its decrease is amplified both by the MNPs cellular uptake and the hyperthermia protocol. Significant increases in DNA double-strand breaks at 6 h are observed in samples exposed to MNP uptake, treated with 0.75 Gy carbon-ion irradiation and hyperthermia. The proposed experimental protocol, based on the combination of hadron irradiation and hyperthermia, represents a first step towards an innovative clinical option for pancreatic cancer.

Keywords: hadron therapy; magnetic nanoparticles; hyperthermia; nanomaterials; magnetic fluid hyperthermia; pancreatic cancer

---

1. Introduction

Pancreatic cancer is the seventh leading cause of cancer deaths [1], being responsible for 6% of all cancer-related deaths. Researchers are tirelessly endeavoring to develop therapies for pancreatic adenocarcinoma, but in spite of their efforts, survival rate remains poor and most patients have an unresectable tumor at the time of the diagnosis [2]. To overcome these limits, conventional radiotherapy, commonly carried out with X-ray beams has been applied in the context of neoadjuvant or adjuvant therapy concepts, but only modest results have been obtained both because pancreatic cancer is radioresistant and because of the radiosensitivity of normal tissues and organs surrounding the tumor [3–6]. Therefore, highly and inherently conformal radiation therapy techniques, e.g., hadron therapy (HT), present promising alternative treatment options. As an emerging approach, the combination of these new modalities with more conventional therapeutical protocols seems to offer a more efficient way to kill cancer tissues and/or to control or possibly inhibit the tumor progression [7–9].

Hadron therapy employs radiation beams consisting of charged particles, like protons, carbon and helium ions. HT offers some important advantages in comparison to X-ray radiotherapy: (i) the damage induced on the tumor tissues is generally higher; (ii) as a consequence of point (i), it allows the treatment of radioresistant tumors; (iii) the surrounding healthy cells are kept safe because most energy is deposited within the tumor site (Bragg peak), maximizing the cancer cells damage and (iv) the particle beam remains more collimated along the full path and, therefore, any side effects to the adjacent normal tissues can be further reduced [10].

Among therapies additional to irradiation, hyperthermia (Hyp) is clinically investigated for its efficacy [11–15]. The main reason is that the cancer cells are intrinsically sensitive to hyperthermia, because of the highly disorganized development of the tumor and the consequent blood perfusion distortion, which leads to low pH and hypoxia; such an environment may favor cell death by temperature increase [16,17]. On the contrary, healthy tissues are kept safe because they are rarely in conditions of hypoxia and high acidity. The local heating of tumor tissue has been historically realized in different ways, e.g., bath heating [18], microwave irradiation [19], radiofrequency waves [20], focused ultrasounds [21,22], capacitance hyperthermia [23], concentrated laser light [24], magnetic fluid hyperthermia (MFH) [25–34] and, more recently, innovative techniques resulting from the coupling of different types of hyperthermia (for example, magnetic and ultrasonic hyperthermia or magnetic hyperthermia and phototherapy [35,36]). In a relatively recent literature-based review, Peeken et al. [37] show the currently used hyperthermia techniques for heat delivery and temperature control, remarking the different modes of action of Hyp. From a different point of view, Datta et al. [38] reviewed the advantages of using multifunctional MNPs in local tumor MFH, and discussed their role as multimodal theranostic vectors.

MFH is used for thermoablation ($T > 50$ °C) or as mild hyperthermia (40–45 °C) and has a minimal invasivity. It is performed by injecting superparamagnetic nanoparticles directly into the tumor and subsequently exposing the patient to an alternating magnetic field (AMF) of a specific intensity H and frequency f. The Brezovich criterion [39] $H{\cdot}f < 4.85 \times 10^8$ $Am^{-1}s^{-1}$, although under discussion [23], is used to save patients' health. MFH has already been successfully applied to treat patients in specialized hospitals, and many studies evaluated its efficacy on animal models too [40–42]. For what concerns clinical cases, MagForce AG [43], a company specialized in nanocancer therapy for treatment of tumors, initiated in 1997 a more than decennial study on materials and treatment methods, finally receiving the

European Conformity (CE) marking for its MFH system in 2010. To date, it can boast the treatment of more than 100 patients affected by different types of cancer, e.g., glioblastoma multiforme and prostate cancer, spreading the MFH protocol to German, Polish and, recently, US hospitals. It is worth noting that in all successful cases (life expectation increased until 1.5 times), MFH is applied in combination with photon radiation therapy and, when proper, chemotherapy. The material injected intratumorally consists of aminosilane-coated magnetic nanoparticles (MNPs) with an iron-oxide magnetite core of a diameter d ≈ 12–15 nm and dispersed in an aqueous solution [44–47].

The patients underwent six semi-weekly hyperthermia sessions for 60 min: after the MNPs injection they were exposed to an alternating magnetic field (f = 100 kHz, μ_0H = 2.5~19 mTesla) combined with radiotherapy [48]. At the preclinical level, for instance MNPs were used for a case of glioblastoma multiforme on Wistar rats, using μ_0H = 20 mTesla and f = 874 kHz (40 min at T = 42 °C) [49]. Here, as in many other in-vitro and in-vivo cases, the Brezovich limit is exceeded, thus allowing the release of more thermal energy [50–52]. Remarkably, in most preclinical cases, superparamagnetic particles based on a Fe_3O_4 or γ-Fe_2O_3 core with d ≈ 20–22 nm are used, properly coated to prevent undesired aggregation and for ensuring biocompatibility [53,54].

In addition to X-ray radiotherapy plus MFH therapy, some cases of hyperthermia combined to HT have been recently reported. Datta et al. [55] report on two patients with unresectable soft-tissue sarcomas in the lower leg. They were treated with local hyperthermia (RF waves) once a week, in combination with a daily proton therapy (for 7 weeks), achieving functional limb preservation with nearly total tumor control. As a second example, Maeda et al. [56] suggested that hyperthermia (water bath at T = 42.5 °C for 1 h), applied immediately after radiation exposure, might induce hypersensitization to hadron radiation (protons and carbon ions). Thirdly, Ahmad et al. [57] collected preliminary data on the sensitization of cells to proton therapy using A549 lung cancer cells subjected to hyperthermia treatment (T = 42 °C using a heating pad) and to proton irradiation. Their results showed that the cell survival fraction dropped on average by 10–15% both at 2 and 4 Gy.

As concerns the combined action of radiotherapy, MNPs and hyperthermia, it should be noted that their effect could be synergistic or additive, as is usual in cancer treatments by means of, e.g., radiotherapy combined with chemotherapy, immunotherapy, surgery, hyperthermia and so on. For example, Dong et al. [58] show the synergistic osteosarcoma therapeutic strategy when hyperthermia is combined with elaborately catalytic Fenton reaction achieved by Fe_3O_4 and CaO_2 NPs. On the other hand, Ito et al. [59] for the treatment of HER2-overexpressing cancer reported the combination of antibody therapy with magnetic hyperthermia showing an additive effect. Thus, being that killing tumor cells is the main goal, both additive and/or synergistic effects have been shown to be able to reduce the survival of tumor cells.

Although sparse data related to radiotherapy plus Hyp are available in the literature, there is a clear lack of systematic studies regarding the possible combination of HT and Hyp where MFH is used for locally increasing temperature. The present work aims to address the lack of investigation in this direction, by presenting a novel and promising approach directed to pancreatic BxPC3 tumor cells treatment. With the use of HT plus Hyp and, for comparison, of X-ray irradiation plus Hyp, we show that the clonogenic survival of BxPC3 cell cultures, decreased at first by HT, is further diminished by MNP uptake and Hyp treatment, the last one giving an additive killing effect of about 15–30% (MNPs action possibly being synergistic when photons are used). Moreover, when a 0.75 Gy carbon ions irradiation is used, we noted a significant increase in DNA double-strand-breaks at 6 h due to MNPs uptake and hyperthermia.

Our experimental results clearly remark the better efficacy of a dual-therapy treatment with respect to single therapy case and pave the way to translate the proposed protocol to clinic.

2. Materials and Methods

2.1. Synthesis and Characterization of Nanoparticles

All the samples were prepared under inert atmosphere using commercially available reagents. Benzyl ether (99%), oleic acid (OA, 90%), oleylamine (OAM, ≥98%), meso-2,3-Dimercaptosuccinic acid (DMSA), toluene (anhydrous, 99.8%), dimethyl sulfoxide (anhydrous, ≥99.9%) and sodium hydroxide (NaOH, ≥98%, pellets) were purchased from Aldrich Chemical Co. Iron(III) acetylacetonate 99% ($Fe(acac)_3$) from Strem Chemicals Inc. Hydrochloric acid (HCl, ≥37%) and absolute ethanol (EtOH) were purchased from Fluka. Ultrapure water was obtained from a Milli-Q® Synthesis system from Millipore, Temecula, CA, USA. All chemicals were used as received.

In brief, $Fe(acac)_3$ (2.83 g, 8 mmol), OAM (8.56 g, 32 mmol), OA (9.04 g, 32 mmol) and benzyl ether (80 mL) were mixed and magnetically stirred under a flow of nitrogen in a 250 mL three-neck round-bottom flask for 15 min. The resulting mixture was heated to reflux (290 °C) at 25 °C/min and kept at this temperature for 90 min under a blanket of nitrogen and vigorous stirring. The black-brown mixture was cooled at room temperature and ethanol (60 mL) was added, causing the precipitation of a black powder. The product was magnetically separated with a permanent magnet, washed several times with ethanol and finally redispersed in toluene.

Afterwards, 400 mg were dispersed in toluene (60 mL), added to a solution of meso-2,3-dimercaptosuccinic acid DMSA (600 mg) in dimethyl sulfoxide (DMSO, 15 mL), sonicated for 1 h and finally incubated at room temperature for 12 h in a rotating agitator. The precipitate was magnetically separated with a permanent magnet, washed several times first with DMSO and then with ethanol and finally redispersed in MilliQ water (80 mL). The suspension was then basified to pH 10 with sodium hydroxide and adjusted to pH 7.4 with hydrochloric acid to make it stable.

The physical characteristics of the MNPs (e.g., size distribution and zeta potential characterized using a dynamic light scattering (DLS) instrument (Malvern Zetasizer ZS, Malvern Instruments Ltd., Malvern, UK). The morphology of the MNPs was determined using transmission electron microscopy, TEM (CM12 PHILIPS Transmission Electron Microscope, 100 kV).

Powder X-ray diffraction (XRD) measurements were carried out using a Bruker D8 Advance diffractometer equipped with a Cu Kα radiation (λ = 1.54178 Å) and operating in θ–θ Bragg Brentano geometry at 40 kV and 40 mA. Lattice parameters, a, and the mean crystallite diameters, d_{XRD}, were evaluated using the TOPAS® software (Bruker) using the method of the fundamental parameter approach considering a cubic space group Fd–3m. The surfactant percentage was determined by elemental analysis on carbon, hydrogen and nitrogen (CHN analysis) performed by a CHN-S Flash E1112 Thermofinnigan Elementary Analyzer.

Magnetic measurements were performed using a SQUID magnetometer (Quantum Design MPMS, San Diego, CA, USA) operating in the 2–350 K temperature range with applied fields up to 5 T. The powder sample was hosted in Teflon tape and then pressed in a pellet to prevent preferential orientation of the nanocrystallites under the magnetic field. The obtained values of magnetization were normalized by the weight of ferrite present in the sample and expressed in Am^2/kg of ferrite.

2.2. Cell Culture

BxPC3 cells were obtained from ICLC (Interlab Cell Line Collection, Genova, Italy). Cells were maintained at 37 °C in a humidified atmosphere containing 5% CO_2 in air as exponentially growing cultures in RPMI 1640 media (Roswell Park Memorial Institute, Sigma-Aldrich, St. Louis, MO, USA) supplemented with 10% fetal bovine serum (FBS, Sigma-Aldrich) and gentamicin (50 mg/mL; Sigma-Aldrich). In these conditions, the doubling time was 35 ± 2 h and the plating efficiency (PE) was about 50%.

2.3. Cell Toxicity

To assess the effect of MNPs on cell proliferation, BxPC3 cells in the logarithmic growth phase were treated with MNPs at a different concentration for 24 and 48 h of incubation. Cell toxicity was assessed by the clonogenic survival quantified after two weeks and was around 50–60% for cells incubated with MNPs of 50 μg/mL (for further data see Tables S1 and S2). Cell toxicity measured by the Trypan Blue assay after 48 h was around 3% for the same concentration. Results obtained from the cell cycle analysis of cells treated with MNPs (50 and 100 μg/mL) for 48 h have not shown alterations in the cell cycle phases (see Table S3).

2.4. Cellular Uptake of MNPs

To quantify the cellular uptake of MNPs (after 48 h of incubation at a concentration of 50 μg/mL), elemental iron (Fe) was quantitatively measured by inductively coupled plasma optical emission spectrometry (ICP-OES) with iCAP 6200 Duo upgrade, Thermofisher. Digestion with nitric acid was carried out at room temperature (T = 22 °C). The mean uptake (computed by averaging the results of all the ICP measurements performed for different experiments) was about 20 pg(Fe)/cell (see Table S4 for more details).

2.5. Irradiation

Cell irradiation with carbon ions was performed using the synchrotron-based clinical scanning beams (fixed horizontal beam line) at the Centro Nazionale di Adroterapia Oncologica (CNAO, Pavia). Since our cells must be placed in appropriate sample holders (that must respect the CNAO beamline geometry) and they deposit in the form of monolayers, to have a sufficient number of cells (1.5×10^6) and irradiate them with the same dose we chose T25 flasks (25 cm^2 surface). The flasks were placed vertically inside a water phantom put at the isocenter on the treatment table, at a depth of 15 cm, corresponding to the mid spread-out Bragg peak (SOBP). The SOBP (6 cm width, from 12 to 18 cm depth in water) was achieved with active beam energy modulation, using 31 different energies (246–312 MeV/u, Linear energy transfer (LET) of about 45 keV/μm). Samples were irradiated at different doses (0–2 Gy).

Photon beam irradiation of cell cultures (dose 0–7 Gy) was performed using a 6 MV linear accelerator (VARIAN Clinac 2100C, Varian Medical Systems, Palo Alto, CA, USA) at the Fondazione IRCCS Istituto Nazionale dei Tumori, Milano, Italy. The flasks containing the cells were irradiated using a vertical beam 20×20 cm^2 field, placing them horizontally at the isocenter in a water phantom at 5 cm depth.

2.6. Clonogenic Assay

The cells were plated in T25 flasks (about 6×10^5 cell/flask) and after two days some cell samples were incubated with MNPs (50 μg/mL) for 48 h at 37 °C. Afterwards, the cells were exposed to radiation types (photons and carbon ions) at different doses and hyperthermia treatment to the determination of cell survival. After irradiation the cells were detached from the flasks (using 0.25% trypsin_EDTA), counted and reseeded in five T25 flasks for each dose at a suitable cell concentration and incubated for 14 days. The cells were then fixed with ethanol and stained with 10% Giemsa solution and colonies consisting of more than 50 cells were scored as survivors. Surviving fractions relative to the untreated cell samples were determined.

2.7. Double Strand Breaks Studies

After irradiation and/or hyperthermic treatment a portion of the BxPC3 cells were seeded in slide flasks (Thermo Fischer Scientific, Waltham, MA, USA) with 3 mL fresh medium and collected 6 and 24 h later. Cells were fixed in 4% paraformaldehyde (PFA), permeabilized with 0.2% Triton X–100 and blocked at 37 °C in BSA 1% (*w/v*) dissolved in phosphate-buffered saline (PBS). Samples were then coimmunostained overnight (ON) at 4 °C, using a rabbit polyclonal anti–53BP1 antibody (Novus Biologicals, Littleton, CO, USA) in combination with a mouse monoclonal anti–γH2AX antibody (Millipore, Temecula, CA, USA). After washes in PBS/BSA 1% samples were incubated for 1 h at 37 °C in the secondary Alexa 546 anti-mouse and Alexa 488 anti-rabbit antibodies (Invitrogen, Life Technologies, Carlsbad, CA, USA). Finally, slides were washed in PBS/BSA 1%, counterstained with DAPI (Sigma-Aldrich, St. Louis, MO, USA) and mounted using the fluorescent mounting medium Vectashiled (Vector Laboratories, Burlingame, CA, USA). Images were acquired with an Axio-Imager.M1 fluorescent microscope (Zeiss, Jena, Germany) and analyzed using ISIS software (Metasystems, Milano, Italy). The frequency of both the DNA damage marker foci per cell were scored in 100 nuclei in four independent experiments.

2.8. Magnetic Fluid Hyperthermia Setup

Magnetic hyperthermia experiments were performed using a MagneTherm™ set-up by Nanotherics, working at 109.8 kHz and amplitude 19.5 mTesla. The temperature of the samples was measured using an Optocon™ optical fiber thermometer positioned at the centre of the sample placed inside an Eppendorf PCR Tube (also called mini-Eppendorf). The sample holder was optimized to accommodate two Eppendorf Tubes (volume 0.2 mL). In fact, the temperature was detected in a "twin-sentinel" sample, placed next to the sample used for the clonogenic assay, since it is necessary, for survival studies, to keep the latter sterile. It should be remarked that, due to the necessary heat insulation and coil geometry (inner diameter 44 mm), and thus to very limited space, we were forced to use Eppendorf tubes in place of much bigger T25 flasks (and thus a pelletization process that followed the irradiation, to transfer the samples from T25 flasks to mini-Eppendorf, see below).

2.9. Magnetic Hyperthermia Treatment

For the hyperthermia treatments the irradiated cells with MNPs were trypsinized, centrifuged (1500 rpm for 10 min), the cellular pellets (about 1.5×10^6 cells in 0.1 mL medium) were transferred into 0.2 mL polypropylene mini-Eppendorf tubes and placed within the MagneTherm™ system. The pelletizing effect, obligatory step to switch from the irradiation configuration (cells in flasks) to that used for hyperthermic treatment (cells in mini-Eppendorf tubes), responsible of an ulterior stress for the cells was taken into account and separated from MFH action in the reported clonogenic survival curves (see paragraph 3.2). This stress is due to all the operations (trypsinization, centrifugation and vortexing) necessary to transfer the irradiated cells from the T25 flask to the mini-Eppendorf. This stress is quantified by a reduced plating efficiency caused by cells that are damaged and therefore no longer proliferate or proliferate less; this effect was taken into account in renormalizing the clonogenic survival curves. When the evaluation of survival rate is given, the eventual stress effect was previously subtracted. A custom thermalization system, based on a Lauda Alpha A thermostat, and a polystyrene sample holder, was placed inside the MagneTherm™ coil to stabilize the initial temperature of the sample to the physiological value of 37 °C, to center the sample in the homogeneity region of the field and to minimize heat dissipation. An AMF (amplitude $\mu_0 H$ = 19.5 mTesla and frequency f = 109.8 kHz) was applied immediately after the irradiation to increase the temperature of the cells up to 42 °C; this temperature was later maintained for 30 min by changing the temperature of the water circulating within the thermalization system. The field and frequency values have been selected not to exceed Brezovich's criterion excessively, as these values reflect those used in the clinic [43].

3. Results and Discussion

3.1. Synthesis and Characterization of MNPs

Since experiments lasted more than 3 years, it was not possible to use the same batch of MNPs. Different batches of MNPs with similar features were then prepared following the same procedure. The characteristics of a representative sample are described in the following. The samples consisted of a spherical Fe_3O_4 core coated with meso-2,3-dimercaptosuccinic acid. The MNPs were synthesized by thermal decomposition of iron acetylacetonate ($Fe(acac)_3$), in benzyl ether in the presence of oleic acid (OA) and oleylamine (OAM) as surfactants.

The core morphodimensional characteristics were studied by means of transmission electron microscopy (TEM), using a CM12 PHILIPS transmission electron microscope. In Figure 1a we report a representative TEM image, which demonstrates the almost spherical shape of the particles; a mean core diameter value of d_{TEM} = 19.2 ± 3.6 nm was extracted from the histogram of the size distribution.

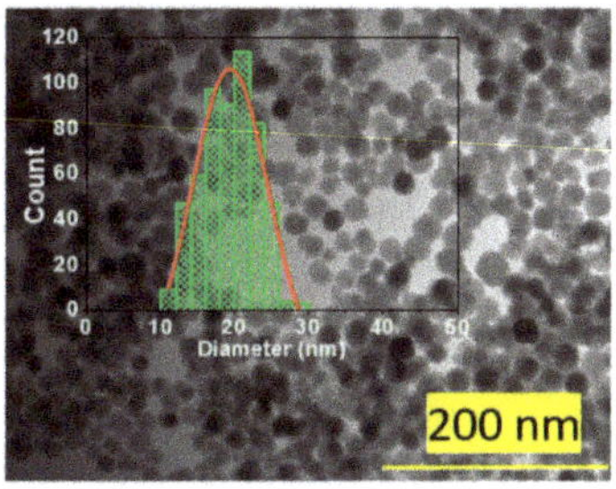

a

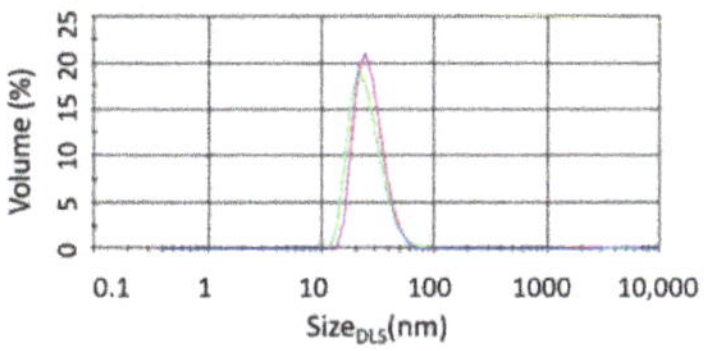

b

Figure 1. (**a**) Representative bright-field TEM image, together with the histogram of the size distribution (d_{TEM} = 19.2 ± 3.6 nm). The solid red line represents the best Gaussian fit. (**b**) Dynamic light scattering (DLS) diameter d_{DLS} measurements at pH 7.4. The green, blue and red lines represent data obtained from three different measurements.

The X-ray diffraction pattern (see Supplementary Materials Figure S1) shows a single crystalline phase with a diffraction pattern compatible with the inverse spinel cubic structure characteristic of magnetite. The peaks all match the reference pattern (JCPDS 19–0629) as regards both position and intensity. The lattice parameter (a = 8.387 Å) is close to the one expected for magnetite (8.396 Å), suggesting a low degree of surface oxidation of the MNPs, which reasonably consist of a magnetite core surrounded by a thin maghemite shell. The crystallite diameter obtained by the Scherrer analysis (d_{XRD} = 18.7 ± 0.4 nm) is comparable to the one obtained by TEM measurement, suggesting that the MNPs are single crystals and present a high degree of crystallinity.

The MNPs were dispersed in pH 7.4 water after coating with DMSA (ca. 2% *w/w* as evaluated by CHN analysis). The successful DMSA functionalization and the stability in water solution were also confirmed by dynamic light scattering (DLS, Figure 1b) and zeta potential (Figure S2) measurements, which provided d_{DLS} = 27 ± 8 nm and Z = −30.5 ± 7.5 mV at pH 7.4.

The hyperthermic efficiency was estimated by evaluating the specific absorption rate (SAR). The SAR of the samples was calculated using the following formula: [60]

$$SAR = \frac{m_{H_2O}\, c_{H_2O} + m_{Fe_3O_4}\, c_{Fe_3O_4}}{m_{Fe_3O_4}} \cdot \frac{\Delta T}{\Delta t} \tag{1}$$

where $c_{H_2O} = 4.18$ J K^{-1} g^{-1} and $c_{Fe_3O_4} = 0.62$ J K^{-1} g^{-1} are the specific heat of water and magnetite in colloidal solutions, while m_{H2O} and m_{Fe3O4} are the respective masses. We neglected the contribution of the DMSA coating because of its small mass fraction. From the experimental temperature kinetics curve, T vs t, the SAR was estimated to be 110 ± 30 W/g_{Fe3O4} under an AMF of frequency f = 109.8 kHz and amplitude $\mu_0 H$ = 19.5 mTesla (for more details see Figure S3).

The field dependence of the sample magnetization was investigated by means of SQUID magnetometry. In Figure 2 we reported the M vs. H curves obtained at 5 K and 300 K.

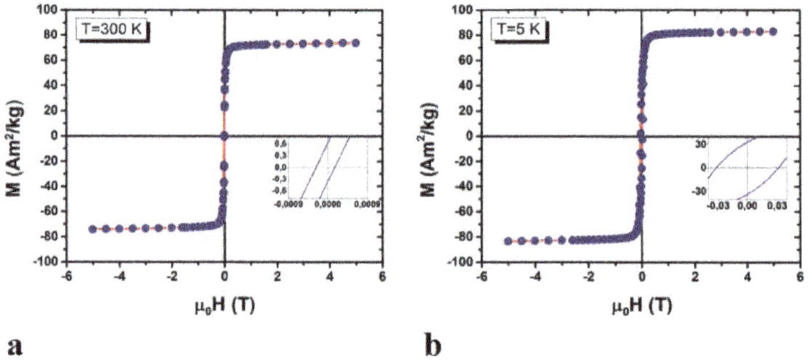

Figure 2. (**a**) Hysteresis loop measured at 300 K and (**b**) 5 K, in the field range ± 5 Tesla. In the insets, we report the low field regions.

At T = 5 K MNPs showed hysteretic behavior with a coercive field $\mu_0 Hc = 30 \pm 2$ mTesla, while a negligible opening-up of the hysteresis loop was observed at T = 300 K (less than 0.3 mTesla, comparable to the remanent field). These results indicate that MNPs display a superparamagnetic behavior, as expected for magnetite based MNPs of this size. At both temperatures, the saturation magnetization, Ms, was very high (83 Am^2/kg at 5 K and 74 Am^2/kg at 300 K) and close to the bulk values (91.7 Am^2/kg_{Fe3O4} at room temperature [61]), confirming the high crystallinity of the inorganic core.

3.2. Experimental Treatment Protocol

As a novel cancer treatment, we investigated the possible synergistic/additive action of hadron therapy and hyperthermia on pancreatic cancer BxPC3 cell culture. The general experimental protocol, sketched in Scheme 1, comprised of three different treatment modes, indicated as 1, 2 and 3 (see next paragraph and Materials and Methods for more details).

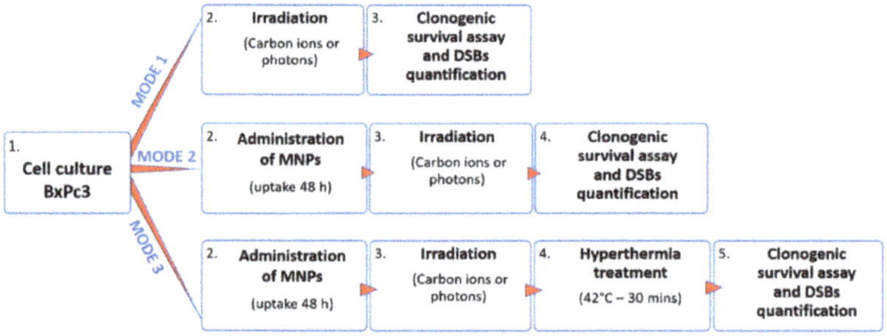

Scheme 1. Description of the 3 different treatment modes used during the experiments. Clonogenic survival has been determined after simple irradiation (mode 1), magnetic nanoparticles (MNPs) administration and irradiation (mode 2) and MNPs administration plus irradiation and subsequent hyperthermia (Hyp; mode 3). DSB stands for a break in double-stranded DNA in which both strands have been cleaved.

The irradiation of samples (cells incubating or not MNPs) have been performed by a C ions beam at the synchrotron-based facility at the National Center for Oncological Hadron Therapy (CNAO) in Pavia, or by means of photons, using a 6 MV linear accelerator at the Fondazione IRCCS Istituto Nazionale dei Tumori in Milano (INT).

The MNPs, used to perform part of the hyperthermic process via MFH, have been administered to human pancreatic adenocarcinoma BxPC3 cells with a concentration 50 μg/mL in the culture medium.

The cells were incubated for 48 h to favor cells' uptake (see Materials and Methods) and thus improve the MFH efficacy.

In the cases of combined therapy (mode 3 in Scheme 1), after the irradiation, we applied Hyp to samples of cells incubating MNPs. The temperature was increased from 37 °C to 42 °C (ΔT = 5 °C) and then kept constant for 30 min, by means of bath heating (through a thermalization system that circulates water around the sample) and MFH, realized by an AMF of 19.5 mTesla amplitude and 109.8 kHz frequency. A sketch of the hyperthermia setup is shown in Figure 3. The amount of uptaken MNPs was estimated by inductively coupled plasma-optical emission spectrometry (ICP-OES) and found on average ~20 pg(Fe)/cell, calculated over different experiments. With this amount, on average about 40% of the heating is due to the application of the magnetic field, while the remaining temperature increase was obtained due to the thermalization system. It should be noted that, due to the small cell volume, the value of 20 pg(Fe)/cell corresponded to filling mostly the inner part and the surface of the cell and, thus, a quantity of heat release near to the highest one.

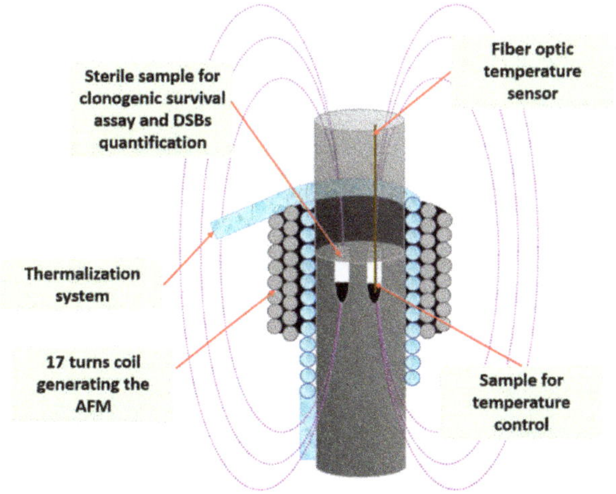

Figure 3. Representation of the hyperthermia setup. A thermalization system (bath heating) circulates water around the two sample vials, one sterile for the clonogenic survival assay and DSBs detection, and one for the temperature registration by a fiber optic probe.

The biological effect of the different treatment modes was finally assessed:

(i) By using a clonogenic assay after two weeks. The clonogenic cell survival assay, i.e., the ability of a cell to produce a viable colony containing at least 50 cells, is considered as a gold standard method for studying cellular sensitivity to irradiation [62];

(ii) By estimating the number of non-repairable double strand breaks (DSBs) per cell after 6 and 24 h, through detection and counting of persistent repair foci, which are DSBs markers. This study was limited to carbon ions irradiation.

3.3. Clonogenic Survival Studies

3.3.1. Carbon Ion Irradiation Experiments

At first, the assessment of the effect of the combination of HT + MNPs + Hyp was performed using the clonogenic assay 2 weeks after the experiment. Figure 4 shows survival data of BxPC3 cells treated with three different modes as described in Scheme 1: (1) only carbon ion (0–2 Gy) irradiation (HT, orange circles); (2) carbon ion irradiation after the administration of magnetic nanoparticles

(HT + MNP, navy triangles) and (3) carbon ion irradiation on culture cells containing MNPs combined with the hyperthermia treatment for 30 min at 42 °C (HT + MNP + Hyp, green stars).

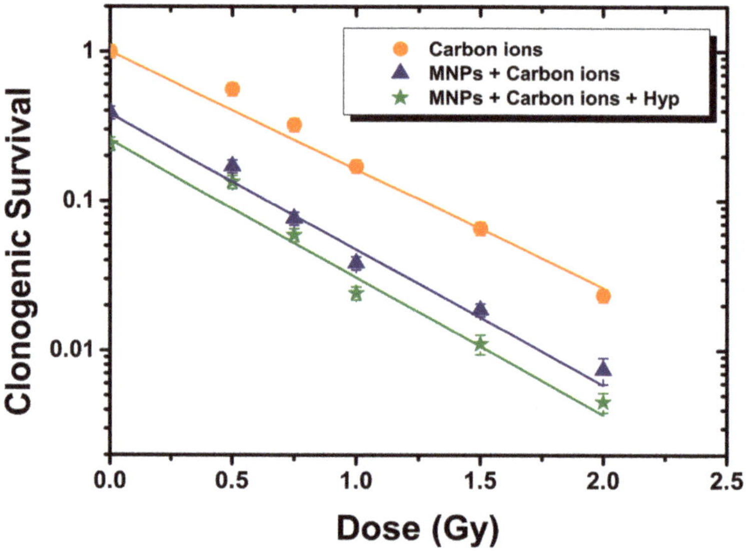

Figure 4. Clonogenic survival of BxPC3 cells culture for 3 different protocols (see text): hadron therapy (HT) only (orange circles), HT + MNPs administration (navy triangles) and HT + MNPs administration + Hyp (green stars). An additive effect of MNPs and Hyp is noted. Solid lines represent the best fit of the clonogenic survival (CS) curves as a function of the radiation dose D according to the law CS $\propto$ exp($-\alpha$D). From the fits: $\alpha_{C\ ions} = 1.82 \pm 0.06\ Gy^{-1}$; $\alpha_{MNPs+C\ ions} = 2.09 \pm 0.15\ Gy^{-1}$ and $\alpha_{MNPs+C\ ions+Hyp} = 2.11 \pm 0.20\ Gy^{-1}$.

The results for each protocol were averaged over four independent experiments.

The following observations were found:

(i) At 0 Gy dose, i.e., unirradiated samples, clonogenic survival (CS) decreased from 1 to 0.4 ± 0.04 when MNPs were added, due to a MNP toxicity at 15 days. A further decrease of CS to 0.24 ± 0.02 was observed when also Hyp was applied (mode 3);

(ii) If carbon ions irradiation alone was applied, the CS decreased on increasing the dose, according to the law: CS $\propto$ exp($-\alpha$D) where D is the dose;

(iii) At all doses, once MNPs were added and irradiation was performed (mode 2), a decrease of CS with respect to irradiation only (mode 1) was observed;

(iv) At all doses, once Hyp was further added (mode 3), the CS dropped further.

It is thus straightforward to conclude that: (i) there is a sizeable MNP toxicity that could result in a potential additive therapeutic effect (although not synergistic) and (ii) hyperthermia grants an additional killing effect on tumor cells with respect to hadron therapy alone.

3.3.2. Photon Irradiation Experiments

To compare the effect of HT plus Hyp treatment with a more conventional one making use of photon irradiation plus Hyp, we performed experiments by using a linear particle accelerator Linac at INT. Figure 5 shows survival data of BxPC3 cells exposed to different doses of 6 MV photons (0–7 Gy) and to Hyp at 42 °C for 30 min, as averaged over two independent experiments.

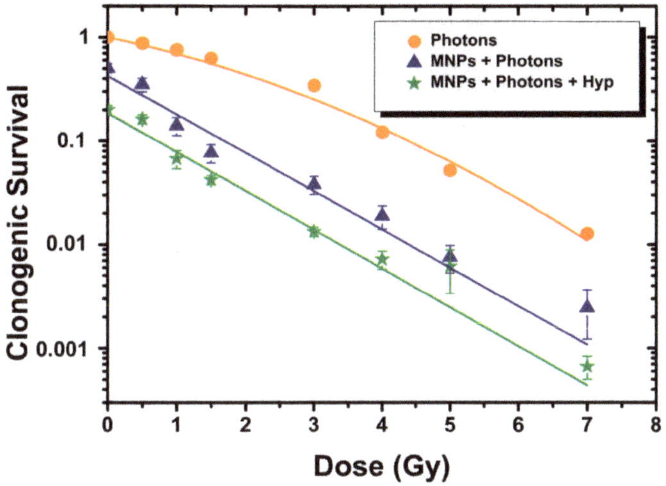

Figure 5. Clonogenic survival of BxPC3 cells incubated with or without MNPs, irradiated with photons (0–7 Gy) and combined (or not) with Hyp. Solid lines represent the best fit of the clonogenic survival (CS) curve as a function of the radiation dose D according to the law CS ∝ exp(−αD-βD^2) for photons alone, and to CS ∝ exp(−αD) for the other two treatments. From the fit: $\alpha_{Photons}$ = 0.32 ± 0.06 Gy^{-1}; $\beta_{Photons}$ = 0.05 ± 0.01 Gy^{-2}; $\alpha_{MNPs+Photons}$ = 0.85 ± 0.09 Gy^{-1} and $\alpha_{MNPs+Photons+Hyp}$ = 0.86 ± 0.06 Gy^{-1}.

In a similar way to the CNAO experiments, three treatment modes were used: (i) photon irradiation alone (orange circles), (ii) photon irradiation after the administration of MNPs for 48 h (navy triangles) and (iii) photon irradiation on cells incubating MNPs combined to 30 min at 42 °C Hyp treatment (green stars).

One could observe the following:

(i) As in the HT case, at 0 Gy dose, i.e., unirradiated samples, clonogenic survival (CS) decreased from 1 to 0.5 ± 0.05 when MNPs were added, and a further decrease of CS to 0.2 ± 0.02 was observed when also Hyp was applied (mode 3);

(ii) If only photon irradiation was applied, the CS decreased once the dose was increased according to the linear quadratic model: CS ∝ exp(−αD−βD^2);

(iii) At all doses, CS decreased once MNPs were added with respect to irradiation with photons only, confirming the results found for HT;

(iv) When MNPs were added, the CS vs. D model changed to CS ∝ exp(−αD), which corresponded to a modification of the cells response to photon irradiation; thus the typical shoulder of the dose–survival curves, found after treatment with radiation alone, was removed;

(v) As in HT, once Hyp was further added, the CS dropped further; also in this case, the dose–survival curve obeyed the law CS ∝ exp(−αD).

Therefore, also in this case we could conclude that: (i) there was a MNP toxicity, comparable to the HT case, and (ii) Hyp gave an additive killing effect on tumor cells with respect to photon irradiation alone.

Let us compare shortly the results of carbon ions and photon irradiation alone. As already observed for other type of tumors [63–65] and other pancreatic tumor cell lines [51], also in the case of BxPC3 cells the survival fractions obtained when carbon ion radiotherapy was used were lower than those for the photon-irradiated samples. This means that carbon ion radiotherapy shows an enhanced efficacy as compared to standard photon radiotherapy. Moreover, as mentioned in the introduction, due to the physical characteristics of high-LET particle radiation (carbon ions) the maximum dose deposition occurs at the so-called Bragg Peak, the depth of which can be adjusted at will within the target tissue. Based on this custom dose profile it should be possible to more accurately administer maximum doses to the tumor with minimum adverse effects to the surrounding tissues. For carbon ion and photon irradiations α values were determined and are given in the captions of Figures 4 and 5. As expected, α values were significantly higher for carbon ions than for photons, reflecting a steeper decline of the initial slope of the survival curves for high-LET beams.

Remarkably, particle radiation has been proven to have a higher relative biological effectiveness (RBE) achieving an increased cytotoxic effect when compared with conventional photon therapy. Additionally, we determined the carbon ions RBE by comparing cell survival data after carbon and photon irradiations as the ratio between the dose of reference radiation (6 MeV photon) and the carbon ions are necessary to produce the same biological effect.

At 10% of the clonogenic survival, the RBE of the CNAO carbon ions beam was equal to about 3.5. This result suggests that the use of hadrons in cancer therapy for the treatment of pancreatic tumors could be a promising modality. Until now the carbon ions RBE values reported in the literature are few, but they seem to agree with our result. El Shafie et al. [66] have found that the RBE of carbon ion irradiation for the BxPC3 cell line ranged from 1.5 to 3.5, depending on the survival level and dose, and data published by Oonishi and colleagues confirms, for the same cell line, the increased RBE of carbon ions irradiation [67].

Remarkably, as the MNPs presence increased the cell mortality rate and, in the case of photons, changed the dose-survival response law, we guess that they have a radiosensitizing (possibly synergistic) effect on BxPC3 cells. Similar results were obtained by Li et al. [68], Liu et al. [69] and Wang et al. [70] with gold nanoparticles and X-rays on 4T1, EMT-6 murine breast carcinoma and HeLa cells, respectively. In another perspective, Goel et al. [71] report progress in nanoparticles-mediated radiosensitization.

3.4. Double Strand Breaks Studies for HT

As a measure of the single/combined effect of hyperthermia and carbon ions irradiations in the induction of DNA damage, the kinetics of DNA double strand breaks rejoining has been evaluated by means of γH2AX and 53BP1 foci formation by immunofluorescence analysis. Both γH2AX (phosphorylation at Ser-139) and 53BP1 are well validated markers of DNA double-strand breaks. [72,73].

Results from four experiments carried out at the CNAO facility were collected and analyzed after exposure of BxPC3 pancreatic tumor cells to 0.75 and 1.5 Gy and harvested at 6 and 24 h from the three different treatment modes. In Figure 6 we report the normalized "foci/cell" values at different doses. It is observed: (i) as expected carbon ions radiation alone (mode 1) increased the number of DSBs with respect to control, i.e., untreated, samples (control, Figure 6a); such a difference was more pronounced at 6 h from treatment, indicating that at 24 h cell repair is more efficient for both DSBs markers used; (ii) at a zero dose, hyperthermia treatment (mode 3) further increased the number of DSBs at 6 h compared to samples without Hyp (mode 1 and 2 treatments, Figure 6b); (iii) at 0.75 Gy, Hyp induced a significant increase in DSBs after 6 h, for both γH2AX and 53BP1 foci, while at 1.5 Gy the differences in the DSBs number between the three different treatment modes was minimal and (iv) at 24 h there were lower DSB values for both doses and for all treatment modes.

These results were similar to those reported by Ma et al. [74] and suggest that the MNPs+Hyp has a radiosensitization effect and inhibits cellular DNA repair mechanisms.

Although the available literature supports the notion that the hyperthermic effect is dependent on the LET of ionizing radiation, the potential molecular mechanism underlying hyperthermia radiosensitization for particle radiation still remain scantly investigated and poorly known. As for low-LET radiations, a major contribution in the untangling of such a mechanism/s was provided by using Chinese hamster ovary (CHO) cell lines specifically defective in the two processes devoted to the repair of radiation-induced DNA DSBs, and namely non homologous end joining (NHEJ) and homologous recombination (HR). In this respect, exposing CHO wild-type cells and a panel of repair defective counterparts to protons (LET = 1 keV/μm 42.5 °C water bath, 1 h) and carbon ions (LET = 13–70 keV/μm; 42.5 °C water bath, 1 h) it was demonstrated a prevalent contribution of HR over NHEJ in radiosensitization, probably affecting the processing of a subset of DNA DSBs lesions [56]. Interestingly, at the molecular level it was shown that Hyp (T > 41 °C) did inhibit HR in human and mouse cells, affecting some of the key players in this process (e.g., delaying recruitment of RAD51 at radiation-induced foci, degrading BRCA2, inactivating RPA, reducing the level of MRN complex, etc.; see [75]). In addition, since heat may explicate the pleiotropic effect on cells, it is generally thought that cytotoxic or sensitizing effects of Hyp cannot be attributed to deactivation of a single DNA repair mechanism, but rather to influencing many pathways (e.g., cell cycle progression and activation of checkpoints) on multiple levels [75,76].

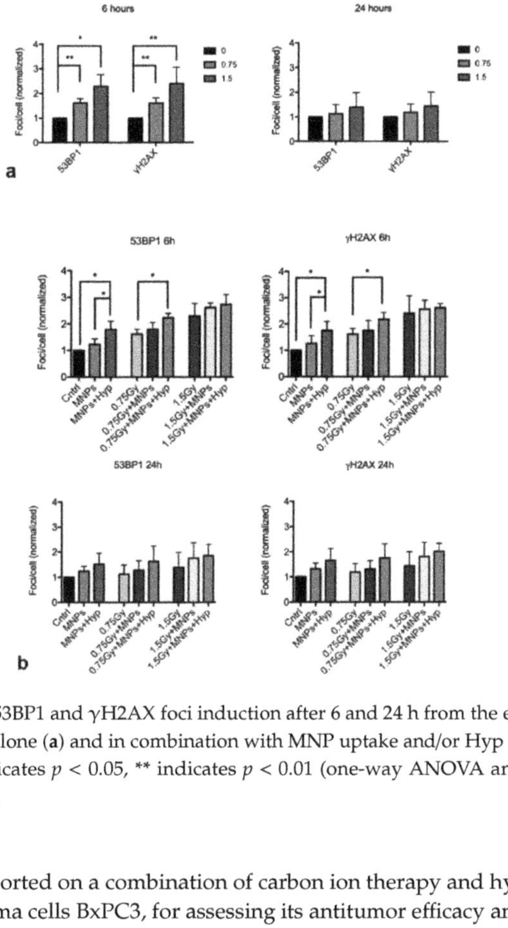

Figure 6. Analysis of 53BP1 and γH2AX foci induction after 6 and 24 h from the exposure to 0.75 and 1.5 Gy of carbon ions alone (**a**) and in combination with MNP uptake and/or Hyp in BxPC3 pancreatic tumor cells (**b**). * indicates $p < 0.05$, ** indicates $p < 0.01$ (one-way ANOVA and Tukey's multiple comparison post-test).

4. Conclusions

In this work, we reported on a combination of carbon ion therapy and hyperthermia applied to pancreatic adenocarcinoma cells BxPC3, for assessing its antitumor efficacy and proposing protocols for future clinical applications. The results obtained were compared to those achieved with a 6 MV photon beam irradiation joint to hyperthermia. A significant part of the heating power was provided through MFH (40% of the total heating), assisted by magnetite nanoparticles with a core diameter of 19 nm and coated with an organic biocompatible ligand, namely DMSA. The hadron carbon ion therapy was shown to have RBE ~3.5, thus confirming a greater efficacy with respect to photon therapy. The clonogenic survival results with respect to a simple irradiation of culture cells, clearly show: (i) at all HT/photon irradiation doses, a further killing (toxicity) additive effect of about 50–60% due to the MNPs cellular uptake and (ii) a significant killing effect of hyperthermia, consisting in an additive 15–30% of total CS, for both irradiation protocols (MNPs action possibly being synergistic when radiation therapy is delivered by photons). Moreover, a significant increase of DNA-DSBs was observed at 6 h after 0.75 Gy-dose HT irradiation plus administered MNPs and Hyp, with respect to the sample exposed to irradiation only.

The increased efficacy of hadron therapy combined with hyperthermia (applied immediately after) lays the foundations for future preclinical studies. Furthermore, these encouraging results point in the direction of further investigating this combination, with a view to finally translate it to a clinical application.

Supplementary Materials: The following are available online at http://www.mdpi.com/2079-4991/10/10/1919/s1, Figure S1: XRD pattern, Figure S2: Zeta potential measurements, Figure S3: SAR estimation method, Tables S1 and S2: determination of optimal time of MNPs uptake and concentration, Table S3: BxPC3 cell cycle analysis after 48 h of treatment with MNPs, Table S4: cellular uptake of MNPs evaluation, Figure S4: dissolution experiments by the radiotracing method.

Author Contributions: F.B. (Francesca Brero), M.A. (Martin Albino), A.A., P.A., M.A. (Matteo Avolio), F.B. (Francesco Berardinelli), D.B., P.C., M.C. (Maurizio Corti), S.G., F.G., A.G., C.I., C.L., S.L., S.M., M.M., F.O., C.S., I.V. and A.L. conceived, designed and performed the experiments; F.B. (Francesca Brero), M.A. (Martin Albino), A.A., P.A., M.A. (Matteo Avolio), F.B. (Francesco Berardinelli), D.B., P.C., F.G., A.G., C.I., C.L., S.M., R.M., F.O., C.S., I.V. and A.L. analyzed the data; F.B. (Francesca Brero) wrote the original draft; A.A., P.C., C.I., C.S., F.O., P.A. and A.L. revised and edited the original draft; E.P., M.C. (Mario Ciocca), A.F. and A.L. supervised the project. All authors have read and agreed to the published version of the manuscript.

Funding: This research received no external funding.

Acknowledgments: The authors thank INFN HADROCOMBI and HADROMAG projects for funding the work, The COST Action TD1402 (RADIOMAG) and the COST Action EURELAX (EU CA15209) are also acknowledged. We are grateful to Marco Cobianchi for his help, especially in the first phase of the work regarding the implementation of the experimental set up. Acknowledgements are also due to A.P. Caricato, P. Cerello and V. Conte, for very useful discussions and suggestions.

Conflicts of Interest: The authors declare no conflict of interest.

References

1. Rawla, P.; Sunkara, T.; Gaduputi, V. Epidemiology of Pancreatic Cancer: Global Trends, Etiology and Risk Factors. *World J. Oncol.* **2019**, *10*, 10–27. [CrossRef]
2. De La Cruz, M.S.D.; Young, A.P.; Ruffin IV, M.T. Diagnosis and Management of Pancreatic Cancer. *Am. Fam. Phys.* **2014**, *89*, 626–632.
3. Rubin, P.; Constine, L.S.; Marks, L.B. *ALERT—Adverse Late Effects of Cancer Treatment*; Springer: Heidelberg, Germany, 2014.
4. Shinoto, M.; Shioyama, Y.; Matsunobu, A.; Okamoto, K.; Suefuji, H.; Toyama, S.; Honda, H.; Kudo, S. Dosimetric Analysis of Upper Gastrointestinal Ulcer after Carbon-Ion Radiotherapy for Pancreatic Cancer. *Radiother. Oncol.* **2016**, *120*, 140–144. [CrossRef]
5. Caivano, D.; Vitolo, V.; Fiore, M.R.; Iannalfi, A.; Vischioni, B.; Bonora, M.; D'ippolito, E.; Ronchi, S.; Molinelli, S.; Ciocca, M.; et al. EP-1421: Carbon Ions In The Treatment of Pancreatic Disease. *Radiother. Oncol.* **2018**, *127*, S773. [CrossRef]
6. Shinoto, M.; Yamada, S.; Terashima, K.; Yasuda, S.; Shioyama, Y.; Honda, H.; Kamada, T.; Tsujii, H.; Saisho, H.; Asano, T.; et al. Carbon Ion Radiation Therapy with Concurrent Gemcitabine for Patients with Locally Advanced Pancreatic Cancer. *Int. J. Radiat. Oncol. Biol. Phys.* **2016**, *95*, 498–504. [CrossRef]
7. NIH, U.S. National Library of Medicine. Available online: https://clinicaltrials.gov/ct2/show/NCT00685763; https://clinicaltrials.gov/ct2/show/NCT03885284; (accessed on 31 May 2020).
8. Nichols, R.C. Proton Therapy for Pancreatic Cancer. *World J. Gastrointest. Oncol.* **2015**, *7*, 141. [CrossRef]
9. Kim, T.H.; Lee, W.J.; Woo, S.M.; Kim, H.; Oh, E.S.; Lee, J.H.; Han, S.S.; Park, S.J.; Suh, Y.G.; Moon, S.H.; et al. Effectiveness and Safety of Simultaneous Integrated Boost-Proton Beam Therapy for Localized Pancreatic Cancer. *Technol. Cancer Res. Treat.* **2018**, *17*. [CrossRef]
10. Available online: www.fondazionecnao.it (accessed on 11 May 2020).
11. Spirou, S.V.; Costa Lima, S.A.; Bouziotis, P.; Vranješ-Djurić, S.; Efthimiadou, E.K.; Laurenzana, A.; Barbosa, A.I.; Garcia-Alonso, I.; Jones, C.; Jankovic, D.; et al. Recommendations for In Vitro and In Vivo Testing of Magnetic Nanoparticle Hyperthermia Combined with Radiation Therapy. *Nanomaterials* **2018**, *8*, 306. [CrossRef]

12. Spirou, S.V.; Basini, M.; Lascialfari, A.; Sangregorio, C.; Innocenti, C. Magnetic Hyperthermia and Radiation Therapy: Radiobiological Principles and Current Practice †. *Nanomaterials* **2018**, *8*, 401. [CrossRef]
13. Datta, N.R.; Ordóñez, S.G.; Gaipl, U.S.; Paulides, M.M.; Crezee, H.; Gellermann, J.; Marder, D.; Puric, E.; Bodis, S. Local hyperthermia combined with radiotherapy and-/or chemotherapy: Recent advances and promises for the future. *Cancer Treat Rev.* **2015**, *41*, 742–753. [CrossRef]
14. Dharmaiah, S.; Zeng, J.; Rao, V.S.; Ouyang, Z.; Ma, T.; Yu, K.; Bhatt, H.; Shah, C.; Godley, A.; Xia, P.; et al. Clinical and dosimetric evaluation of recurrent breast cancer patients treated with hyperthermia and radiation. *Int. J. Hyperth.* **2019**, *36*, 985–991. [CrossRef]
15. Elming, P.B.; Sørensen, B.S.; Oei, A.L.; Franken, N.A.; Crezee, J.; Overgaard, J.; Horsman, M.R. Hyperthermia: The optimal treatment to overcome radiation resistant hypoxia. *Cancers* **2019**, *11*, 60. [CrossRef]
16. van der Zee, J. Heating the Patient: A Promising Approach? *Ann. Oncol.* **2002**, *13*, 1173–1184. [CrossRef]
17. Baronzio, G.F.; Hager, E.D. *Hyperthermia in Cancer Treatment: A Primer*; Springer Science & Business Media: Berlin/Heidelberg, Germany, 2006.
18. Overgaard, J. Simultaneous and Sequential Hyperthermia and Radiation Treatment of an Experimental Tumor and Its Surrounding Normal Tissue in Vivo. *Int. J. Radiat. Oncol. Biol. Phys.* **1980**, *6*, 1507–1517. [CrossRef]
19. Lin, J.C.; Lin, M.F. Microwave Hyperthermia-Induced Blood-Brain Barrier Alterations. *Radiat. Res.* **1982**, *89*, 77–87. [CrossRef]
20. Prasad, B.; Kim, S.; Cho, W.; Kim, J.K.; Kim, Y.A.; Kim, S.; Wu, H.G. Quantitative Estimation of the Equivalent Radiation Dose Escalation Using Radiofrequency Hyperthermia in Mouse Xenograft Models of Human Lung Cancer. *Sci. Rep.* **2019**, *9*, 3942. [CrossRef]
21. Mari, A.D.I.; Giuliano, S.R.; Lanteri, E.; Pumo, V.; Romano, F.; Trombatore, G.; Bucolo, A.; Tralongo, P. Clinical use of high- intensity focused ultra- sound in the management of different solid tumors. *WCRJ* **2014**, *1*, e295.
22. Liang, X.; Gao, J.; Jiang, L.; Luo, J.; Jing, L.; Li, X.; Jin, Y.; Dai, Z. Nanohybrid Liposomal Cerasomes with Good Physiological Stability and Rapid Temperature Responsiveness for High Intensity Focused Ultrasound Triggered Local Chemotherapy of Cancer. *ACS Nano* **2015**, *9*, 1280–1293. [CrossRef]
23. Abe, M.; Hiraoka, M.; Takahashi, M.; Egawa, S.; Matsuda, C.; Onoyama, Y.; Morita, K.; Kakehi, M.; Sugahara, T. Multi-institutional Studies on Hyperthermia Using an 8-MHz Radiofrequency Capacitive Heating Device (Thermotron RF-8) in Combination with Radiation for Cancer Therapy. *Cancer* **1986**, *58*, 1589–1595. [CrossRef]
24. Müller, G.J.; Roggan, A. (Eds.) *Laser-Induced Interstitial Thermotherapy*; SPIE Press: Bellingham, WA, USA, 1995.
25. Guardia, P.; Di Corato, R.; Lartigue, L.; Wilhelm, C.; Espinosa, A.; Garcia-Hernandez, M.; Gazeau, F.; Manna, L.; Pellegrino, T. Water-Soluble Iron Oxide Nanocubes with High Values of Specific Absorption Rate for Cancer Cell Hyperthermia Treatment. *ACS Nano* **2012**, *6*, 3080–3091. [CrossRef]
26. Ortega, D.; Pankhurst, Q.A. Magnetic Hyperthermia. *Nanoscience* **2013**, *1*, e88.
27. Périgo, E.A.; Hemery, G.; Sandre, O.; Ortega, D.; Garaio, E.; Plazaola, F.; Teran, F.J. Fundamentals and Advances in Magnetic Hyperthermia. *Appl. Phys. Rev.* **2015**, *2*, 041302. [CrossRef]
28. Tay, Z.W.; Chandrasekharan, P.; Chiu-Lam, A.; Hensley, D.W.; Dhavalikar, R.; Zhou, X.Y.; Yu, E.Y.; Goodwill, P.W.; Zheng, B.; Rinaldi, C.; et al. Magnetic Particle Imaging-Guided Heating in Vivo Using Gradient Fields for Arbitrary Localization of Magnetic Hyperthermia Therapy. *ACS Nano* **2018**, *12*, 3699–3713. [CrossRef]
29. Cabrera, D.; Coene, A.; Leliaert, J.; Artés-Ibáñez, E.J.; Dupré, L.; Telling, N.D.; Teran, F.J. Dynamical Magnetic Response of Iron Oxide Nanoparticles Inside Live Cells. *ACS Nano* **2018**, *12*, 2741–2752. [CrossRef]
30. Pan, J.; Hu, P.; Guo, Y.; Hao, J.; Ni, D.; Xu, Y.; Bao, Q.; Yao, H.; Wei, C.; Wu, Q.; et al. Combined Magnetic Hyperthermia and Immune Therapy for Primary and Metastatic Tumor Treatments. *ACS Nano* **2020**, *14*, 1033–1044. [CrossRef]
31. Niculaes, D.; Lak, A.; Anyfantis, G.C.; Marras, S.; Laslett, O.; Avugadda, S.K.; Cassani, M.; Serantes, D.; Hovorka, O.; Chantrell, R.; et al. Asymmetric Assembling of Iron Oxide Nanocubes for Improving Magnetic Hyperthermia Performance. *ACS Nano* **2017**, *11*, 12121–12133. [CrossRef]
32. Espinosa, A.; Di Corato, R.; Kolosnjaj-Tabi, J.; Flaud, P.; Pellegrino, T.; Wilhelm, C. Duality of Iron Oxide Nanoparticles in Cancer Therapy: Amplification of Heating Efficiency by Magnetic Hyperthermia and Photothermal Bimodal Treatment. *ACS Nano* **2016**, *10*, 2436–2446. [CrossRef]
33. Xu, C.; Zheng, Y.; Gao, W.; Xu, J.; Zuo, G.; Chen, Y.; Zhao, M.; Li, J.; Song, J.; Zhang, N.; et al. Magnetic Hyperthermia Ablation of Tumors Using Injectable Fe3O4/Calcium Phosphate Cement. *ACS Appl. Mater. Interfaces* **2015**, *7*, 13866–13875. [CrossRef]

34. El Hajj Diab, D.; Clerc, P.; Serhan, N.; Fourmy, D.; Gigoux, V. Combined Treatments of Magnetic Intra-Lysosomal Hyperthermia with Doxorubicin Promotes Synergistic Anti-Tumoral Activity. *Nanomaterials* **2018**, *8*, 468. [CrossRef]
35. Kaczmarek, K.; Hornowski, T.; Antal, I.; Rajnak, M.; Timko, M.; Józefczak, A. Sono-magnetic heating in tumor phantom. *J. Magn. Magn. Mater.* **2020**, *500*, 166396. [CrossRef]
36. Curcio, A.; Silva, A.; Cabana, S.; Espinosa, A.; Baptiste, B.; Menguy, N.; Wilhelm, C.; Abou-Hassan, A. Iron Oxide Nanoflowers @ CuS Hybrids for Cancer Tri-Therapy: Interplay of Photothermal Therapy, Magnetic Hyperthermia and Photodynamic Therapy. *Theranostics* **2019**, *9*, 1288–1302. [CrossRef] [PubMed]
37. Peeken, J.C.; Vaupel, P.; Combs, S.E. Integrating Hyperthermia into Modern Radiation Oncology: What Evidence Is Necessary? *Front. Oncol.* **2017**, *7*, 132. [CrossRef] [PubMed]
38. Datta, N.R.; Krishnan, S.; Speiser, D.E.; Neufeld, E.; Kuster, N.; Bodis, S.; Hofmann, H. Magnetic Nanoparticle-Induced Hyperthermia with Appropriate Payloads: Paul Ehrlich's "Magic (Nano)Bullet" for Cancer Theranostics? *Cancer Treat. Rev.* **2016**, *50*, 217–227. [CrossRef] [PubMed]
39. Brezovich, I.A. Low Frequency Hyperthermia: Capacitive and Ferromagnetic Thermoseed Methods. In *Medical Physics Monograph No 16: Biological, Physical, and Clinical Aspects of Hyperthermia*; American Institute of Physics: University Park, MD, USA, 1988.
40. Johannsen, M.; Jordan, A.; Scholz, R.; Koch, M.; Lein, M.; Deger, S.; Roigas, J.; Jung, K.; Loening, S. Evaluation of Magnetic Fluid Hyperthermia in a Standard Rat Model of Prostate Cancer. *J. Endourol.* **2004**, *18*, 495–500. [CrossRef]
41. Johannsen, M.; Thiesen, B.; Jordan, A.; Taymoorian, K.; Gneveckow, U.; Waldöfner, N.; Scholz, R.; Koch, M.; Lein, M.; Jung, K.; et al. Magnetic Fluid Hyperthermia (MFH) Reduces Prostate Cancer Growth in the Orthotopic Dunning R3327 Rat Model. *Prostate* **2005**, *64*, 283–292. [CrossRef]
42. Jordan, A.; Scholz, R.; Wust, P.; Fähling, H.; Krause, J.; Wlodarczyk, W.; Sander, B.; Vogl, T.; Felix, R. Effects of Magnetic Fluid Hyperthermia (MFH) on C3H Mammary Carcinoma in Vivo. *Int. J. Hyperth.* **1997**, *13*, 587–605. [CrossRef]
43. Available online: www.magforce.com (accessed on 31 August 2020).
44. Jordan, A.; Scholz, R.; Maier-Hauff, K.; Johannsen, M.; Wust, P.; Nadobny, J.; Schirra, H.; Schmidt, H.; Deger, S.; Loening, S.; et al. Presentation of a new magnetic field therapy system for the treatment of human solid tumors with magnetic fluid hyperthermia. *J. Magn. Magn. Mater.* **2001**, *225*, 118–126. [CrossRef]
45. Jordan, A.; Scholz, R.; Wust, P.; Fähling, H.; Roland, F. Magnetic fluid hyperthermia (MFH): Cancer treatment with AC magnetic field induced excitation of biocompatible superparamagnetic nanoparticles. *J. Magn. Magn. Mater.* **1999**, *201*, 413–419. [CrossRef]
46. Maier-Hauff, K.; Rothe, R.; Scholz, R.; Gneveckow, U.; Wust, P.; Thiesen, B.; Feussner, A.; von Deimling, A.; Waldoefner, N.; Felix, R.; et al. Intracranial thermotherapy using magnetic nanoparticles combined with external beam radiotherapy: Results of a feasibility study on patients with glioblastoma multiforme. *J. Neuro-Oncol.* **2007**, *81*, 53–60. [CrossRef]
47. Johannsen, M.; Thiesen, B.; Wust, P.; Jordan, A. Magnetic nanoparticle hyperthermia for prostate cancer. *Int. J. Hyperth.* **2010**, *26*, 790–795. [CrossRef]
48. Maier-Hauff, K.; Ulrich, F.; Nestler, D.; Niehoff, H.; Wust, P.; Thiesen, B.; Orawa, H.; Budach, V.; Jordan, A. Efficacy and Safety of Intratumoral Thermotherapy Using Magnetic Iron-Oxide Nanoparticles Combined with External Beam Radiotherapy on Patients with Recurrent Glioblastoma Multiforme. *J. Neurooncol.* **2011**, *103*, 317–324. [CrossRef]
49. de Rego, G.N.A.; Mamani, J.B.; Souza, T.K.F.; Nucci, M.P.; da Silva, H.R.; Gamarra, L.F. Therapeutic Evaluation of Magnetic Hyperthermia Using Fe_3O_4-Aminosilane-Coated Iron Oxide Nanoparticles in Glioblastoma Animal Model. *Einstein (Sao Paulo)* **2019**, *17*. [CrossRef] [PubMed]
50. Mannucci, S.; Ghin, L.; Conti, G.; Tambalo, S.; Lascialfari, A.; Orlando, T.; Benati, D.; Bernardi, P.; Betterle, N.; Bassi, R.; et al. Magnetic Nanoparticles from Magnetospirillum Gryphiswaldense Increase the Efficacy of Thermotherapy in a Model of Colon Carcinoma. *PLoS ONE* **2014**, *9*, e0108959. [CrossRef] [PubMed]
51. Dutz, S.; Hergt, R. Magnetic Particle Hyperthermia—A Promising Tumour Therapy? *Nanotechnology* **2014**, *25*, 452001. [CrossRef] [PubMed]
52. Alphandéry, E.; Faure, S.; Seksek, O.; Guyot, F.; Chebbi, I. Chains of Magnetosomes Extracted from AMB-1 Magnetotactic Bacteria for Application in Alternative Magnetic Field Cancer Therapy. *ACS Nano* **2011**, *5*, 6279–6296. [CrossRef]

53. Jordan, A.; Wust, P.; Scholz, R.; Tesche, B.; Fähling, H.; Mitrovics, T.; Vogl, T.; Cervós-Navarro, J.; Felix, R. Cellular Uptake of Magnetic Fluid Particles and Their Effects on Human Adenocarcinoma Cells Exposed to AC Magnetic Fields in Vitro. *Int. J. Hyperth.* **1996**, *12*, 705–722. [CrossRef]
54. Oh, Y.; Lee, N.; Kang, H.W.; Oh, J. In Vitro Study on Apoptotic Cell Death by Effective Magnetic Hyperthermia with Chitosan-Coated MnFe2O4. *Nanotechnology* **2016**, *27*, 115101. [CrossRef]
55. Datta, N.R.; Schneider, R.; Puric, E.; Ahlhelm, F.J.; Marder, D.; Bodis, S.; Weber, D.C. Proton Irradiation with Hyperthermia in Unresectable Soft Tissue Sarcoma. *Int. J. Part. Ther.* **2016**, *3*, 327–336. [CrossRef]
56. Maeda, J.; Fujii, Y.; Fujisawa, H.; Hirakawa, H.; Cartwright, I.M.; Uesaka, M.; Kitamura, H.; Fujimori, A.; Kato, T.A. Hyperthermia-Induced Radiosensitization in CHO Wild-Type, NHEJ Repair Mutant and HR Repair Mutant Following Proton and Carbon-Ion Exposure. *Oncol. Lett.* **2015**, *10*, 2828–2834. [CrossRef]
57. Ahmad, S.; Jin, H.; Sahoo, K.; Griffin, R.J.; Herman, T.S.; Ranjan, A. Proton Therapy in Combination With Mild Hyperthermia Enhances Killing of Radio-Resistant Hypoxic Tumor Cells. *Int. J. Radiat. Oncol.* **2017**, *99*, E574. [CrossRef]
58. Dong, S.; Chen, Y.; Yu, L.; Lin, K.; Wang, X. Magnetic Hyperthermia–Synergistic H_2O_2 Self-Sufficient Catalytic Suppression of Osteosarcoma with Enhanced Bone-Regeneration Bioactivity by 3D-Printing Composite Scaffolds. *Adv. Funct. Mater.* **2020**, *30*, 1907071. [CrossRef]
59. Ito, A.; Kuga, Y.; Honda, H.; Kikkawa, H.; Horiuchi, A.; Watanabe, Y.; Kobayashi, T. Magnetite Nanoparticle-Loaded Anti-HER2 Immunoliposomes for Combination of Antibody Therapy with Hyperthermia. *Cancer Lett.* **2004**, *212*, 167–175. [CrossRef] [PubMed]
60. Cervadoro, A.; Giverso, C.; Pande, R.; Sarangi, S.; Preziosi, L.; Wosik, J.; Brazdeikis, A.; Decuzzi, P. Design Maps for the Hyperthermic Treatment of Tumors with Superparamagnetic Nanoparticles. *PLoS ONE* **2013**, *8*, e57332. [CrossRef] [PubMed]
61. Soeya, S.; Hayakawa, J.; Takahashi, H.; Ito, K.; Yamamoto, C.; Kida, A.; Asano, H.; Matsui, M. Development of Half-Metallic Ultrathin Fe3O4 Films for Spin-Transport Devices. *Appl. Phys. Lett.* **2002**, *80*, 823–825. [CrossRef]
62. Hall, E.J.; Giaccia, A.J. (Eds.) *Radiobiology for the Radiologist*, 6th ed.; Lippincott Williams & Wilkins: Philadelphia, PA, USA, 2006.
63. Habermehl, D.; Ilicic, K.; Dehne, S.; Rieken, S.; Orschiedt, L.; Brons, S.; Haberer, T.; Weber, K.J.; Debus, J.; Combs, S.E. The Relative Biological Effectiveness for Carbon and Oxygen Ion Beams Using the Raster-Scanning Technique in Hepatocellular Carcinoma Cell Lines. *PLoS ONE* **2014**, *9*, e113591. [CrossRef] [PubMed]
64. Combs, S.E.; Zipp, L.; Rieken, S.; Habermehl, D.; Brons, S.; Winter, M.; Haberer, T.; Debus, J.; Weber, K.J. In Vitro Evaluation of Photon and Carbon Ion Radiotherapy in Combination with Chemotherapy in Glioblastoma Cells. *Radiat. Oncol.* **2012**, *7*, 1–6. [CrossRef] [PubMed]
65. Cui, X.; Oonishi, K.; Tsujii, H.; Yasuda, T.; Matsumoto, Y.; Furusawa, Y.; Akashi, M.; Kamada, T.; Okayasu, R. Effects of Carbon Ion Beam on Putative Colon Cancer Stem Cells and Its Comparison with X-Rays. *Cancer Res.* **2011**, *71*, 3676–3687. [CrossRef]
66. El Shafie, R.A.; Habermehl, D.; Rieken, S.; Mairani, A.; Orschiedt, L.; Brons, S.; Haberer, T.; Weber, K.J.; Debus, J.; Combs, S.E. In Vitro Evaluation of Photon and Raster-Scanned Carbon Ion Radiotherapy in Combination with Gemcitabine in Pancreatic Cancer Cell Lines. *J. Radiat. Res.* **2013**, *54* (Suppl. 1), i113–i119. [CrossRef]
67. Oonishi, K.; Cui, X.; Hirakawa, H.; Fujimori, A.; Kamijo, T.; Yamada, S.; Yokosuka, O.; Kamada, T. Different Effects of Carbon Ion Beams and X-Rays on Clonogenic Survival and DNA Repair in Human Pancreatic Cancer Stem-like Cells. *Radiother. Oncol.* **2012**, *105*, 258–265. [CrossRef]
68. Li, M.; Zhao, Q.; Yi, X.; Zhong, X.; Song, G.; Chai, Z.; Liu, Z.; Yang, K. Au@MnS@ZnS Core Shell/Shell Nanoparticles for Magnetic Resonance Imaging and Enhanced Cancer Radiation Therapy. *ACS Appl. Mater. Interfaces* **2016**, *8*, 9557–9564. [CrossRef]
69. Liu, C.J.; Wang, C.H.; Chen, S.T.; Chen, H.H.; Leng, W.H.; Chien, C.C.; Wang, C.L.; Kempson, I.M.; Hwu, Y.; Lai, T.C.; et al. Enhancement of Cell Radiation Sensitivity by Pegylated Gold Nanoparticles. *Phys. Med. Biol.* **2010**, *55*, 931. [CrossRef] [PubMed]
70. Wang, X.; Zhang, C.; Du, J.; Dong, X.; Jian, S.; Yan, L.; Gu, Z.; Zhao, Y. Enhanced Generation of Non-Oxygen Dependent Free Radicals by Schottky-Type Heterostructures of Au-Bi_2S_3 Nanoparticles via X-Ray-Induced Catalytic Reaction for Radiosensitization. *ACS Nano* **2019**, *13*, 5947–5958. [CrossRef] [PubMed]
71. Goel, S.; Ni, D.; Cai, W. Harnessing the Power of Nanotechnology for Enhanced Radiation Therapy. *ACS Nano* **2017**, *11*, 5233–5237. [CrossRef] [PubMed]

72. Rogakou, E.P.; Boon, C.; Redon, C.; Bonner, W.M. Megabase Chromatin Domains Involved in DNA Double-Strand Breaks in Vivo. *J. Cell Biol.* **1999**, *146*, 905–916. [CrossRef]
73. Schultz, L.B.; Chehab, N.H.; Malikzay, A.; Halazonetis, T.D. P53 Binding Protein 1 (53BP1) Is an Early Participant in the Cellular Response to DNA Double-Strand Breaks. *J. Cell Biol.* **2000**, *151*, 1381–1390. [CrossRef]
74. Ma, J.; Zhang, Z.; Zhang, Z.; Huang, J.; Qin, Y.; Li, X.; Liu, H.; Yang, K.; Wu, G. Magnetic Nanoparticle Clusters Radiosensitise Human Nasopharyngeal and Lung Cancer Cells after Alternating Magnetic Field Treatment. *Int. J. Hyperth.* **2015**, *31*, 800–812. [CrossRef]
75. Oei, A.L.; Vriend, L.E.; Crezee, J.; Franken, N.A.; Krawczyk, P.M. Effects of hyperthermia on DNA repair pathways: One treatment to inhibit them all. *Radiat. Oncol.* **2015**, *10*, 1–13. [CrossRef]
76. Oei, A.L.; Kok, H.P.; Oei, S.B.; Horsman, M.R.; Stalpers, L.J.A.; Franken, N.A.P.; Crezee, J. Molecular and biological rationale of hyperthermia as radio- and chemosensitizer. *Adv. Drug Deliv. Rev.* **2020**. [CrossRef]

nanomaterials

MDPI

Article

Analysis of Fluorescence Decay Kinetics of Indocyanine Green Monomers and Aggregates in Brain Tumor Model In Vivo

Dina Farrakhova [1,*], Igor Romanishkin [1], Yuliya Maklygina [1], Lina Bezdetnaya [2,3] and Victor Loschenov [1,4]

1 Prokhorov General Physics Institute of the Russian Academy of Science, 119991 Moscow, Russia; igor.romanishkin@nsc.gpi.ru (I.R.); us.samsonova@physics.msu.ru (Y.M.); loschenov@nsc.gpi.ru (V.L.)
2 Centre de Recherche en Automatique de Nancy, CNRS, Université de Lorraine, 54519 Vandoeuvre-lès-Nancy, France; l.bolotine@nancy.unicancer.fr
3 Institut de Cancérologie de Lorraine, 54519 Vandoeuvre-lès-Nancy, France
4 Institute of Engineering Physics for Biomedicine, National Research Nuclear University MEPhI, 115409 Moscow, Russia
* Correspondence: farrakhova.dina@mail.ru; Tel.: +7-968-587-52-75

Abstract: Spectroscopic approach with fluorescence time resolution allows one to determine the state of a brain tumor and its microenvironment via changes in the fluorescent dye's fluorescence lifetime. Indocyanine green (ICG) is an acknowledged infra-red fluorescent dye that self-assembles into stable aggregate forms (ICG NPs). ICG NPs aggregates have a tendency to accumulate in the tumor with a maximum accumulation at 24 h after systemic administration, enabling extended intraoperative diagnostic. Fluorescence lifetime analysis of ICG and ICG NPs demonstrates different values for ICG monomers and H-aggregates, indicating promising suitability for fluorescent diagnostics of brain tumors due to their affinity to tumor cells and stability in biological tissue.

Keywords: fluorescent diagnosis; fluorescent lifetime; near-infrared range; fluorescent dyes; H-aggregates; indocyanine green; brain cancer; glioma C6

Citation: Farrakhova, D.; Romanishkin, I.; Maklygina, Y.; Bezdetnaya, L.; Loschenov, V. Analysis of Fluorescence Decay Kinetics of Indocyanine Green Monomers and Aggregates in Brain Tumor Model In Vivo. *Nanomaterials* **2021**, *11*, 3185. https://doi.org/10.3390/nano11123185

Academic Editors: Fiore P. Nicoletta and Francesca Iemma

Received: 30 October 2021
Accepted: 23 November 2021
Published: 24 November 2021

1. Introduction

Any surgical intervention in the central nervous system for tumor resection requires high accuracy and selectivity of its effect on tissue. Currently, the main problem is the lack of a rapid, objective, and comprehensive intraoperative assessment of the boundaries of tumor tissue during surgical resection or a laser-induced therapy session [1]. The laser spectroscopic methods provide a unique opportunity to determine noninvasively the most significant parameters characterizing the condition of tissues and the prognosis of its possible evolutionary pathological changes, which may occur in the absence of timely treatment. Optical spectroscopy allows one to obain a wide range of information about physiological and morphological parameters [2,3]. This approach identifies a relationship between data from pathological tissue such as absorption, fluorescence, scattering caused by substances originally inherent to tissues and cells, and externally introduced markers. Laser spectroscopic methods with time resolution allow one to determine the state of a tumor and its microenvironment based on the changes in its photosensitizer (PS) fluorescence lifetime. The difference between the fluorescence lifetime is due to photosensitizer accumulation in cells with different phenotypes [4]. Immunocompetent cells, including macrophages, are a predominant component of brain tumors bioenvironment and allow an enhanced uptake of the photosensitizer molecules compared with cancer cells [5]. The change of fluorescence lifetime of ICG in cancer cells is due to a modification in the refractive index of biological tissues [6]. This approach has the advantage of being less sensitive to changes in the irradiation intensity and the dye concentration than the (relative) fluorescence intensities [7–9]. Noninvasive conditions for assessment of brain tumor tissue and surrounding tissues is important for performing a relapse-free operation without reducing the patient's quality of life. It is important to register small and fast functional changes to assess the relationship between surface metabolic and structural changes that occur during the formation and growth of brain cancer [10].

Exogenous fluorophores with selective accumulation in tumor cells and successive photo-cytotoxic effects can significantly increase the delineation of cancer boundaries and enhance their therapeutic effect. Indocyanine green (ICG) is a well-known infra-red fluorescent dye that is used extensively for vascular system imaging [11], tumor angiogenesis [11,12], and internal organs [13] imaging in vivo. ICG has an absorption in the far-red and near-infrared ranges, where biological components such as water and hemoglobin have low coefficients of absorption and scattering. ICG monomers and H- and J-aggregates have absorption peaks at 780 nm, 715 nm, and 895 nm, respectively, where penetration depth fluctuates from 0.5 cm to 0.8 cm. Unfortunately, ICG is not preferable for tumor diagnosis due to limited accumulation and a high photobleaching effect [14]. There is a lot of research related to ICG embedded in different delivery system such as liposomes [15], polymers [16], lipid nanoparticles [17,18], albumin nanoparticles [19,20], and nanofibers [21], which significantly improve limited accumulation in vivo and photostability. High temperatures enable ICG to form stable H- and J-aggregates [22,23]. At present, there are numerous studies that use ICG colloidal solutions (ICG NPs) for cancer treatment [24–27]. ICG aggregates have a tendency to accumulate in the tumor, enabling extended intraoperative diagnostic. Moreover, after internalization of the aggregates by cells, it could be disassociated into the monomeric form of ICG [28], which can lead to the destruction of tumor tissue via photodynamic action.

2. Materials and Methods

Colloidal solution of ICG NPs and molecular solution of ICG were used as fluorescent dyes. ICG molecular form was heated at 65 °C for 20 h to form ICG NPs according to a previously published method [28]. After forming ICG NPs, the solution was filtered through 0.40 μm syringe filters to remove large aggregates. BALB/c mice with glioma C6 xenograft were used as experimental models. The experimental group consisted of 18 female mice with 20–23 g body weight in the age range of 6–8 weeks. The mice were kept in individually ventilated cages with controlled environmental values.

Animal care guidelines were used under the protocol of European Convention for the Protection of Vertebrate Animals used for Experimental and Other Scientific Purposes (Strasbourg, 18.III.1986). For spectroscopic research, the mice were restrained in tube restrainer configurations. Animals' supervision was conducted daily and efforts were made to prevent mice from suffering.

Fluorescent dyes of ICG and ICG NPs were administered intravenously into the tail vein at 10 mg/kg dose in 0.2 μL volume for 30 s. The concentrations of ICG and ICG NPs were selected corresponding to the doses used in clinical practice.

The fluorescence spectra of ICG and ICG NPs accumulated in tumor xenograft at different time intervals after fluorescent dyes administration were obtained via LESA-01-Biospec spectroscopic system (Moscow, Russia). The excitation wavelength was 633 nm with 5 mW/cm^2 power density and a corresponding longpass filter was set at 635 nm for fluorescence registration. The filter used is convenient for fluorescence registration of ICG monomers and H-aggregates. A diffusely reflected signal for fluorescence curves was detected by optical fiber in contact of the distal tip to the tumor area and nearby healthy normal tissue (norma). At least three spectra for each measurement were obtained for statistical analysis. Excretory organs study was carried out after sacrificing the mice according to the protocol.

Fluorescence kinetics of ICG and ICG NPs were recorded by a system based on a Hamamatsu C10627-13 streak-camera (Iwata City, Japan) with a time resolution of 15 ps and laser with 637 nm excitation wavelength and 65 ps pulse. The system works on a time-correlated single photon counting approach. The description of the system is reported in the [29]. The fiber was fixed on a tripod to maintain the working distance between the distal end and the tumor tissue. A 637 nm laser is preferable for excitation of ICG H-aggregates, since it falls on the left shoulder of the absorption peak with the maximum at 715 nm. The absorption peaks corresponding to ICG monomers and H- and J-aggregates are presented in our recent paper [30].

Statistical analysis was conducted with the paired Student's *t*-test for demonstration of significance of ICG monomers and H-aggregates, accumulated in tumor tissue. The empirical value is $t = 4.7$, greater than the critical one ($t_{0.01} = 2.78$, $t_{0.05} = 4.6$).

3. Results

The accumulation kinetics of ICG and ICG NPs accumulation in tumor xenografts was assessed according to the fluorescence intensity. Fluorescence spectra were also obtained in the skin, which served as normal tissue sample (Figure 1a). The distribution of the fluorescent dye in excretory organs was obtained at times corresponding to the maximum of its accumulation (Figure 1b,c).

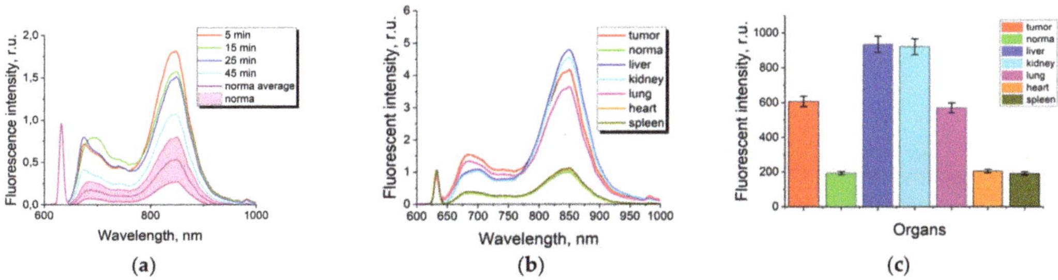

Figure 1. (**a**) Fluorescence spectra of ICG molecular solution in tumor xenografts at different times ($\lambda_{ex} = 633$ nm). The data obtained from the study of normal skin samples at all time points fluctuate in the range between two spectral curves (pink color); (**b**) fluorescence spectra of ICG molecular solution in excretory organs 5 min after intravenous administration ($\lambda_{ex} = 633$ nm); (**c**) integral dependence of the spectral curve of ICG molecular solution in tumor xenografts 5 min after intravenous administration ($\lambda_{ex} = 633$ nm).

The maximum accumulation of ICG in tumor tissue was observed within the first 5 min after intravenous injection and then the fluorescent dye was rapidly eliminated (Figure 1a). To assess the accumulation of ICG in the excretory organs at the time corresponding to ICG maximum accumulation in the tumor, we registered the fluorescence spectra and the distribution of the fluorescence intensity in the organs and in the skin (taken as normal tissue) (Figure 1b,c). Thus, the absence of significant contrast of ICG is demonstrated since the level of accumulation in the tumor slightly exceeds the level of total accumulation in the skin, while the level of ICG accumulation in the liver and kidneys is the highest, which confirms the presence of rapid elimination processes. The obtained results are consistent with other studies where ICG is used in the clinic as a contrast agent for visualization of the vascular network, which is permissible due to its rapid elimination [11–13].

The spectral analysis showed that ICG NPs accumulation in tumor tissue was maximal at 24 h after systemic administration (Figure 2a). Only H-aggregates of ICG were observed at 633 nm excitation after ICG NPs administration corresponding to the fluorescent maximum at 705 nm. The obtained dynamics allow one to conclude that the aggregates enable the fluorescent dye to accumulate and stay in the tumor for a long time (up to 24 h post-administration). It should be noted that the comparison analysis of the fluorescence of ICG and ICG NP at the same time points was not possible. The molecular form of ICG has the rapid clearance by the liver within 1 h after intravenous administration with a rapid successive decline, while the ICG NP fluorescent signal peaks at 5 h after intravenous administration. There is an obvious mismatch between peaks of ICG molecular form and ICG NPs accumulation. The excretory organs analysis demonstrated that H-type aggregates of ICG NP noticeably accumulate in tumor tissue compared with ICG molecular form (Figure 2b,c). As a result of the study, it is worth emphasizing that the aggregated form of ICG NPs is promising from the point of view of its spectral-fluorescent properties, as well as the local contrast accumulation of the nanoform in the tumor.

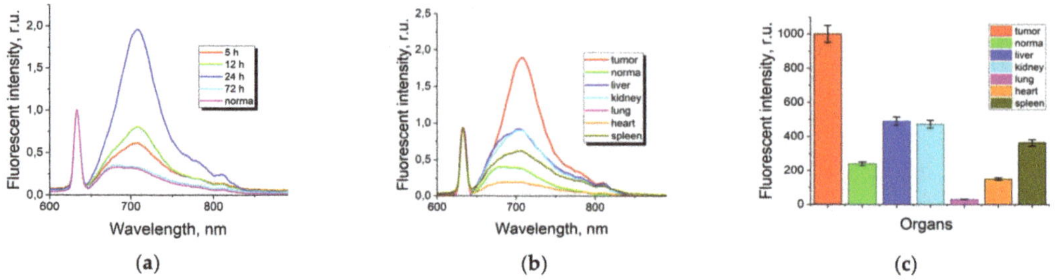

Figure 2. (**a**) Fluorescence spectra of ICG NPs colloidal solution in tumor xenografts at different times (λ_{ex} = 633 nm); (**b**) fluorescence spectra of ICG NPs colloidal solution in excretory organs 24 h after intravenous administration (λ_{ex} = 633 nm); (**c**) integral dependence of the spectral curve of ICG NPs colloidal solution in tumor xenografts under 24 h after intravenous administration (λ_{ex} = 633 nm).

The fluorescence lifetimes of ICG and ICG NPs were obtained via the streak camera at times matching the maximal dyes' accumulation in the tumor (Figure 3).

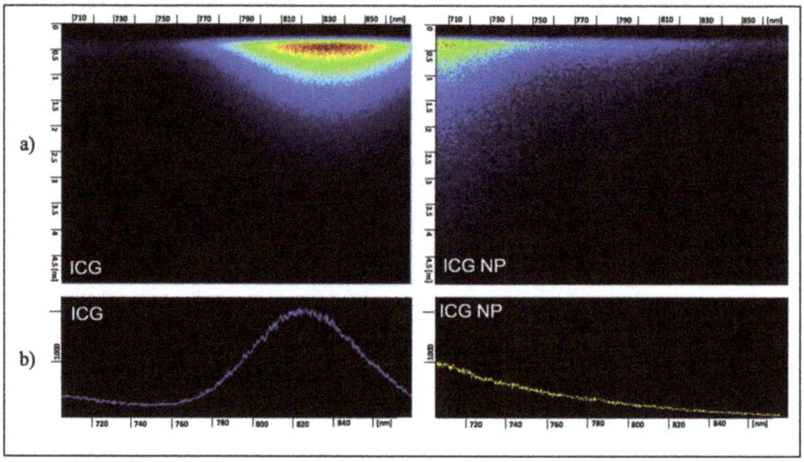

Figure 3. (**a**) Fluorescence spectra of fluorescence of ICG and ICG NPs accumulated in tumor xenograft; (**b**) profiles of fluorescence spectra of ICG and ICG NPs accumulated in tumor xenograft.

The spectra of fluorescence lifetime illustrate a fluorescence shoulder of ICG at 700–730 nm, which corresponds to H-aggregates, and an intense fluorescent peak of ICG at 790–860 nm, which corresponds to monomers. For ICG NPs, the fluorescence signal was registered at 700–730 nm, which corresponded to ICG H-aggregates.

The fitting of the obtained fluorescence decay kinetics after the laser pulse excitation was approximated by an exponential function:

$$I(t) = A_1 e^{-\frac{x}{t1}} + A_2 e^{-\frac{x}{t2}} + \cdots \quad (1)$$

where $A_1, A_2, \ldots$ are the amplitude components of the fluorescent lifetime, directly proportional to the contribution of each exponential component; $t_1, t_2, \ldots$ are the corresponding fluorescent lifetime indicators, measured in nanoseconds.

The fluorescence lifetimes of ICG accumulated in tumor xenografts were obtained for H-aggregates and monomers separately, while ICG NPs fluorescence lifetimes were obtained for H-aggregates. Mathematical fitting of fluorescence kinetic spectra demonstrated the availability of two fluorescence lifetimes for each form of ICG. For each sample five spectra were recorded, which were averaged to perform statistical analysis (Table 1). The second column of the Table presents two components of ICG fluorescence lifetimes (ns).

Table 1. Fluorescence lifetime of ICG in monomeric and aggregate forms in tumor and healthy tissue.

Sample	Fluorescence Lifetime τ, ns	Amplitude of Fluorescence Lifetime, %
ICG monomers in tumor	0.55 ± 0.09 0.82 ± 0.03	21% 79%
ICG monomers in norma	0.37 ± 0.09 0.75 ± 0.05	64% 36%
ICG H-aggregates in tumor	0.28 ± 0.04 1.10 ± 0.09	83% 17%
ICG H-aggregates in norma	0.27 ± 0.08 1.21 ± 0.11	82% 18%
ICG NPs H-aggregates in tumor	0.26 ± 0.03 1.34 ± 0.07	91% 9%
ICG NPs H-aggregates in norma	0.20 ± 0.04 1.00 ± 0.11	97% 3%

The obtained data show the distinction between fluorescence lifetime values of ICG and different forms of ICG NPs accumulated in tumor tissue. The short component of the fluorescence lifetime of the monomeric form is significantly different from the short component of the H-aggregates. The statistical analysis shows statistically significant differences between the monomers and H-aggregates in the molecular solution. Additionally, the long component of monomers is lower in comparison with the long component of H-aggregates. The obtained data were compared with fluorescence lifetimes in normal tissue after intravenous administration. The fluorescence decay kinetics in tumor tissue are slightly different compared to intact tissue. The fluorescence lifetime of ICG monomeric form in the tumor for both components is slightly higher compared to normal tissue. Meanwhile, H-aggregate values vary within the error margin in tumor and normal tissue.

4. Discussion

ICG has significant impact for intraoperative fluorescence navigation in the Near-Infrared-II "window" (1000–1700 nm), especially in brain malignances [10,31,32]. However it has drawbacks like elimination from circulation and low photostability. In addition, once in circulation ICG is rapidly captured by liver, thus limiting its delivery to other sides. This is why the colloidal solution of ICG should be considered due to its stability. The results of the dynamics of the fluorescence signal under 633 nm laser excitation in the tumor after intravenous administration of ICG molecular form demonstrated the maximum fluorescence signal immediately after administration (0–5 min), followed by its uniform elimination within an hour. The studied fluorescent dye ICG in molecular form has promising spectral-fluorescent characteristics necessary for the excitation of deep layers of biological tissues. The fluorescence spectra of ICG NPs in xenografted mice at different time intervals after systemic administration were recorded using $\lambda = 633$ nm, which effectively excite H-aggregates for visualization. We demonstrated the kinetics of accumulation of the fluorescent dye ICG NPs, and at the same time, the processes of its interaction with the biological tissue were controlled. The development and testing of ICG NPs colloidal solution is promising in order to increase its specificity to the tumor. ICG molecular form and ICG NPs have various accumulation times in tumor tissue. The fluorescence of ICG molecular form is observed only in the liver within one hour after intravenous administration, while ICG NPs fluorescent signal is detectable in the tumor from 5 h after intravenous administration. The presence of the ICG aggregates is not detectable immediately, but on average, 24 h after intravenous administration demonstrating an intense fluorescent signal in tumor tissue. ICG NPs are expected to be preferentially accumulated by the enhanced permeability and retention effect (EPR) similar to protein-bound ICG molecular form [33]. We hypothesize that aggregates are not internalized by tumor cells but are retained in the extracellular matrix due to surface negative charge of aggregated forms [28]. It was found that ICG NPs tend to be eliminated from the body by the reticuloendothelial system. As we showed in our recent study [30], a high concentration of ICG colloidal solution is observed in the liver and kidneys, indicating the metabolic pathway.

Fluorescence lifetime analysis of ICG and ICG NPs demonstrate different lifetime components for ICG monomers and H-aggregates. The first fluorescence lifetime component of ICG H-aggregate is twice as short as that of the monomer, while the second component of H-type aggregate is longer compared to monomers. According to these data, fluorescent decay kinetics allow to get information on tumor tissue and its bioenvironment. ICG H-aggregates selectively accumulate in tumor tissue, resulting in a clear advantage for long-lasting fluorescent diagnosis. The monomeric form of ICG is mostly used as a contrast agent for visualization of the vascular network, enabling predicting of metastasis ways. While ICG nanoparticles are very promising for fluorescent diagnostics of brain tumors due to their spectroscopic properties and a high selectivity, thus sparing healthy brain tissue.

5. Conclusions

To conduct a comparative analysis of ICG in molecular and colloidal solutions, the interstitial distribution of the fluorescent dye was studied in pre-clinical models. The accumulation maximum for ICG and ICG NPs were established in the tumor tissue, in normal tissue, and in the excretory organs. The obtained results showed different accumulation times of the studied fluorescent dyes in the pathological tissue: the molecular form of ICG accumulates within 5 min, while ICG NPs accumulation takes 24 h. It was demonstrated that ICG NPs are promising for fluorescence diagnostics due to their unique optical properties and selective accumulation. The analysis of fluorescence lifetime illustrates the difference between lifetime components allowing to separate ICG monomers and H-aggregates in biological tissue. ICG colloidal solutions have significant prospects for fluorescent diagnostics of brain tumors due to their affinity to tumor cells and stability in biological tissue.

Author Contributions: Conceptualization, D.F. and Y.M.; methodology, V.L.; software, I.R.; validation, D.F. and I.R.; formal analysis, L.B.; investigation, V.L.; resources, D.F.; data curation, I.R.; writing—original draft preparation, Y.M.; writing—review and editing, L.B.; visualization, D.F.; supervision, V.L.; project administration, V.L.; funding acquisition, Y.M. All authors have read and agreed to the published version of the manuscript.

Funding: The reported study was funded by RFBR according to the research project No. 21-52-15025 and partly by IEA (International Emerging Action) CNRS grant 00534.

Institutional Review Board Statement: Animal care guidelines were used under the protocol of European Convention for the Protection of Vertebrate Animals used for Experimental and Other Scientific Purposes (Strasbourg, 18.III.1986).

Informed Consent Statement: Not applicable.

Data Availability Statement: Not applicable.

Conflicts of Interest: The authors declare no conflict of interest.

References

1. Nagaya, T.; Nakamura, Y.A.; Choyke, P.L.; Kobayashi, H. Fluorescence-guided surgery. *Front. Oncol.* **2017**, *7*, 314. [CrossRef] [PubMed]
2. Endo, H.; Owada, S.; Inagaki, Y.; Shida, Y.; Tatemichi, M. Metabolic reprogramming sustains cancer cell survival following extracellular matrix detachment. *Redox Biol.* **2020**, *36*, 101643. [CrossRef] [PubMed]
3. Muller, P.J.; Wilson, B.C. *Photodynamic Therapy//Neurooncology: The Essentials*; Thieme: New York, NY, USA, 2000; pp. 249–256.
4. Ma, H.; Gao, Z.; Yu, P.; Shen, S.; Liu, Y.; Xu, B. A dual functional fluorescent probe for glioma imaging mediated by blood-brain barrier penetration and glioma cell targeting. *Biochem. Biophys. Res. Commun.* **2014**, *449*, 44–48. [CrossRef] [PubMed]
5. Korbelik, M.; Krosl, G. Photofrin accumulation in malignant and host cell populations of various tumours. *Br. J. Cancer* **1996**, *73*, 506–513. [CrossRef]
6. Suhling, K.; French, P.M.; Phillips, D. Time-resolved fluorescence microscopy. *Photochem. Photobiol. Sci.* **2005**, *4*, 13–22. [CrossRef]
7. Grabolle, M.; Kapusta, P.; Nann, T.; Shu, X.; Ziegler, J.; Resch-Genger, U. Fluorescence lifetime multiplexing with nanocrystals and organic labels. *Anal. Chem.* **2009**, *81*, 7807–7813. [CrossRef]
8. Raymond, S.B.; Boas, D.A.; Bacskai, B.J.; Kumar, A.T.N. Lifetime-based tomographic multiplexing. *J. Biomed. Opt.* **2010**, *15*, 046011. [CrossRef] [PubMed]
9. Ehlert, O.; Thomann, R.; Darbandi, M.; Nann, T. A four-color colloidal multiplexing nanoparticle system. *ACS Nano* **2008**, *2*, 120–124. [CrossRef] [PubMed]
10. Cho, S.S.; Salinas, R.; Lee, J.Y.K. Indocyanine-Green for Fluorescence-Guided Surgery of Brain Tumors: Evidence, Techniques, and Practical Experience. *Front. Surg.* **2019**, *6*, 11. [CrossRef]
11. Zhang, H.F.; Maslov, K.; Stoica, G.; Wang, L.V. Functional Photoacoustic Microscopy for High-Resolution and Noninvasive In Vivo Imaging. *Nat. Biotechnol.* **2006**, *24*, 848–851. [CrossRef]
12. Diot, G.; Metz, S.; Noske, A.; Liapis, E.; Schroeder, B.; Ovsepian, S.V.; Meier, R.; Rummeny, E.; Ntziachristos, V. Multispectral Optoacoustic Tomography (MSOT) of Human Breast Cancer. *Clin. Cancer Res.* **2017**, *23*, 6912–6922. [CrossRef] [PubMed]
13. Song, K.H.; Wang, L.V. Deep Reflection-Mode Photoacoustic Imaging of Biological Tissue. *J. Biomed. Opt.* **2007**, *12*, 060503. [CrossRef] [PubMed]
14. Desmettre, T.; Devoisselle, J.; Mordon, S. Fluorescence Properties and Metabolic Features of Indocyanine Green (ICG) as Related to Angiography. *Surv. Ophthalmol.* **2000**, *45*, 15–27. [CrossRef]
15. Yan, F.; Wu, H.; Liu, H.; Deng, Z.; Liu, H.; Duan, W.; Liu, X.; Zheng, H. Molecular imaging-guided photothermal/photodynamic therapy against tumor by iRGD-modified indocyanine green nanoparticles. *J. Control Release* **2016**, *224*, 217–228. [CrossRef] [PubMed]
16. Ma, Y.; Tong, S.; Bao, G.; Gao, C.; Dai, Z. Indocyanine green loaded SPIO nanoparticles with phospholipid-PEG coating for dual-modal imaging and photothermal therapy. *Biomaterials* **2013**, *34*, 7706–7714. [CrossRef]
17. Zheng, M.; Yue, C.; Ma, Y.; Gong, P.; Zhao, P.; Zheng, C.; Sheng, Z.; Zhang, P.; Wang, Z.; Cai, L. Single-step assembly of DOX/ICG loaded lipid–polymer nanoparticles for highly effective chemo-photothermal combination therapy. *ACS Nano* **2013**, *7*, 2056–2067. [CrossRef]
18. Navarro, F.P.; Berger, M.; Guillermet, S.; Josserand, V.; Guyon, L.; Neumann, E.; Vinet, F.; Texier, I. Lipid nanoparticle vectorization of indocyanine green improves fluorescence imaging for tumor diagnosis and lymph node resection. *J. Biomed. Nanotechnol.* **2012**, *8*, 730–741. [CrossRef] [PubMed]
19. Sheng, Z.; Hu, D.; Zheng, M.; Zhao, P.; Liu, H.; Gao, D.; Gong, P.; Gao, G.; Zhang, P.; Ma, Y.; et al. Smar human serum albumin-indocyanine green nanoparticles generated by programmed assembly for dual-modal imaging-guided cancer synergistic phototherapy. *ACS Nano* **2014**, *8*, 12310–12322. [CrossRef]
20. Chen, Q.; Liang, C.; Wang, X.; He, J.; Li, Y.; Liu, Z. An albumin-based theranostic nano-agent for dual-modal imaging guided photothermal therapy to inhibit lymphatic metastasis of cancer post surgery. *Biomaterials* **2014**, *35*, 9355–9362. [CrossRef] [PubMed]
21. Huang, P.; Gao, Y.; Lin, J.; Hu, H.; Liao, H.-S.; Yan, X.; Tang, Y.; Jin, A.; Song, J.; Niu, G.; et al. Tumor-specific formation of enzyme-instructed supramolecular self-assemblies as cancer theranostics. *ACS Nano* **2015**, *9*, 9517–9527. [CrossRef]
22. Belfield, K.D.; Bondar, M.V.; Hernandez, F.E.; Przhonska, O.V.; Yao, S. Two-photon absorption of a supramolecular pseudoisocyanine J-aggregate assembly. *Chem. Phys.* **2006**, *320*, 118–124. [CrossRef]
23. Yao, H.; Domoto, K.; Isohashi, T.; Kimura, K. In situ detection of birefringent mesoscopic H and J aggregates of thiacarbocyanine dye in solution. *Langmuir* **2005**, *21*, 1067–1073. [CrossRef] [PubMed]
24. Lovell, J.F.; Jin, C.S.; Huynh, E.; Jin, H.; Kim, C.; Rubinstein, J.L.; Chan, W.C.W.; Cao, W.; Wang, L.V.; Zheng, G. Porphysome nanovesicles generated by porphyrin bilayers for use as multimodal biophotonic contrast agents. *Nat. Mater.* **2011**, *10*, 324–332. [CrossRef] [PubMed]
25. Shakiba, M.; Ng, K.K.; Huynh, E.; Chan, H.; Charron, D.M.; Chen, J.; Muhanna, N.; Foster, F.S.; Wilsom, B.C.; Zheng, G. Stable J-aggregation enabled dual photoacoustic and fluorescence nanoparticles for intraoperative cancer imaging. *Nanoscale* **2016**, *8*, 12618–12625. [CrossRef] [PubMed]
26. Song, X.; Gong, H.; Liu, T.; Cheng, L.; Wang, C.; Sun, X.; Liang, C.; Liu, Z. J-aggregates of organic dye molecules complexed with iron oxide nanoparticles for imaging-guided photothermal therapy under 915-nm light. *Small* **2014**, *10*, 4362–4370. [CrossRef]

27. Song, X.; Zhang, R.; Liang, C.; Chen, Q.; Gong, H.; Liu, Z. Nano-assemblies of J-aggregates based on a NIR dye as a multifunctional drug carrier for combination cancer therapy. *Biomaterials* **2015**, *57*, 84–92. [CrossRef] [PubMed]
28. Liu, R.; Tang, J.; Xu, Y.; Zhou, Y.; Dai, Z. Nano-sized indocyanine green J-aggregate as a one-component theranostic agent. *Nanotheranostics* **2017**, *1*, 430. [CrossRef] [PubMed]
29. Bystrov, F.G.; Makarov, V.I.; Pominova, D.V.; Ryabova, A.V.; Loschenov, V.B. Analysis of photoluminescence decay kinetics of aluminum phthalocyanine nanoparticles interacting with immune cells. *Biomed. Photonics* **2016**, *5*, 3–8. [CrossRef]
30. Farrakhova, D.; Maklygina, Y.; Romanishkin, I.; Yakovlev, D.; Plyutinskaya, A.; Bezdetnaya, L.; Loschenov, V. Fluorescence imaging analysis of distribution of indocyanine green in molecular and nanoform in tumor model. *Photodiagnosis Photodyn. Ther.* **2021**, in press. [CrossRef] [PubMed]
31. Zhu, S.; Tian, R.; Antaris, A.L.; Chen, X.; Dai, H. Near-infrared-II molecular dyes for cancer imaging and surgery. *Adv. Mater.* **2019**, *31*, 1900321. [CrossRef]
32. Teng, C.W.; Cho, S.S.; Singh, Y.; De Ravin, E.; Somers, K.; Buch, L.; Bream, S.; Singhal, S.; Delikatny, J.; Lee, J.Y. Second Window ICG Predicts Gross Total Resection and Progression Free Survival During Brain Metastasis Surgery. *Neurosurgery* **2020**, *67*, 1026–1035. [CrossRef] [PubMed]
33. Belykh, E.; Shaffer, K.V.; Lin, C.; Byvaltsev, V.A.; Preul, M.C.; Chen, L. Blood-brain barrier, blood-brain tumor barrier, and fluorescence-guided neurosurgical oncology: Delivering optical labels to brain tumors. *Front. Oncol.* **2020**, *10*, 739. [CrossRef] [PubMed]

nanomaterials

MDPI

Article

Combining Augmented Radiotherapy and Immunotherapy through a Nano-Gold and Bacterial Outer-Membrane Vesicle Complex for the Treatment of Glioblastoma

Mei-Hsiu Chen [1,2,†], Tse-Ying Liu [3,†], Yu-Chiao Chen [3] and Ming-Hong Chen [4,5,*]

1 Department of Internal Medicine, Far Eastern Memorial Hospital, New Taipei 220, Taiwan; michelle8989@gmail.com
2 Department of Biomedical Engineering, Ming Chuang University, Taoyuan 333, Taiwan
3 Department of Biomedical Engineering, National Yang Ming Chiao Tung University, Taipei 112, Taiwan; tyliu5@ym.edu.tw (T.-Y.L.); ian53011@gmail.com (Y.-C.C.)
4 Graduate Institute of Nanomedical and Medical Engineering, Taipei Medical University, Taipei 110, Taiwan
5 Department of Neurosurgery, Wang Fang Hospital, Taipei Medical University, Taipei 116, Taiwan
* Correspondence: chen.minghong@gmail.com
† These authors contributed equally to this work.

Abstract: Glioblastoma, formerly known as glioblastoma multiforme (GBM), is refractory to existing adjuvant chemotherapy and radiotherapy. We successfully synthesized a complex, Au–OMV, with two specific nanoparticles: gold nanoparticles (AuNPs) and outer-membrane vesicles (OMVs) from *E. coli*. Au–OMV, when combined with radiotherapy, produced radiosensitizing and immunomodulatory effects that successfully suppressed tumor growth in both subcutaneous G261 tumor-bearing and in situ (brain) tumor-bearing C57BL/6 mice. Longer survival was also noted with in situ tumor-bearing mice treated with Au–OMV and radiotherapy. The mechanisms for the successful treatment were evaluated. Intracellular reactive oxygen species (ROS) greatly increased in response to Au–OMV in combination with radiotherapy in G261 glioma cells. Furthermore, with a co-culture of G261 glioma cells and RAW 264.7 macrophages, we found that GL261 cell viability was related to chemotaxis of macrophages and TNF-α production.

Keywords: glioblastoma; gold nanoparticles; outer membrane vesicles; immunotherapy; radioenhancer

Citation: Chen, M.-H.; Liu, T.-Y.; Chen, Y.-C.; Chen, M.-H. Combining Augmented Radiotherapy and Immunotherapy through a Nano-Gold and Bacterial Outer-Membrane Vesicle Complex for the Treatment of Glioblastoma. *Nanomaterials* **2021**, *11*, 1661. https://doi.org/10.3390/nano11071661

Academic Editors: Fiore P. Nicoletta and Francesca Iemma

Received: 26 May 2021
Accepted: 21 June 2021
Published: 24 June 2021

1. Introduction

Glioblastoma (Grade IV), previously known as glioblastoma multiforme (GBM), is the most malignant brain tumor. Despite the great technological advances in imaging, surgery, and adjuvant therapies, the median survival is 14–16 months after diagnosis, and the 5-year overall survival is only 9.8% [1]. In addition, its infiltrative nature is one of the main reasons for tumor recurrence. Current glioblastoma therapy represents a combination of surgery, radiation, and chemotherapy. There are two FDA-approved glioblastoma drugs: one is temozolomide, a DNA alkylating agent, and the other is bevacizumab, a humanized monoclonal antibody IgG1. Both of them are not effective enough; therefore, there is a constant search for new treatments with improved efficiency and less adverse effects [2,3]. For non-surgical treatment of glioblastoma, enhancement of radiotherapy and better targeting of pharmaceutical agents to tumors through immunotherapy were attempted to improve survival. Metal nanoparticles have been widely used in clinical practice, including diagnostic, therapeutic (radiation dose enhancers, hyperthermia inducers, drug delivery vehicles, vaccine adjuvants, photosensitizers, and enhancers of immunotherapy), and theranostic (combining both diagnostic and therapeutic) applications [4]. Gold nanoparticles (AuNPs) are one of the promising agents. They have several advantages: biocompatibility, well-established methods for synthesis in a wide range of sizes, and the possibility of surface coating with many different molecules to provide, for example, surface charge or interactivity with serum proteins [5]. Safety is the most important concern for all clinical applications. Radiotherapy cannot spare 100% of the healthy tissues neighboring cancerous ones. In order to overcome these limitations, combinations of therapeutic modalities have been proposed, and have already been showing promising results with AuNPs [6,7]. In addition, AuNPs are able to target the tumor by passive targeting (enhanced permeability and retention (EPR) effect and mononuclear phagocyte system (MPS) escape) and active targeting (tumor cell targeting and stimuli-response) [8]. Furthermore, photoexcitation of the AuNPs would lead to the generation of reactive oxygen species (ROS), which are known to play a central role in the photodymamic therapy of cancer [9]. Bacterial outer-membrane vesicles (OMVs) are naturally produced from all Gram-negative bacteria and have nano-sized, lipid-bilayered vesicular structures composed of various immunostimulatory components. An OMV-based vaccine is being clinically used as a meningococcal group B vaccine under the trade name Bexsero. Furthermore, studies have revealed the remarkable ability of OMVs to effectively induce long-term antitumor immune responses without notable adverse effects [10]. In this study, we mixed AuNPs and OMVs into a co-suspension of Au–OMV and tested its safety and effects on glioblastoma through the combination of augmented radiotherapy and immunotherapy.

2. Materials and Methods

2.1. Preparation and Characterization of Au–OMV

E. coli (MAX Efficiency^{TM} DH5α Competent Cells, Thermo Fisher, Waltham, MA, USA), were cultured for 16 h on lysogeny broth (1% tryptone, 0.5% yeast extract, 1% NaCl, pH 7.0) at 37 °C with shaking (180 rpm) until the OD600 reached 0.8–1. The cultured cells were pelleted twice at 4000× *g* for 10 min. The supernatant was filtered with a filter having 0.45 μm pore size and was concentrated using a 100 kDa hollow fiber membrane (Amicon, Millipore, Kenilworth, NJ, USA). The concentrate was filtered again using a 0.22 μm filter and was pelleted by ultracentrifugation at 150,000× *g* in a SW 41 Ti rotor (Beckman Coulter Inc., Brea, CA, USA) for 3 h. The pellet was suspended in 50% iodixanol, and we used buoyant density gradients of 10%, 40%, and 50% iodixanol layers at 200,000× *g* for 2 h. The fractions containing bacterial OMVs and extracellular vesicles were collected from the third fraction from the top layer. The purified OMVs were filtered with 0.22 μm filters to avoid any bacteria or cell debris contamination. Protein concentration was determined using Bradford assay. The sample was aliquoted and stored at −80 °C until use [11]. For AuNPs synthesis, we dissolved 1 mg $HAuCl_4$ in 90 mL deionized water and heated the mixture to boil. The reducing solution, 500 μL 250 mM sodium citrate, was added until the light-yellow solution turned to red. We kept stirring for another 30 min and measured the particle size. The solution containing particles was stored at 4 °C. For the preparation of Au–OMV, solutions containing Au and OMV particles were mixed (at concentrations of 200 $\mu g\ mL^{-1}$ and 2 $\mu g\ mL^{-1}$ for Au and OMV, respectively) thoroughly in a homogenizer to form a stable co-suspension without forming aggregations of particles.

The morphologies of Au nanoparticles, OMV, and Au–OMV were examined using surface (JEOL, JSM-7600F, Tokyo, Japan) and transmission electron microscopy (JEOL, JEM-2000EX II, Tokyo, Japan). Size distribution was measured using dynamic light scattering at a scattering angle of 90° using a Zetasizer nano ZS90 (Malvern Instruments, Worcestershire, UK) at 25 °C.

2.2. Cell Culture

GL261 mouse glioma cells, C8D1A mouse astrocytes, B.end3 mouse endothelial cells, and RAW264.7 mouse macrophages were grown in 90% Dulbecco's modified eagle medium (DMEM) supplemented with 10% fetal bovine serum (FBS) and 4 mM glutamine in 75T culture plates at 37 °C in a 5% CO_2-containing atmosphere.

2.3. Cell Viability Test

Cells were seeded in 98-well plates and incubated for 24 h. After washing three times with phosphate-buffered saline (PBS), OMV and AuNPs at various concentrations (diluted with FBS-containing culture medium) were added to the plates and incubated for 24 h. After washing with PBS, 20X diluted PrestoBlue® reagent was added and reacted with cells for 20 min. Viable cells were evaluated using TECAN Sunrise ELISA Reader (TECAN, Zurich, Switzerland) at excitation/emission (Ex/Em) 560/590 nm.

2.4. The Combination of Au–OMV and Radiotherapy Applied to the Co-Culture of GL261 Glioma Cells and RAW264.7 Macrophages

2.4.1. Cell Viability

In total, 3×10^5 GL261 mouse glioma cells were seeded in 6 wells until well attached. After washed with PBS, cells were incubated with 5 μg mL^{-1} OMV and 400 μg mL^{-1} AuNP for 5 h. X-ray radiation (6 MeV, a dose rate of 100 rad/min; and the total dose of X-ray radiation was 2 Gy in a single fraction) was given to the cells, and the 0.4 μm pored upper chambers of transwells with 1×10^6 RAW264.7 mouse macrophages were inserted and co-cultured for 24 h. PrestoBlue® Cell Viability Reagent was used to evaluate the viability of the cells using TECAN Sunrise ELISA Reader at Ex/Em 560/590 nm.

2.4.2. Macrophage Migration (Chemotaxis Assay)

GL261 mouse glioma cells were seeded in 12 wells with a density of 5×10^5/well. After well attached, cells were incubated with 2 μg mL^{-1} OMV and 400 μg mL^{-1} AuNP for 5 h. X-ray radiation (6 MeV, a dose rate of 100 rad/min) was given to the cells, and the total dose of X-ray radiation was 2 Gy in a single fraction. On the other hand, 1×10^6 RAW264.7 mouse macrophages in 1 mL PBS were labeled with 5 μL Vybrant™ DiD (red for macrophages) Cell Labeling Solution at 37 °C for 20 min. After being centrifuged at 15,000 rpm for 5 min, RAW264.7 cell lysates were resuspended in PBS and were placed in 3 μm pored upper chambers of transwells (inserts) with a density of 5×10^5/well at 37 °C and co-cultured for 24 h. After washing with PBS for 3 times, cells were fixed with 4% formalin for 10 min and then 0.1% Triton X-100 was added. After washing with PBS for 3 times, Alexa Fluor™ 488 phalloidin (Ex/Em 493/519, green for actin) was added and the solutions were incubated for 30 min. Furthermore, H33342/DAPI (Ex/Em 358/519, blue for nucleus) was added, and then all were observed with the ZEISS LSM 880 (Zeiss, Oberkochen, Germany) confocal microscope.

2.5. Intracellular ROS Detection Using Flow Cytometry

In total, 3×10^5 GL261 mouse glioma cells were seeded in 6-well plates until well attached. After washed with PBS, cells were incubated with 5 μg mL^{-1} OMV and 400 μg mL^{-1} AuNP for 5 h. After washed with PBS, Trypsin was added for the detachment of cells. The cell lysate was centrifuged with 1000 rpm for 5 min. After discarding the supernatant, 100 μL CellROX® Deep Red Reagent was added to the centrifuged cells. X-ray (6 MeV, a dose rate of 100 rad/min) was given to the cells and the total dose of X-ray radiation was 2 Gy in a single fraction. After washed with 900 μL PBS, cells were dissolved in 1 μL medium and subjected to flow cytometry (Beckman Coulter CytoFLEX, Beckman Coulter Inc., Brea, CA, USA) for the detection of ROS fluorescence at Ex/Em 495/529 nm.

2.6. Confocal Microscopic ROS Detection

GL261 mouse glioma cells were seeded in a Ultra-Low Attachment Surface 6-well plate with a density of 1×10^6 cells/well and cultured with Au–OMV complex overnight. X-ray radiation (6 MeV, a dose rate of 100 rad/min) was given to the cells, and the total dose of X-ray radiation was 2 Gy in a single fraction. The live cells were incubated with 100 μL of CellROX™ Deep Red Reagent (5 μM; Ex/Em 640/665, pink for ROS), 100 μL of H33342/DAPI (Ex/Em 358/519, blue for nucleus), and Alexa Fluor™ 488 phalloidin (Ex/Em 493/519, green for actin) for 1 h and observed under a Zeiss LSM 880 confocal microscope.

2.7. Western Blot

Cells were washed three times using PBS at 24 h after treatment and subsequently added to lysis buffer for 5 min. The protein concentration in each cell lysate was then quantified and adjusted depending on the concentration. The proteins were further separated using sodium dodecyl sulfate polyacrylamide gel electrophoresis (SDS-PAGE) and transferred to a polyvinylidene fluoride (PVDF) membrane. The samples were then blocked with 5% bovine serum albumin (BSA) for 15 min followed by incubating with according primary antibodies at 4 °C overnight. Subsequently, the samples were rinsed using PBS three times and incubated with respective secondary antibodies at 4 °C for 2 h. The samples were then washed with PBS and observed using a luminescence/fluorescence imaging system, and the bands were recorded.

2.8. Animal Models

Six to ten-week old of C57BL/6 mice were obtained from BioLASCO Taiwan Co., Ltd., Taipei, Taiwan.

2.8.1. The Subcutaneous Tumor Models

In total, 5×10^6 GL261 glioma cells in 100 µL PBS were injected into the subcutaneous tissues of the left legs of the C57BL/6 mice. Successful inductions of 100 mm^3 subcutaneous tumors were seen in 14 days. The tumor-bearing mice were randomly divided into different treatment groups. On day 14, the first treatment included intraperitoneal injection of 2 µg mL^{-1} OMV and/or 200 µg mL^{-1} Au–OMV (10 µL in total) with or without 2 Gy of radiotherapy. The treatments were given every 3 days for 5 times. The tumor sizes were measured before and after each treatment every 4 days.

2.8.2. The Brain Tumor Models (In Situ)

In total, 5×10^6 GL261 glioma cells were injected into 25 brains of C57BL/6 mice through a specific locator. On day 5, after confirming successful induction of brain tumors using a non-invasive in vivo imaging system (IVIS) or MRI, mice were randomly divided and treated with 2 µg mL^{-1} OMV and/or 200 µg mL^{-1} Au–OMV (10 µL in total) using in situ injection. On the other day, 2 Gy of radiotherapy was given with an interval of 2–3 days, 5 times. Tumor sizes were evaluated using IVIS and MRI on days 12, 15, and 18.

3. Results

3.1. Au–OMV Complex

The average diameter of synthetic Au-nanoparticles was 18 nm, and the sizes of these particles were homogeneous with a polydispersity index (PDI) equal to 0.073 (Table 1). The average diameter of OMVs was 126 nm. During the preparation of Au–OMV, solutions containing Au and OMV particles were mixed thoroughly in a homogenizer. As both Au and OMV particles were negatively charged, they could be mixed well to form a stable co-suspension solution without forming aggregations of particles (Figure 1).

3.2. Selective Cytotoxicity of OMV to Glioma Cells at Concentration ≥ 2 µg mL^{-1}

OMVs with concentrations ranging from 0 to 10 µg mL^{-1} were incubated with B.end3 mouse brain endothelial cells, C8D1A mouse astrocytes, and GL261 mouse glioma cells at 37 °C for 24 h. When the cells were cultured with OMV concentrations greater than or equal to 2 µg mL^{-1}, the survival rate of GL261 glioma cells reduced to 60%, whereas the survival rates of B.end3 endothelial cells and C8D1A astrocytes remained higher than 80% (Figure 2A).

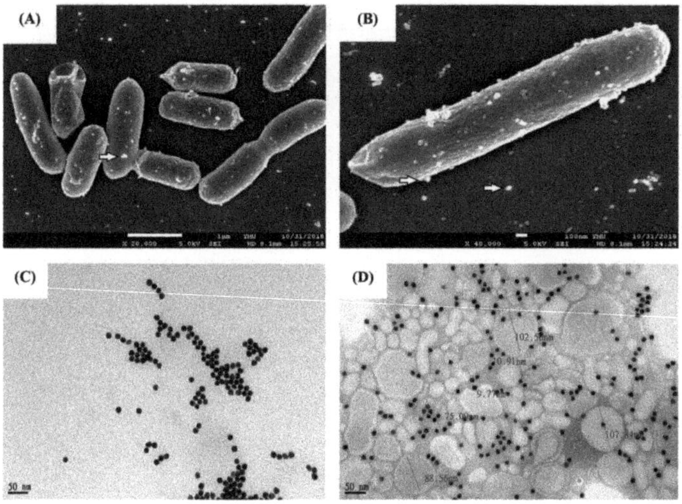

Figure 1. Characterizations of Au–OMV nanoparticles. (**A**,**B**) Scanning electron microscope image (SEM) of OMV production by *E. coli*. (**C**,**D**) TEM images of Au and Au–OMV.

Table 1. The size distributions of Au and OMV particles.

Nanoparticles	Size (nm)	PDI
Au	17.85	0.073
OMV	126	0.238
Au–OMV	42	0.521

3.3. AuNPs Were Nontoxic to Cells Unless Exposed to Radiotherapy at a Concentration of 200 μg mL^{-1}

AuNPs with concentrations ranging from 0 to 800 μg mL^{-1} were incubated with B.end3 mouse brain endothelial cells, C8D1A mouse astrocytes, and GL261 mouse glioma cells at 37 °C for 24 h. The survival rates of all three types of cells were above 80% (Figure 2B). If AuNps-treated GL261 cells were exposed to 2 Gy of radiotherapy, the survival rate of cells was reduced by 30% when the AuNPs' concentration reached 200 μg mL^{-1}. However, with the concentration increased up to 800 μg mL^{-1}, the cytotoxic effect remained at 30%, which showed the limitation of the augmented radiotherapy effect of AuNPs (Figure 2C). The cytotoxicity analysis showed that AuNPs were nontoxic to cells (Figure 2B) unless exposed to radiotherapy at a concentration of 200 μg mL^{-1} (Figure 2C; i.e., showing selective toxicity to the GL261 cells). Therefore, a dose of 2 μg mL^{-1} OMV and/or 200 μg mL^{-1} Au–OMV was chosen for the following animal studies.

3.4. GL261 Cell Viability Reduced Significantly by Combination of Radiotherapy and Au–OMV Treatment When Co-Cultured with Macrophages

Cultured GL261 glioma cells were treated with various treatments and co-cultured with RAW 264.7 macrophages using 0.4 μm pored transwells which allowed only cytokines to move, not macrophages (Figure 3A). GL261 cell viability did not show significant change if they were not co-cultured with macrophages. However, the survival rate of glioma cells reduced to 60% when the co-culture systems were treated with OMV only, and further reduced to 30% if the treatments combined both OMV and AuNPs with 2 Gy of radiotherapy (Figure 3B).

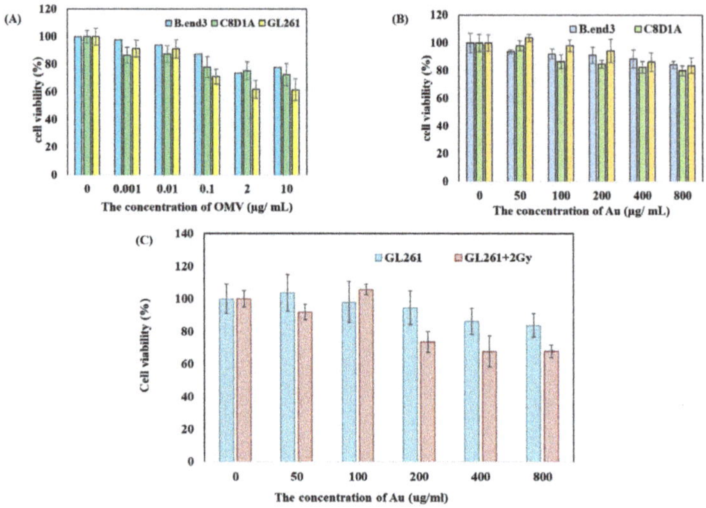

Figure 2. Toxicity of (**A**) OMV and (**B**) Au to B.end3 mouse brain endothelial cells, C8D1A mouse astrocytes, and GL261 mouse glioma cells. (**C**) Au augmented the radiotoxic effect towards GL261 mouse glioma cells.

3.5. Chemotaxis of Macrophages to Glioma Cells Was Induced by Au–OMV in Combination with Radiotherapy

GL261 glioma cells were treated with various treatments and co-cultured with RAW 264.7 macrophages for 24 h using 3 μm transwells, which allowed the migration of macrophages. The migrations of macrophages towards glioma cells (chemotaxis of macrophages) were observed under confocal microscopy using VybrantTM DiD-stained (red) macrophages (Figure 3C). The chemotaxis of macrophages largely increased when glioma cells were treated with Au–OMV combined with 2-Gy radiotherapy (Figure 4). The degrees of chemotaxis of macrophages in response to various treatments seemed correlated with the changes of glioma cell viability to the same treatment.

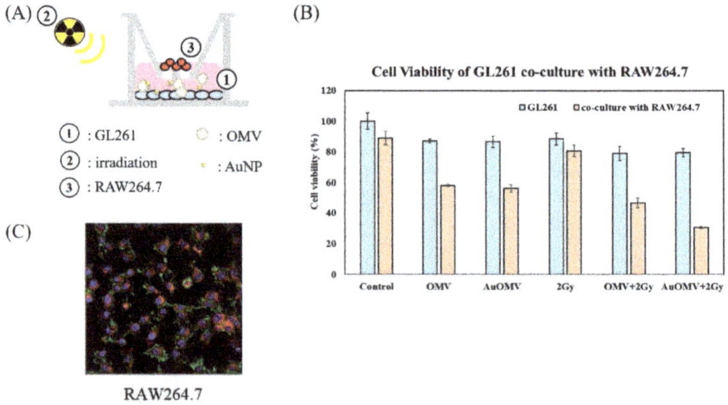

Figure 3. (**A**) GL261 glioma cells (in lower well) were treated with various treatments and co-cultured with RAW 264.7 Macrophages (in upper well) using 0.4 μm pored transwells. (**B**) GL261 cell viability in response to various treatments when co-cultured for 24 h with RAW 264.7 cells. (**C**) Co-culture in transwell system with a 3 μm porous membrane at the bottom of the upper chamber. Migrations of RAW 264.7 mouse macrophages towards glioma cells (chemoatxis of microphages) were noted and were labeled with VybrantTM DiD (red) in the lower chamber.

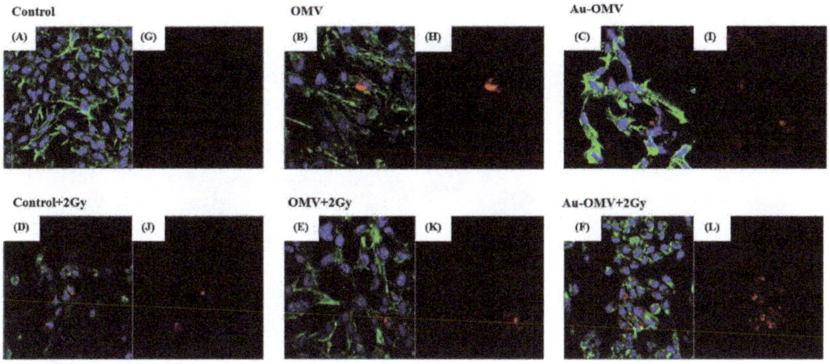

Figure 4. Confocal microscopy of immunofluorescence imaging of the GL 261 glioma cells co-cultured with RAW 264.7 macrophages. (**A**,**G**) Control and treatments with (**B**,**H**) OMV, (**C**,**I**) Au–OMV, (**D**,**J**) 2 Gy X-ray, (**E**,**K**) OMV + 2 Gy X-ray, and (**F**,**L**) Au–OMV + 2 Gy X-ray. (immunofluorescences: H33342/DAPI (blue for nucleus), phalloidin (green for actin), and VybrantTM DiD (red for macrophages).

3.6. ROS Production of Glioma Cells Greatly Increased with Au–OMV in Combination with Radiotherapy

It is well known that metal-based nanoparticles' induction of cancer cell death is related to the generation of ROS [12]. However, ROS production of cancer cells related to OMV has not been explored. In this study, ROS production in GL261 glioma cells was induced by OMV administration, and the reaction was dose-dependent (Figure 5A). With the combination of Au–OMV and radiotherapy, ROS production increased as much as five times the control (Figure 5B).

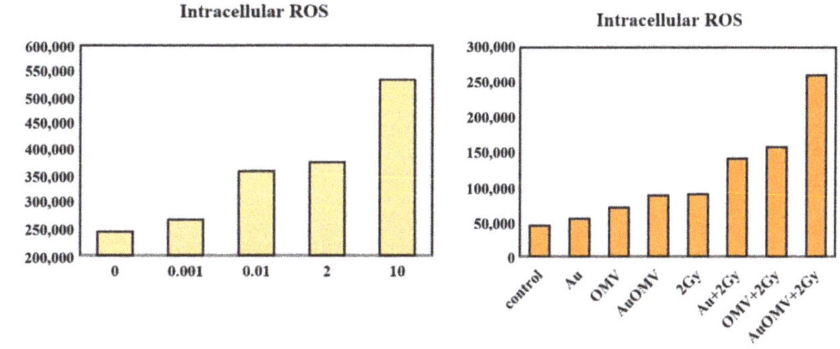

(**A**) (**B**)

Figure 5. Intracellular ROS production of G261 glioma cells in response to (**A**) different concentrations of OMV and (**B**) various treatments.

To simulate 3D tumor behavior, we built a 3D sphere grown with GL261 glioma cells. Briefly, GL261 glioma cells were grown in Nunclon Sphera microplates at 10,000 cells/well, and the live cells were labeled with H33342/ DAPI (blue for nucleus), phalloidin (green for actin), and CellROX Deep Red Reagent for oxidative stress detection (pink). Under confocal microscopy (Figure 6), a mild increase in ROS production was observed when the glioma cells were treated with Au–OMV (Figure 6B) and 2-Gy radiotherapy (Figure 6C). The greatest increase of ROS production was detected while 3D spherical glioma cells were treated with Au–OMV and 2-Gy radiotherapy (Figure 6D). The results in 3D spheres were correlated with those in the culture discs.

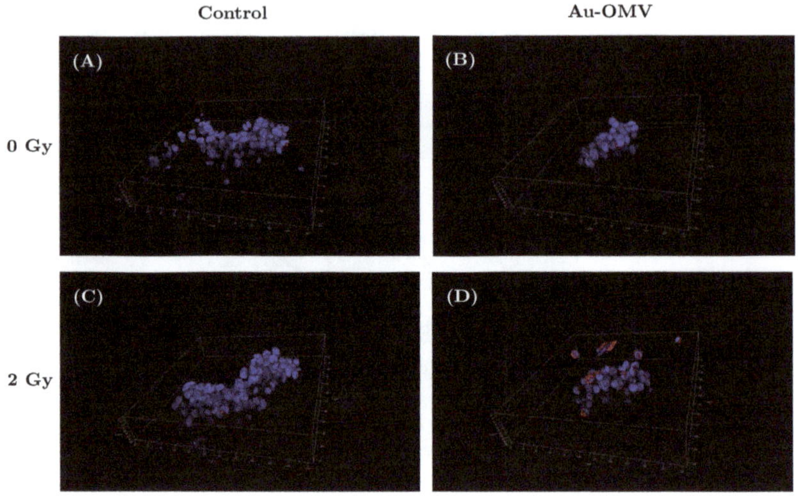

Figure 6. ROS production of live G261 glioma cells grown on 3D spheres in control (**A**), in response to (**B**) Au–OMV, (**C**) 2-Gy radiotherapy, and (**D**) the combination of Au–OMV and radiotherapy under confocal microscopy. The live cells were labeled with H33342/DAPI (blue for nucleus), phalloidin (green for actin), and CellROXTM Deep Red Reagent for oxidative stress detection (pink for ROS).

3.7. TNF-α Expression in the GL261 and RAW 264.7 Co-Culture Was Increased by Combining Radiotherapy with Au–OMV

GL261 glioma cells were treated with various treatments and co-cultured with RAW 264.7 macrophages for 24 h using 0.4 μm transwells, which allowed only cytokines and not macrophages to move. TNF-*α* protein expression in cell lysates was evaluated by Western blots. Increased production of TNF-*α* protein after various treatments was found in the lysate of cells treated with Au–OMV combined with radiotherapy (Figure 7).

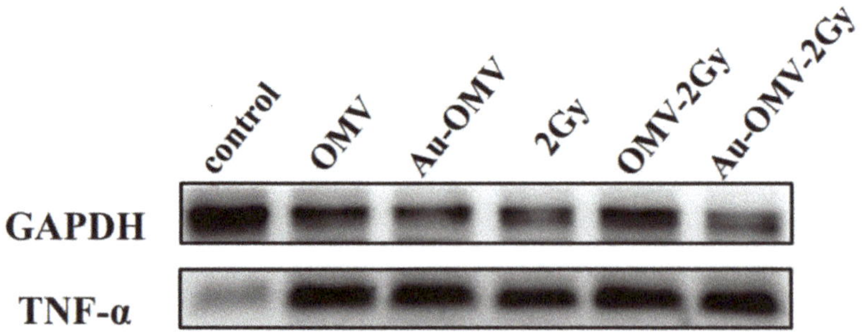

Figure 7. Western blot of TNF-*α* released from RAW 264.7 macrophages co-cultured with GL261 glioma cells. GAPDH was used as a loading control.

3.8. In Vivo Glioblastoma Animal Models Successfully Treated with Intraperitoneal Injection and In Situ Injection of Au–OMV Combined with 2 Gy Radiotherapy

3.8.1. The Subcutaneous Tumor Models

In this step, 5×10^6 GL261 glioma cells were injected into the subcutaneous tissues of the left legs of the C57BL/6 mice. The successful induction of 100 mm^3 subcutaneous tumors was seen in 14 days. The tumor-bearing mice were randomly divided into different treatment groups. Treatment groups were given intraperitoneal injections of 2 μg OMV or 200 μg Au–OMV with or without 2-Gy radiotherapy every 3 days, five times (Figure 8A).

The size of each tumor was measured before and after treatment (Figure 8B). The tumor volume steadily increased with time in the control group, but was successfully suppressed by treatments, especially in the group treated with Au–OMV with 2-Gy radiotherapy (Figure 8C). Growth of tumors was suppressed in all treatment groups, especially in the group treated with Au–OMV + 2 Gy radiotherapy, whose tumors did not grow even at the end of the experiment.

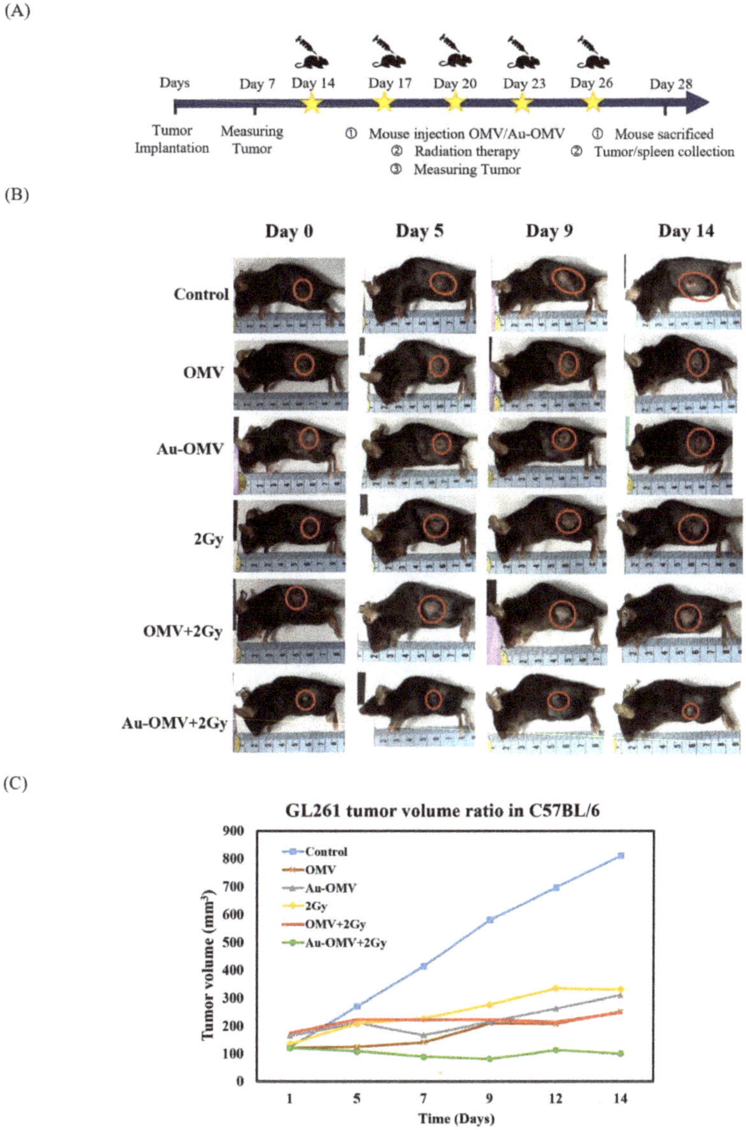

Figure 8. (**A**) Schematic diagram of subcutaneous OMV or Au–OMV injections with radiation therapy for GL261 glioma tumor-bearing mice. (**B**,**C**) Tumor volume kept growing in the control group, but the growth was suppressed in all treatment groups, especially in the group treated with Au–OMV + 2 Gy radiotherapy, whose tumors did not grow, even at the end of the experiment.

3.8.2. The In Situ Brain Tumor Models

In addition, 5×10^6 GL261 glioma cells were injected into the brains of C57BL/6 mice through a specific locator. On day 5, after confirming successful induction of brain tumors using a noninvasive in vivo imaging system (IVIS), mice were randomly divided into groups and treated with Au–OMV in a 10 μL solution using in situ injection on day 7. To some groups, 2-Gy radiotherapy was given. Tumor sizes were evaluated using IVIS on days 12, 15, and 18 (Figure 9A). Only when mice were treated with Au–OMV combined with 2-Gy radiotherapy was tumor growth suppressed at the beginning of the treatment. If tumor size was small, eradication of the tumor was noted after complete treatment. Only when tumor-bearing mice were treated with Au–OMV combined with 2-Gy radiotherapy could they live more than 18 days after tumor implantation (Figure 9B). The survival rate after 18 days was 35% for the Au–OMV with 2-Gy radiotherapy group (Figure 9C).

(A)

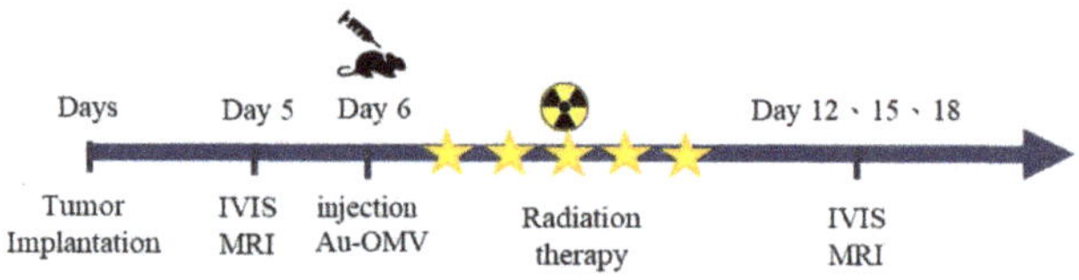

(B)

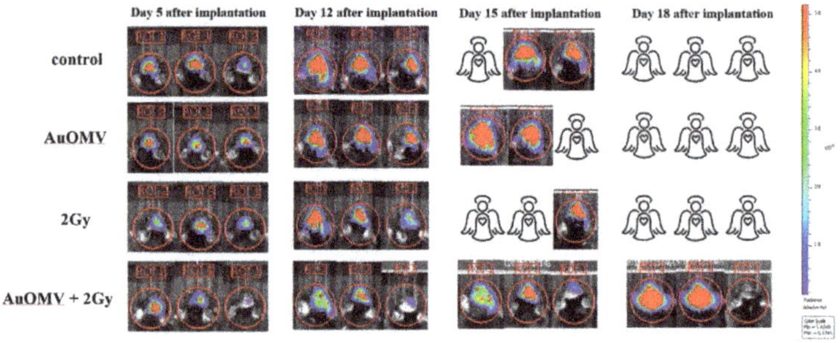

(C)

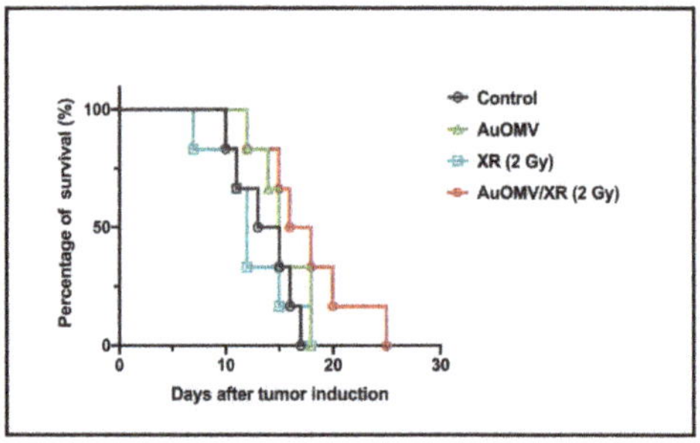

Figure 9. (**A**) Schematic diagram of intracranial Au–OMV injection with radiation therapy for GL261 glioma tumor-bearing mice. (**B**) Assessment of the IVIS imaging system as a method for monitoring tumor growth in a GL261 in situ tumor mode. IVIS imaging of mouse brain tumors on days 5, 12, 15, and 18 after implantation of tumor cells; imaging results of three mice in each group. All images were at the same scale. (**C**) Kaplan–Meier survival curves of control and treatment groups (N > 5 per arm).

4. Discussion

Glioblastoma is known as the most difficult brain tumor to remove because of its infiltrating nature. In addition, the difficulty of reaching the tumor with chemotherapeutic drugs due to the blood brain barrier (BBB) and the limitations of radiotherapy in eradicating radio-resistant glioblastoma cells further prevent the development of effective treatment. To overcome these hurdles, the use of nano-particulate anti-GBM drugs has been suggested [13]. Before our study, Au-related nanoparticles had been used to radio-sensitize glioma stem cells [14], and OMVs had been evaluated for antitumor treatments in multiple tumors, but never been studied for the treatment of glioblastoma.

In this study, we successfully synthesized a complex of two specific nanoparticles, AuNPs and OMVs, to form Au–OMVs (Figure 1). Au–OMVs improve the GBM treatment via several different mechanisms. The efficacy of radiotherapy against GBM tumor is increased due to its radio-sensitizing effect, and the immune mechanisms relying on activation of anti-tumor cytokines and immune cells and local production of ROS are enhanced. At first, we evaluated our synthesized AuNPs and OMVs individually. OMVs had specific cytotoxicity to mouse glioma cells and were relatively safe to endothelial cells and astrocytes. At concentrations above 2 $\mu g\ mL^{-1}$, GL261 cells' survival rate reduced to 60% as compared to 80% of B.end3 and C8D1A cells (Figure 2A). Up to 800 $\mu g\ mL^{-1}$ of AuNPs, glioma cells, and normal endothelial cells and astrocytes had survival rates of at least 80% (Figure 2B). AuNPs had a radio-sensitizing effect at the concentration of 200 $\mu g\ dL^{-1}$, causing a 30% reduction of GL261 cell survival, and the augmentation was not dose-related (Figure 2C). In general, both OMV and Au particles were relatively safe to normal endothelial cells and astrocytes and had mild cytotoxic effects on glioma cells at specific concentrations.

The histopathology of glioblastoma is characterized by significant infiltration of resident microglia and peripheral macrophages in the tumor and pervasive infiltration of tumor cells into the healthy surroundings of the tumor. Microglia and macrophages are the main innate immune cells of the central nervous system. These macrophage populations infiltrate the brain tumor area and constitute up to 50% of non-neoplastic cells, presenting an opening for therapeutic strategies [15]. It has been speculated that the recruitment of tumor-associated microglia and macrophages by tumor cells could be a potential approach for drug delivery [16,17]. Before our study, it was clear that macrophages play a significant role in the pathophysiology of glioblastomas. In our study, we co-cultured RAW 264.7 macrophages and G261 glioma cells using transwells with different pore sizes to simulate the microenvironment of the glioblastoma and evaluate the immunological responses to different treatments, especially our new Au–OMVs. Using 0.4 μm transwells, macrophages cultured in the transwell-insert influenced glioma cell growth without migration to the lower compartment (Figure 3A). The survival of GL261 glioma cells reduced to 30% when the co-cultured systems were treated with Au–OMV combined with 2-Gy radiotherapy (Figure 3B).

Reactive oxygen species (ROS) produced in eukaryotic cells through aerobic metabolism have evolved as regulators of important signaling pathways. ROS are generated after exposure to physical agents and after chemotherapy and radiotherapy [18]. It was not surprising that intracellular ROS increased after radiotherapy. However, in spite of limited literature, there was still evidence showing that OMVs dramatically increased the production of MCP-1 (macrophage/monocyte chemoattractant protein-1) and MCP-2 (macrophage/monocyte chemoattractant protein-2) cytokines from neurons [19]. The elevated secretion of MCP-1 was then associated with changes in oxidative stress [20,21]. In our study, intracellular ROS production also increased with OMV treatment alone in a dose-dependent way (Figure 5A). OMVs are known to have immunomodulatory function [22]. The ROS production of GL261 glioma cells in response to OMV treatment might have been related to the immunomodulation. To our knowledge, this is the first study to demonstrate OMVs' effect on ROS production in glioma cells. Furthermore, treatment with Au–OMV caused more ROS production than OMV alone. Both Au and OMVs increased the intracellular ROS production in response to 2-Gy radiotherapy. The complex of Au–OMV in combination with 2-Gy radiotherapy induced the greatest intracellular ROS production (Figure 5B). The ROS production was mostly induced by the combination of Au–OMV with 2 Gy radiotherapy, not only in 2D culture, but also in 3D culture (Figure 6). ROS have been implicated as mediators of cell survival and cell death triggered by TNF-α signaling [23,24].

TNF-α was thought to be produced primarily by macrophages [25]. Using 3 μm transwells co-culture system, we labeled macrophages with red fluorescence (Figure 3C) and observed macrophage migration (chemotaxis) under a confocal microscope. The complex of Au–OMV in combination with 2-Gy radiotherapy made the most macrophages migrate from the upper compartment to the lower compartment (Figure 4). The reduction of GL261 glioma cell survival seemed correlated with macrophage chemotaxis to glioma cells. Not only did macrophage chemotaxis increase with Au–OMV and 2-Gy radiotherapy treatment, but macrophage-related cytokine also increased. Using a 0.4 μm transwells co-culture system and Western blotting, we also found that the protein levels of TNF-α increased after the system was treated with Au–OMV with 2-Gy radiotherapy (Figure 7).

Based on the successful in vitro results, we tested the Au–OMV complex on two in vivo animal models with C57BL/6 mice. First, for easier tumor observation and more convenient drug administration, we created the subcutaneous tumor mass using GL261 glioma cells were implanted on the left thighs of C57BL/6 mice. Treatment groups were given intraperitoneal injections of 2 μg OMV or 200 μg Au–OMV with or without 2-Gy radiotherapy every 3 days, five times (Figure 8A). The tumor volume steadily increased with time in the control group, but was successfully suppressed by treatments, especially in the group treated with Au–OMV and 2-Gy radiotherapy. After 2 weeks, tumor volume barely changed in tumor-bearing mice treated with Au–OMV and 2-Gy radiotherapy in comparison to an eight-fold increase in tumor volume of the control mice (Figure 8C).

Encouraged by the successful subcutaneous model results, we created an in situ brain tumor animal model by implanting GL261 cells into the brain using a specific locator. The in situ brain tumor model was more challenging for tumor size measurements and drug administration. For tumor measurements, we injected D-luciferin preparation into peritoneum and observed tumor size by IVIS. Successful tumor reduction was observed in the in situ tumor-bearing mice treated with Au–OMV and 2-Gy radiotherapy; control mice or mice treated with only Au–OMV or 2-Gy radiotherapy experienced tumor growth or death. If tumor size was small, eradication of the tumor was noted at the end of the experiment. However, tumor regrowth was noted in some mice treated with Au–OMV and 2-Gy radiotherapy, whereas non-treated mice or mice that received a single treatment died (Figure 9B). Non-treated, in situ, GL261 tumor-bearing mice had very short lives (ranged from 15–18 days). Although tumor regrowth was noted, tumor-bearing mice treated with Au–OMV and 2-Gy radiotherapy had longer survival (Figure 9C). More experimentation is needed to evaluate whether higher dosages or repeated treatments will help to prolong the disease-free time.

5. Conclusions

In this study, the combination of Au–OMV and radiotherapy had a specific cytotoxic effect on GL261 glioma cells which may have been related to intracellular ROS production, chemotaxis of macrophages, and TNF-α production. We also created two animal models using GL261 cells implanted in the subcutaneous tissues or brains of C57BL/6 mice. As a result, suppression of tumor growth was noted in all tumor-bearing mice treated with both Au–OMV and radiotherapy. Au–OMV opened up a new avenue for the treatment of glioblastoma using low-dose combination radiotherapy. However, further studies on the detailed mechanisms of OMVs are important for more understanding. Such understanding could be crucial for the future development of chemically conjugated Au–OMV preparations. As Au–OMV-based therapy uses different mechanisms from temozolomide and bevacizumab in the treatment of glioblastoma, more studies are necessary to compare these therapeutic agents and investigate the feasibility of combinations of them in chemoradiation strategies of glioblastoma management.

Author Contributions: Conceptualization, M.-H.C. (Ming-Hong Chen) and T.-Y.L.; methodology, Y.-C.C.; validation, M.-H.C. (Mei-Hsiu Chen) and T.-Y.L.; formal analysis, M.-H.C. (Ming-Hong Chen); investigation, Y.-C.C.; data curation, Y.-C.C.; writing—original draft preparation, M.-H.C. (Mei-Hsiu Chen); writing—review and editing, M.-H.C. (Mei-Hsiu Chen) and M.-H.C. (Ming-Hong Chen); supervision, M.-H.C. (Ming-Hong Chen) and T.-Y.L.; project administration, M.-H.C. (Ming-Hong Chen); funding acquisition, M.-H.C. (Mei-Hsiu Chen), T.-Y.L. and M.-H.C. (Ming-Hong Chen). All authors have read and agreed to the published version of the manuscript.

Funding: This research was funded by Ministry of Science and Technology, Taiwan, grant number MOST-108-2314-B010-035-MY3; and the Far Eastern Memorial Hospital, grant number FEMH-2020-C-035 and FEMH-2021-C-018.

Institutional Review Board Statement: The study was conducted according to the guidelines of the Declaration of Helsinki, and approved by the Ethics Committee of National Yang Ming Chiao Tung University. Number: IACUC No. 1080316.

Acknowledgments: The authors would like to thank Taipei Veterans General Hospital for providing facilities.

Conflicts of Interest: The authors declare that they have no known competing financial interests or personal relationships that could have appeared to influence the work reported in this paper. The funders had no role in the design of the study; in the collection, analyses, or interpretation of data; in the writing of the manuscript, or in the decision to publish the results.

References

1. Nørøxe, D.S.; Poulsen, H.S.; Lassen, U. Hallmarks of glioblastoma: A systematic review. *ESMO Open* **2016**, *1*, e000144. [CrossRef]
2. Mujokoro, B.; Adabi, M.; Sadroddiny, E.; Adabi, M.; Khosravani, M. Nano-structures mediated co-delivery of therapeutic agents for glioblastoma treatment: A review. *Mater. Sci. Eng.* **2016**, *69*, 1092–1102. [CrossRef]
3. Šamec, N.; Zottel, A.; Videtič Paska, A.; Jovčevska, I. Nanomedicine and Immunotherapy: A Step Further towards Precision Medicine for Glioblastoma. *Molecules* **2020**, *25*, 490. [CrossRef] [PubMed]
4. Schuemann, J.; Bagley, A.F.; Berbeco, R.; Bromma, K.; Butterworth, K.T.; Byrne, H.L.; Chithrani, B.D.; Cho, S.H.; Cook, J.R.; Favaudon, V.; et al. Roadmap for metal nanoparticles in radiation therapy: Current status, translational challenges, and future directions. *Phys. Med. Biol.* **2020**, *65*, 21RM02. [CrossRef] [PubMed]
5. Haume, K.; Rosa, S.; Grellet, S.; Śmiałek, M.A.; Butterworth, K.T.; Solov'yov, A.V.; Prise, K.M.; Golding, J.; Mason, N.J. Gold nanoparticles for cancer radiotherapy: A review. *Cancer Nanotechnol.* **2016**, *7*, 8. [CrossRef] [PubMed]
6. Kareliotis, G.; Tremi, I.; Kaitatzi, M.; Drakaki, E.; Serafetinides, A.A.; Makropoulou, M.; Georgakilas, A.G. Combined radiation strategies for novel and enhanced cancer treatment. *Int. J. Radiat. Biol.* **2020**, *96*, 1087–1103. [CrossRef]
7. Spyratou, E.; Makropoulou, M.; Efstathopoulos, E.P.; Georgakilas, A.G.; Sihver, L. Recent Advances in Cancer Therapy Based on Dual Mode Gold Nanoparticles. *Cancers* **2017**, *9*, 173. [CrossRef] [PubMed]
8. Iyer, A.K.; Khaled, G.; Fang, J.; Maeda, H. Exploiting the enhanced permeability and retention effect for tumor targeting. *Drug Discov. Today* **2006**, *11*, 812–818. [CrossRef] [PubMed]
9. Lucky, S.S.; Soo, K.C.; Zhang, Y. Nanoparticles in photodynamic therapy. *Chem. Rev.* **2015**, *115*, 1990–2042. [CrossRef]
10. Kim, O.Y.; Park, H.T.; Dinh, N.T.H.; Choi, S.J.; Lee, J.; Kim, J.H.; Lee, S.W.; Gho, Y.S. Bacterial outer membrane vesicles suppress tumor by interferon-γ-mediated antitumor response. *Nat. Commun.* **2017**, *8*, 626. [CrossRef]
11. Klimentová, J.; Stulík, J. Methods of isolation and purification of outer membrane vesicles from gram-negative bacteria. *Microbiol. Res.* **2015**, *170*, 1–9. [CrossRef]
12. Liu, Y.; Zhang, P.; Li, F.; Jin, X.; Li, J.; Chen, W.; Li, Q. Metal-based NanoEnhancers for Future Radiotherapy: Radiosensitizing and Synergistic Effects on Tumor Cells. *Theranostics* **2018**, *8*, 1824. [CrossRef]
13. Alphandéry, E. Nano-Therapies for Glioblastoma Treatment. *Cancers* **2020**, *12*, 242. [CrossRef] [PubMed]
14. Kunoh, T.; Shimura, T.; Kasai, T.; Matsumoto, S.; Mahmud, H.; Khayrani, A.C.; Seno, M.; Kunoh, H.; Takada, J. Use of DNA-generated gold nanoparticles to radiosensitize and eradicate radioresistant glioma stem cells. *Nanotechnology* **2018**, *30*, 055101. [CrossRef] [PubMed]
15. Hambardzumyan, D.; Gutmann, D.H.; Kettenmann, H. The role of microglia and macrophages in glioma maintenance and progression. *Nat. Neurosci.* **2016**, *19*, 20. [CrossRef]
16. Phillips, W.T.; Bao, A.; Brenner, A.J.; Goins, B.A. Image-guided interventional therapy for cancer with radiotherapeutic nanoparticles. *Adv. Drug Deliv. Rev.* **2014**, *76*, 39–59. [CrossRef] [PubMed]
17. Poon, C.C.; Sarkar, S.; Yong, V.W.; Kelly, J.J.P. Glioblastoma-associated microglia and macrophages: Targets for therapies to improve prognosis. *Brain* **2017**, *140*, 1548–1560. [CrossRef] [PubMed]

18. Perillo, B.; Di Donato, M.; Pezone, A.; Di Zazzo, E.; Giovannelli, P.; Galasso, G.; Castoria, G.; Migliaccio, A. ROS in cancer therapy: The bright side of the moon. *Exp. Mol. Med.* **2020**, *52*, 192–203. [CrossRef]
19. Wawrzeniak, K.; Gaur, G.; Sapi, E.; Senejani, A.G. Effect of Borrelia burgdorferi Outer Membrane Vesicles on Host Oxidative Stress Response. *Antibiotics* **2020**, *9*, 275. [CrossRef]
20. Kolattukudy, P.E.; Niu, J. Inflammation, endoplasmic reticulum stress, autophagy, and the monocyte chemoattractant protein-1/CCR2 pathway. *Circ. Res.* **2012**, *110*, 174–189. [CrossRef]
21. Kim, W.K.; Choi, E.K.; Sul, O.J.; Park, Y.K.; Kim, E.S.; Yu, R.; Suh, J.H.; Choi, H.S. Monocyte chemoattractant protein-1 deficiency attenuates oxidative stress and protects against ovariectomy-induced chronic inflammation in mice. *PLoS ONE* **2013**, *8*, e72108. [CrossRef] [PubMed]
22. Jan, A.T. Outer Membrane Vesicles (OMVs) of Gram-negative Bacteria: A Perspective Update. *Front. Microbiol.* **2017**, *8*, 1053. [CrossRef] [PubMed]
23. Bae, Y.S.; Oh, H.; Rhee, S.G.; Yoo, Y.D. Regulation of reactive oxygen species generation in cell signaling. *Mol. Cells* **2011**, *32*, 491–509. [CrossRef] [PubMed]
24. Kim, J.J.; Lee, S.B.; Park, J.K.; Yoo, Y.D. TNF-alpha-induced ROS production triggering apoptosis is directly linked to Romo1 and Bcl-X(L). *Cell Death Differ.* **2010**, *17*, 1420–1434. [CrossRef] [PubMed]
25. Olszewski, M.B.; Groot, A.J.; Dastych, J.; Knol, E.F. TNF trafficking to human mast cell granules: Mature chain-dependent endocytosis. *J. Immunol.* **2007**, *178*, 5701–5709. [CrossRef] [PubMed]

nanomaterials

MDPI

Article

Combining Dextran Conjugates with Stimuli-Responsive and Folate-Targeting Activity: A New Class of Multifunctional Nanoparticles for Cancer Therapy

Manuela Curcio [1], Alessandro Paolì [1], Giuseppe Cirillo [1], Sebastiano Di Pietro [2], Martina Forestiero [1], Francesca Giordano [1], Loredana Mauro [1], Diana Amantea [1], Valeria Di Bussolo [2], Fiore Pasquale Nicoletta [1,*] and Francesca Iemma [1]

[1] Department of Pharmacy, Health and Nutritional Sciences, University of Calabria, 87036 Rende, Italy; manuela.curcio@unical.it (M.C.); alessandro.paoli28@gmail.com (A.P.); giuseppe.cirillo@unical.it (G.C.); marti.forestiero@libero.it (M.F.); francesca.giordano@unical.it (F.G.); loredana.mauro@unical.it (L.M.); diana.amantea@unical.it (D.A.); francesca.iemma@unical.it (F.I.)

[2] Department of Pharmacy, University of Pisa, Via Bonanno Pisano 33, 56126 Pisa, Italy; sebastiano.dipietro@unipi.it (S.D.P.); valeria.dibussolo@unipi.it (V.D.B.)

* Correspondence: fiore.nicoletta@unical.it; Tel.: +39-0984-493194

Citation: Curcio, M.; Paolì, A.; Cirillo, G.; Di Pietro, S.; Forestiero, M.; Giordano, F.; Mauro, L.; Amantea, D.; Di Bussolo, V.; Nicoletta, F.P.; et al. Combining Dextran Conjugates with Stimuli-Responsive and Folate-Targeting Activity: A New Class of Multifunctional Nanoparticles for Cancer Therapy. *Nanomaterials* **2021**, *11*, 1108. https://doi.org/10.3390/nano11051108

Academic Editor: Manuel Arruebo

Received: 30 March 2021
Accepted: 23 April 2021
Published: 25 April 2021

Abstract: Nanoparticles with active-targeting and stimuli-responsive behavior are a promising class of engineered materials able to recognize the site of cancer disease, targeting the drug release and limiting side effects in the healthy organs. In this work, new dual pH/redox-responsive nanoparticles with affinity for folate receptors were prepared by the combination of two amphiphilic dextran (DEX) derivatives. DEXFA conjugate was obtained by covalent coupling of the polysaccharide with folic acid (FA), whereas DEXssPEGCOOH derived from a reductive amination step of DEX was followed by condensation with polyethylene glycol 600. After self-assembling, nanoparticles with a mean size of 50 nm, able to be destabilized in acidic pH and reducing media, were obtained. Doxorubicin was loaded during the self-assembling process, and the release experiments showed the ability of the proposed system to modulate the drug release in response to different pH and redox conditions. Finally, the viability and uptake experiments on healthy (MCF-10A) and metastatic cancer (MDA-MB-231) cells proved the potential applicability of the proposed system as a new drug vector in cancer therapy.

Keywords: dextran conjugate; folic acid; pH/redox responsive nanoparticles; targeted release; cystamine; PEG diacid

1. Introduction

In the last decades, the acquisition of ever more complete information about the physiopathological features of cancer tissues coupled with the tremendous progresses in nanotechnology applied in the biomedical field, and in cancer therapy in particular, has led to the development of several nanoparticles for the targeted release of anticancer drugs as an alternative approach to overcome the well-known limits of conventional chemotherapy [1,2].

By virtue of the Enhanced Permeation and Retention (EPR) effect, a typical condition of the tumor tissues characterized by angiogenesis and lack of lymphatic drainage, nanoparticles can efficiently accumulate at the tumor site [3,4]. In addition, this kind of system can be designed to recognize specific elements in cancer cells (i.e., overexpressed membrane receptors) [5,6] and/or to respond to specific signals (variation of temperature, pH, redox potential) from the tumor microenvironment [7], enhancing, in both cases, the amount of drug released in the target site. More in detail, it is well known that folate receptors are overexpressed in many solid tumors [8,9], that the extracellular environment is more acidic (pH 6.5) in tumors than in blood and in normal tissues (pH 7.4) [10,11], and that pH values of endosomes/lysosomes are even lower (5.0–5.5) [12,13]. Moreover, the glutathione (GSH) concentration in cancer cells (approximately 2–10 mM) is almost 1000-fold higher than the extracellular matrix (approximately 2–20 μM), generating a high redox potential across cell membranes [14]. Taken together, all these pieces of evidence can serve as ideal triggers for the vectorization of anticancer drugs in tumor cells mediated by nanoparticle systems [15].

Natural polymers, and polysaccharides in particular, were extensively investigated as base materials for the preparation of targeted nanocarriers because of their non-toxicity, cost-effectivity, and physico-chemical features [16,17].

Among others, dextran (DEX), a bacterial-deriving glucose homopolysaccharide, is widely employed in drug delivery due to its high-water solubility, biocompatibility, biodegradability, resistance to protein adsorption, and ease of chemical modification due to the presence of reactive hydroxyl groups [18–20].

The functionalization of DEX with hydrophobic moieties endowed with targeting activity is a useful approach to obtain self-assembling nanocarriers able to vectorize the payload in cancer cells [21]. As an example, the derivatization of DEX with folic acid (FA) generated actively targeted nanoparticle structures for the release of Doxorubicin (DOX) in breast cancer [22,23]. Similarly, the conjugation with hydrophobic chemical species endowed with disulfide bridges, hydrazone, or imine bonds carried out to self-assembling materials for the pH and redox responsive release of anticancer drugs [18,24,25].

In this work, we prepared two new amphiphilic DEX derivatives, DEXFA and DEXssPEGCOOH, with targeted and stimuli-responsive activity, respectively. DEXFA conjugate derived from the covalent coupling of the polysaccharide with FA, whereas DEXssPEGCOOH was obtained from the reaction of a cystamine-modified DEX with polyethylene glycol 600 diacid ($PEG_{600}COOH$). The combination of both derivatives allowed obtaining multifunctional self-assembling nanoparticles (DFNPs) with active-targeted and pH/redox responsive activities. The nanoparticles were characterized by Dynamic Light Scattering (DLS) and Transmission Electron Microscopy (TEM), whereas pH/redox-triggered destabilization assays were performed by measuring the variation of the nanoparticles mean diameter in reducing media and acidic pH. DFNPs were loaded with Doxorubicin hydrochloride (DOX), a DNA topoisomerase II inhibitor with a broad-spectrum antineoplastic activity [26], and in vitro release experiments from DOX-loaded DFNPs were performed varying the pH and redox potential of the surrounding medium. Cytotoxicity and cellular uptake experiments were performed on healthy (MCF-10A) and cancer (MDA-MB-231) cells to evaluate the safety and potential suitability of the system in cancer therapy. Ultimately, cell cycle analysis confirmed the efficacy of the drug delivery system tested in MDA-MB-231 cells.

2. Materials and Methods

2.1. Synthesis of DEXcys

DEXcys was obtained by reductive amination [27,28]. Briefly, after 0.2 g (1.2 mmol glucose repeating units) of DEX (40 kDa) was dissolved in 20 mL of a H_2O:DMSO (3:7 *v*/*v*) mixture, 0.78 g (12.3 mmol) of sodium cyanoborohydride and 2.78 g (12.3 mmol) of cystamine dihydrochloride (cysHCl) were added, and the mixture was stirred for 24 h at room temperature. The resulting solution was purified by dialysis (MWCO 12–14 kDa) against water at 20 °C for 72 h and finally freeze-dried (98% yield). ^{1}H-NMR and 2D-HSQC spectra were recorded on a Bruker Avance III 400 MHz (Bruker Italy, Milan, Italy) at 25 °C using a DMSO/D_2O (1:1 *v*/*v*) mixture as solvent. Dialysis membranes were purchased from Medicell International LTD (London, UK).

All chemicals were purchased from Merck/Sigma Aldrich, Darmstadt, Germany.

2.2. Synthesis of DEXssPEGCOOH Conjugate

PEG_{600}diacid (0.063 g), 1-etil-3-(3-dimetilamminopropil) carbodiimide (EDC) (0.04 g, 0.21 mmol), and *N*-hydroxy succinimide (0.024 g, 0.21 mmol) were dissolved in 3 mL of

DMSO and left to react for 1 h at room temperature under magnetic stirring. Then, DEXcys (0.017 g) dissolved in 2 mL DMSO was added. The mixture was magnetically stirred for 24 h at room temperature, purified by dialysis (MWCO 12–14 kDa) against water at 20 °C for 72 h, and finally freeze-dried (98% yield). Dialysis membranes were purchased from Medicell International LTD (London, UK). ^{1}H-NMR and 2D-HSQC spectra were recorded on a Bruker Avance III 400 MHz (Bruker Italy, Milan, Italy) at 25 °C using DMSO/D_2O (1:1 v/v) mixture as solvent. All chemicals were purchased from Merck/Sigma Aldrich, Darmstadt, Germany.

2.3. Synthesis of DEXFA

FA (0.5 g, 1.13 mmol) was dissolved in 10 mL DMSO, then 0.66 g (3.44 mmol) of EDC, 0.44 g (3.47 mmol) of NHS, and 0.5 g (2.98 mmol) of DEX were added, and the mixture was left to react for 48 h at 45 °C under magnetic stirring. The resulting solution was purified by dialysis (MWCO 12–14 kDa) against a phosphate buffer (0.01 M, pH 7.4) and water for 24 and 48 h, respectively, and finally freeze-dried (98% yield). ^{1}H-NMR and 2D-HSQC spectra were recorded on a Bruker Avance III 400 MHz (Bruker Italy, Milan, Italy) at 25 °C using DMSO as solvent. Dialysis membranes were purchased from Medicell International LTD (London, UK). All chemicals were purchased from Merck/Sigma Aldrich, Darmstadt, Germany.

2.4. Determination of the Critical Aggregation Concentration (CAC)

The CAC of DEXssPEGCOOH, DEXFA, and their combination in the aqueous phase were measured by fluorescence analysis using pyrene as a nonpolar probe [29,30]. In separate experiments, 20.0 μL pyrene solution at a concentration of 3.0×10^{-5} M in acetone was evaporated in vials. Meanwhile, each conjugate was dissolved at concentrations ranging from 1.6×10^{-7} to 1 mg mL^{-1} in phosphate buffer (0.01 M, pH 7.4) under magnetic stirring, and 1 mL of each solution was added to the pyrene vials. The content of the vials was mixed for 12 h, thereby leading to solutions with pyrene concentration of ca. 6.0×10^{-7} M. Then, the intensity ratios (I_3/I_1) of the third vibronic band at 385 nm to the first one at 373 nm of the fluorescence emission spectra of pyrene were recorded at 25 °C. Pyrene fluorescence emission spectra (λ_{exc} = 336 nm; λ_{em} = 350–500 nm) were recorded on Hitachi F-2500 spectrometer (Tokyo, Japan). All chemicals were purchased from Merck/Sigma Aldrich, Darmstadt, Germany.

2.5. Preparation of Nanoparticles and DOX Loading

In the same vial, DEXssPEGCOOH and DEXFA were dispersed in phosphate buffer solution (0.01 M, pH 7.4) at a final concentration of 1 mg mL^{-1} and magnetically stirred for 2 h at 25 °C. DOX-loaded nanoparticles (DOX@DFNPs) were prepared by dispersing each conjugate (final concentration 1 mg mL^{-1}) in a 58.8 μM DOX hydrochloride solution in phosphate buffer (0.01 M, pH 7.4) and magnetically stirred for 12 h at room temperature [31]. The dispersion was used as such in the next release experiments. The DOX content in DOX@DFNPs was confirmed by diluting 1 mL of DOX@DFNPs dispersion in 25 mL of methanol, in order to disrupt nanoparticle structure [32], followed by the measurement of the fluorescence of the solution (λ_{exc} = 480 nm; λ_{em} = 590 nm).

Size distributions were determined using a 90 Plus Particle Size Analyzer DLS equipment (Brookhaven Instruments Corporation, New York, NY, USA) at 25 °C. The autocorrelation function was measured at 90° and the laser beam operated at 658 nm. The polydispersity index (PDI) was directly obtained from the instrumental data fitting procedures by the inverse Laplace transformation and Contin methods. PDI values ≤ 0.3 indicate homogeneous and mono-disperse populations [33]. Morphological analysis of nanoparticles was carried out using transmission electron microscopy (TEM; HRTEM/Tecnai F30 [80 kV] FEI company, Hillsboro, OR, USA). A drop of the vesicle dispersion was placed on a Cu TEM grid (200 mesh, Plano GmbH, Wetzlar, Germany), and the sample in excess was removed using a piece of filter paper. A drop of 2% (w/v) phosphotungstic acid solution was then deposited on the carbon grid for 2 min. Once the excess of staining agent was removed with filter paper, the samples were air-dried, and the thin film of stained nanoparticles was observed.

2.6. Destabilization Experiments

Destabilization experiments of empty nanoparticles in reductive environments were performed by the dialysis method. Briefly, in separate experiments, 4 mL of freshly prepared empty nanoparticles (final concentration 1 mg mL^{-1}) were loaded in a dialysis bag (MWCO 12–14 kDa) and dialyzed against 30 mL phosphate (0.01 M, pH 7.4) and an acetate buffer (0.01 M, pH 5.5) containing GSH at different concentrations (0 and 10 mM) at 37 °C in a beaker with constant stirring. After 24 h, the mean diameter and PDI were measured by DLS. Each analysis was performed in triplicate.

2.7. Release Experiments

Release experiments were carried out by means of a dialysis method under sink conditions. In two different experiments, 2 mL DOX@DFNPs dispersion were loaded in a dialysis bag (cut-off molecular weight of 3.5 kDa) and dialyzed against 10 mL phosphate (0.01 M, pH 7.4) and an acetate (0.01 M, pH 5.5) buffer containing GSH at different concentrations (0 and 10 mM) at 37 °C in a beaker with constant stirring. At pre-established times, samples (0.5 mL) of release medium were withdrawn, replaced with fresh medium, and quantified by a fluorescence spectrometer (F-2500 Hitaki, Tokyo, Japan). Experiments were performed in triplicate.

2.8. Stability in Plasma Simulating Medium

The stability of DOX@DFNPs dispersion in plasma simulating fluid was performed by measuring the drug content in PBS solution containing 90% FBS [34]. In particular, 2 mL DOX@DFNPs dispersion was inserted in a dialysis bag (cut-off molecular weight of 3.5 kDa) and dialyzed against 10 mL of phosphate buffer solution (0.01 M, pH 7.4) containing 90% FBS at 37 °C in a beaker with constant stirring. At pre-established times, samples (0.5 mL) of release medium were withdrawn, replaced with fresh medium, and quantified by a fluorescence spectrometer (F-2500 Hitaki, Tokyo, Japan). Experiments were performed in triplicate.

2.9. Cell Culture

MCF-10A and MDA-MB-231 cell lines were from American Type Culture Collection (Manassas, VA, USA). All cell lines were authenticated and stored according to the supplier's instructions. Cells were used within four months after recovery of frozen aliquots and regularly tested for mycoplasma-negativity (MycoAlert Mycoplasma Detection Assay, Lonza, Basel, Switzerland). The MCF-10A cell line, a non-tumorigenic human epithelial breast cell line, was cultured in DMEM/F-12 supplemented with 5% horse serum (HS), L-glutamine (1%), penicillin/streptomycin (1%), 100 ng mL^{-1} cholera toxin, hydrocortisone (0.5 mg mL^{-1}), insulin (10 mg mL^{-1}), and epidermal growth factor (EGF) (20 ng mL^{-1}). MDA-MB-231 human breast cancer cells were cultured in Dulbecco's modified Eagle's medium (DMEM)/Nutrient Mixture F-12 Ham (DMEM/F12) supplemented with 5% fetal bovine serum (FBS) containing L-glutamine (1%) and penicillin/streptomycin (1%). The cells were maintained at 37 °C in a 5% CO_2 humidified incubator. Before each experiment, cells were grown in a phenol red-free medium, containing 5% charcoal-stripped FBS (cs-FBS), for at least 24 h.

2.10. Cell Viability Assay

The effect of DOX, DFNPs, and DOX@DFNPs was tested in MCF-10A and MDA-MB-231 cells using the 3-(4,5-dimethylthiazol-2-yl)-2,5-diphenyltetrazolium bromide (MTT) cell viability assay. Cells were plated in the appropriate medium (1×10^4) in 96-well tissue culture plates and incubated for 24 h at 37 °C and 5% CO_2, to allow cell adhesion. After 24 h, the culture medium was replaced with medium supplemented with 5% cs-FBS, and the cells were treated with DOX (0.08–1.28 μg mL^{-1}) and/or DFNPs (0.625–1.0 mg mL^{-1}) for 72 h. The MTT assay was performed adding 100 μL MTT stock solution in PBS (2 mg mL^{-1}) to each well and incubated for 4 h to allow the formation of violet formazan crystals. Then, the culture medium was removed, and 100 μL of DMSO was added to solubilize the formazan crystals. The absorbance was measured with the Multiskan EX Microplate Reader (Thermo Scientific, Waltham, MA, USA) at the wavelength of 570 nm.

2.11. Cellular Uptake Experiments

MDA-MB-231 and MCF-10A cells were treated or untreated with DOX and DOX@DFNPs for 24 h. After incubation, the cells were washed twice with phosphate-buffered saline (PBS, pH 7.4) and fixed in 3.7% formaldehyde solution in PBS for 10 min at room temperature and then washed again twice in PBS. The cells were permeabilized with a solution of 0.1% Triton X-100 in PBS for 3–5 min, followed by two washes in PBS. 4′,6-diamidino-2-phenylindole dihydrochloride (DAPI, 2 g mL^{-1}) staining was used for nuclei detection. To confirm the folate receptor-mediated uptake efficacy of DOX@DFNPs, both cell lines were pretreated with free FA (20 μg) for 1 h [35]. The cellular uptake of DOX and DOX@DFNPs (excitation 480 nm, emission 590–610 nm), was quantified in MDA-MB-231 cells using a confocal laser scanning microscope (Fluoview FV300, Olympus, London, UK). Fluorescence intensity was measured in three different (40×) optic fields of three different wells per each experimental condition.

2.12. DNA Flow Cytometry

To perform cell cycle analysis, MDA-MB-231 cells were treated or untreated with DOX or DOX@DFNPs for 48 h. The cells were then fixed with a solution of propidium iodide (100 μg mL^{-1}), and then RNase A (20 μg mL^{-1}) was added. Cellular cycle was measured using a FACScan flow cytometer (Becton Dickinson, Mountain View, CA, USA) and the data acquired using CellQuest software. Cell cycle profiles were determined using Mod-Fit LT.

3. Results and Discussion

The design of drug delivery systems able to recognize characteristic elements or respond to specific signals of the tumor tissues is the main strategy to target the drug release to the site of interest, optimizing the drug efficacy and minimizing the insurgence of undesirable side-effects. In this work, two different self-assembling specimens endowed with actively targeted (DEXFA) and pH/redox-responsive (DEXssPEGCOOH) elements were combined to obtain the DFNPs nanoparticles with affinity to folate receptors (through folate moieties) and the ability to trigger the drug release in response to different pH and GSH concentrations due to the presence of COOH functionalities and cystamine residues, respectively.

3.1. Synthesis and Characterization of DEXssPEGCOOH

As depicted in Figure 1, DEXssPEGCOOH conjugate was obtained by a two-step procedure consisting of (i) reductive amination of DEX in presence of cysHCl and sodiumcyano-borohydride and (ii) covalent coupling of DEXcys and $PEG_{600}COOH$ via carbodiimide chemistry.

DEXcys and DEXssPEGCOOH were characterized by FT-IR and ^1H-NMR analyses (Figure 2). In FT-IR spectrum of DEXcys (Figure 2a), no significant variation of the absorption bands was observed compared to native DEX, because the signals of cystamine were overshadowed by the more intense absorption bands of the polysaccharide. In the ^1H-NMR spectrum (Figure 2b), the signal of DEX anomeric proton (β) was evident at around 4.9 ppm, while the presence of cystamine residues was confirmed by the enhancement of the relative intensity of aliphatic methylene protons in the region from 3.1 to 4.2 ppm with respect to DEX spectrum. In particular, from the ratio between the intensity of aliphatic to anomeric protons of DEX and DEXcys, a derivatization degree of 9.2% was estimated. The successful covalent attachment of $PEG_{600}COOH$ on DEX was confirmed by the appearance, in FT-IR spectrum (Figure 2a), of new absorption bands at 1732 cm^{-1}, ascribable to C=O stretching of carboxyl functions, respectively. The further increase of the aliphatic region in ^{1}H-NMR spectrum of DEXssPEGCOOH conjugate allowed estimating that a 38% amount of cystamine residues in DEXcys was coupled with $PEG_{600}COOH$.

For both conjugates, 2D-HSQC spectra were recorded to further characterize the system (Figures S1 and S2 in Supplementary Material). DEXcys spectrum showed a 64 ppm resonance signal, ascribable to the cystamine protons, in addition to that attributed to the dextran carbon backbone, whereas in DEXssPEGCOOH conjugate an additional 70 ppm resonance, due to the -O-CH_2CH_2-O- repetitive PEG fragment, was observed.

In order to achieve further information about the conjugates structures, their diffusion coefficients were measured by ^{1}H-DOSY analyses [36] (see Supplementary Materials). A slight decrease of the diffusion coefficients was found moving from DEX (1.60×10^{-10} m^2/s), DEXcys (1.52×10^{-10} m^2/s), and DEXssPEGCOOH (1.33×10^{-10} m^2/s), as a consequence of the functionalization process enhancing the molecular weight of the resulting conjugate (Figures S4–S6 in Supplementary Materials). In addition, the DOSY spectra can be rationalized as a further confirmation of the purity of the conjugates: no small molecule signal was envisaged superimposed under the polymer proton resonances, with the exponential fittings of the decay curves for the different signals of the proton spectrum obtained with very small errors, and in line with each other (Figures S5 and S6 in Supplementary Materials).

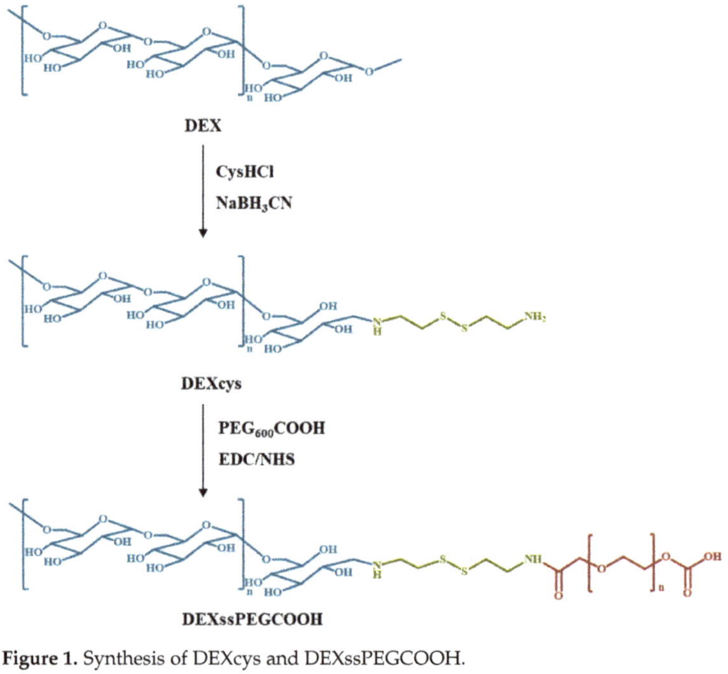

Figure 1. Synthesis of DEXcys and DEXssPEGCOOH.

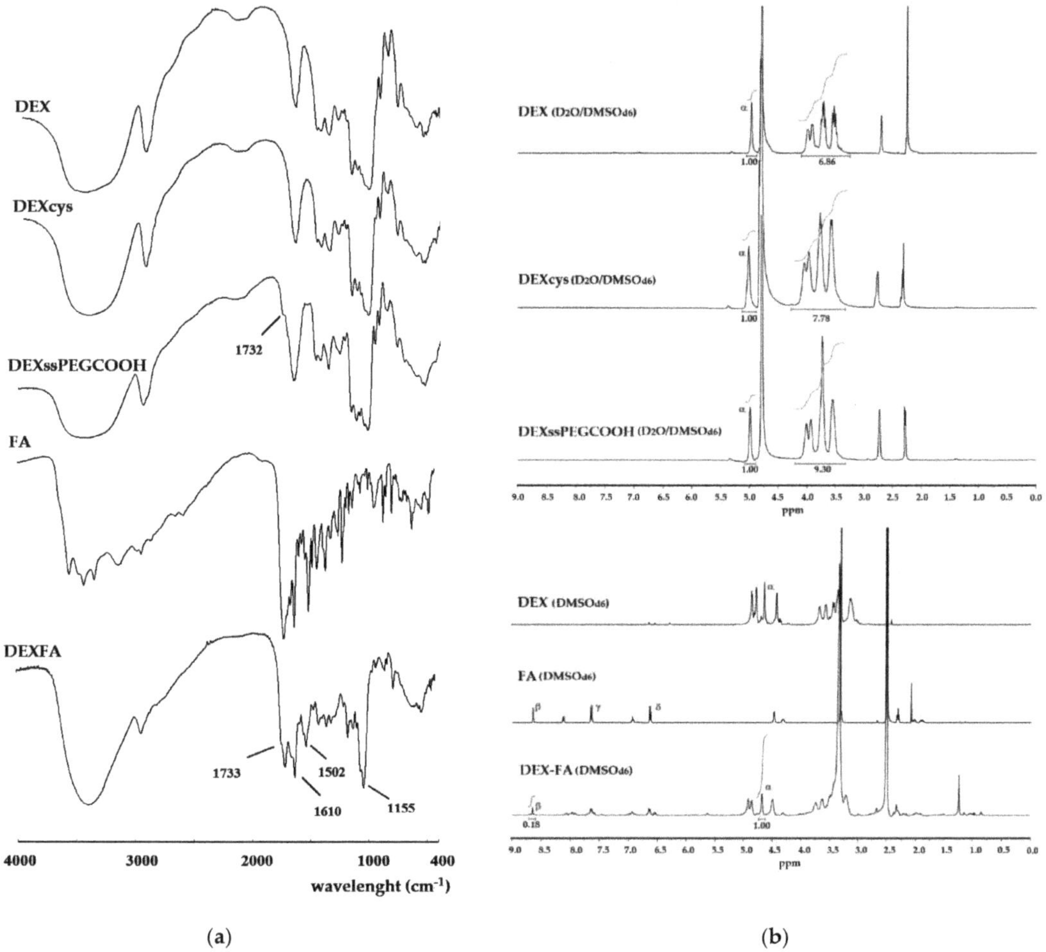

Figure 2. (**a**) FT-IR and (**b**) ^{1}H-NMR spectra of DEX and DEX conjugates.

3.2. Synthesis and Characterization of DEXFA Conjugate

DEXFA conjugate was prepared through esterification reaction between the γ-carboxylic acid group of FA and the hydroxyl groups of DEX (Figure 3).

In the FT-IR spectrum of DEXFA (Figure 2a), new absorption bands at 1733 and 1610 cm^{-1}, ascribable to C=O and -C=C- stretching vibration of FA, respectively, were observed. In addition, the characteristic bands of $-NH_2$ and -CONH- stretching vibrations at 1502 and 1155 cm^{-1} indicated the successful grafting of FA onto DEX. In the ^{1}H-NMR spectrum of DEXFA (Figure 2b), the signal of the DEX anomeric proton (α) was observed at 4.68 ppm, while the presence of FA residues was evidenced by the characteristic resonance signals of FA at 8.5 (pteridine proton, β), 7.6, and 6.8 ppm (γ and δ aromatic protons). The relative integration of pteridine proton peak (β) and anomeric proton (α) allowed calculating a derivatization degree (DD) of 18%, expressed as FA moieties with respect to DEX repeating units. Additionally, for DEXFA, a 2D-HSQC spectrum (Figure S3 in Supplementary Materials) was recorded, and it is in accordance with the reported proton spectrum.

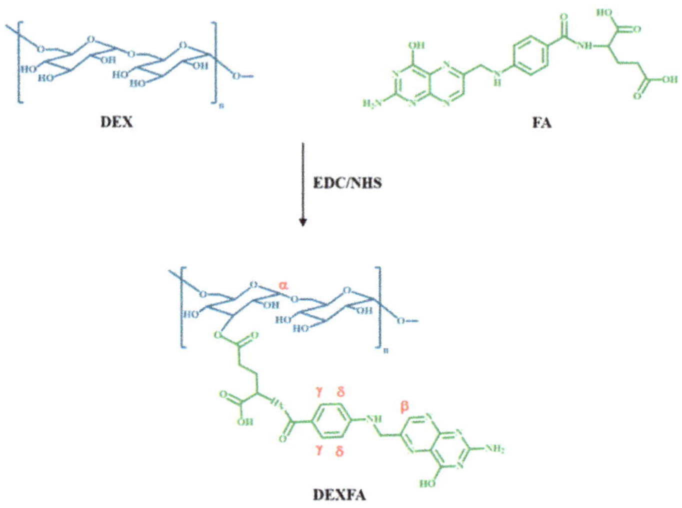

Figure 3. Synthesis of DEXFA conjugate.

3.3. Determination of CAC of DEXssPEGCOOH and DEXFA

The amphiphilic properties of DEXssPEGCOOH, DEXFA, and the combination of two conjugates were evaluated by measuring their CAC, defined as the concentration value above which the conjugates can organize in micellar structures. The determination of this parameter is crucial in view of a potential application in vivo, where the system is highly diluted in the systemic circulation: the lower the CAC value, the greater the stability of the micellar structures formed at low conjugate concentrations. For this measurement, pyrene was used as a probe in virtue of the ability to modify its fluorescent properties when located inside or in the proximity of the micellar hydrophobic domains. In Figure 4, the dependence of pyrene fluorescence spectra (I_{385}/I_{373} ratio) on the logarithm of conjugate concentration was reported. At low concentrations, the intensity values remained almost unchanged; in contrast, when the amount of conjugate increased, a sharp change of the intensity was observed, indicating the onset of self-assembly. From the crossover points, CAC values of 1.66 and 5.60 μg mL^{-1} were calculated for DEXssPEGCOOH and DEXFA, respectively. The lower CAC of DEXssPEGCOOH could be ascribable to the amphiphilic behavior of PEG enhancing the ability of the conjugate to self-assembly in water media. When the experiments were performed on the combination of the two DEX conjugates, in according with literature data [37], a further increase of the CAC value was recorded (8.9 μg mL^{-1}) as a consequence of the enhanced solubility of the whole system.

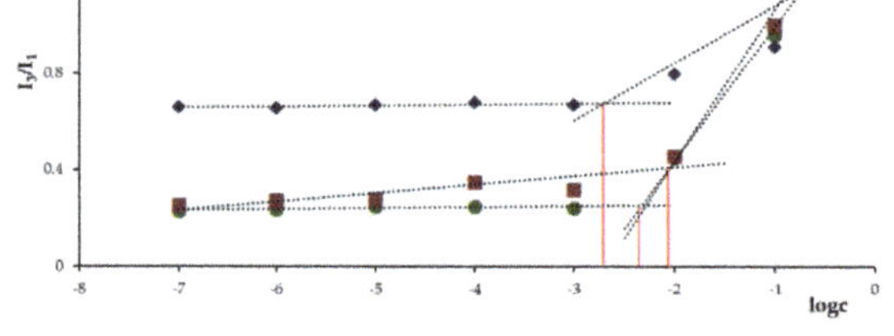

Figure 4. Dependence of pyrene fluorescence spectrum signals on (•) DEXFA, (♦) DEXssPEGCOOH, and (■) DEXFA/DEXssPEGCOOH concentration at pH 7.4.

3.4. Preparation of DFNPs

DFNPs were prepared in a straightforward procedure exploiting the amphiphilic character of the systems by dispersing the DEXssPEGCOOH and DEXFA conjugates in phosphate buffer solution at pH 7.4 (Figure 5).

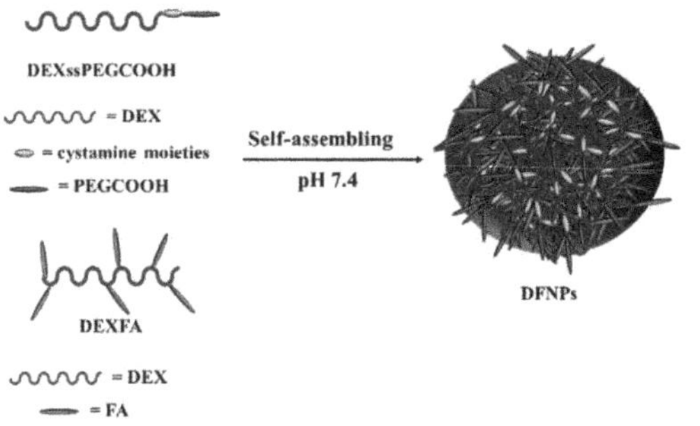

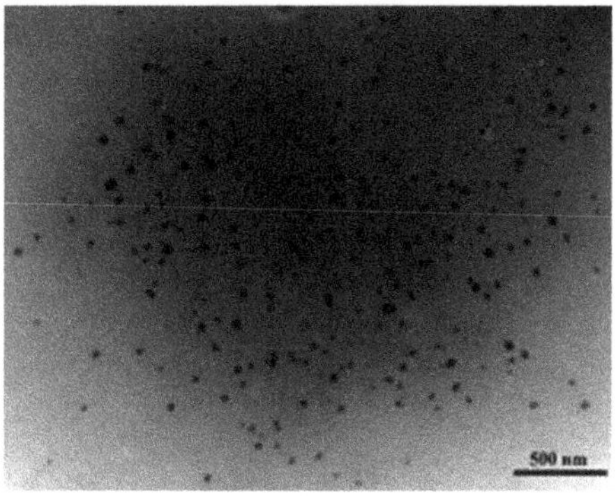

(**a**) (**b**)

Figure 5. (**a**) Representation of self-assembling process and (**b**) TEM image of DFNPs.

We can hypothesize that FA and PEG moieties in DEXFA and DEXssPEGCOOH conjugates constitute the hydrophobic core of the system, respectively, while the polysaccharide represents the hydrophilic nanoparticle corona.

DLS analysis showed a unimodal particle size distribution, with a mean hydrodynamic diameter of 50 ± 5 nm and a PDI of 0.1, while TEM micrographs revealed that the DFNPs were characterized by spherical shape and a mean diameter value close to that recorded by DLS (Figure 5b).

DFNPs present an ideal size to accumulate in tumor tissues via the EPR effect and allow an extensive tumor penetration [38].

3.5. Destabilization Experiments

The main feature of a stimuli-responsive nanocarrier is the ability to undergo specific structural modifications in response to internal or external stimuli in order to enhance cellular internalization and drug release in the diseased site, while remaining stable in the systemic circulation [39,40]. Due to the presence of pH and redox-responsive functionalities, it is expected that DFNPs can be destabilized, varying the pH and the GSH concentration in the surrounding medium. For this determination, the nanoparticles size changes in different pH and redox conditions were evaluated by DLS analyses after 24 h incubation.

In phosphate buffer pH 7.4, mimicking the extracellular environment, the mean diameter of DFNPs remained almost unchanged, while, when 10 mM GSH was added, the mean particle size and the PDI increased to 90 nm and 0.41, respectively, because of the breakage of crosslinking points (disulfide bonds) in the nanoparticle structure [41]. At pH 5.0 (close to the pKa value of PEGCOOH moieties of DEXssPEGCOOH conjugate), the particle diameter rose to 150 nm, probably as a consequence of the nanoparticle aggregation in larger structures due to the formation of intermolecular hydrogen bonds [42]. Finally, the incubation of the nanoparticles in acetate buffer pH 5.0 containing 10 mM GSH resulted in a further increase of the mean diameter and PDI value (nearly to 170 nm and 0.42, respectively), as a consequence of the simultaneous breakage of disulfide bonds and the aggregation phenomena that carried out to the destabilization of the nanoparticles.

3.6. DOX Loading and Release Experiments

DOX hydrochloride was loaded into the nanoparticles in a one-step procedure, adding the drug at a concentration of 32 mg per g of conjugate during the self-assembling of the conjugates (Figure 6), obtaining DOX@DFNPs systems showing similar morphological and dimensional behavior of the unloaded DFNPs. DOX@DFNPs dispersion was used as such in the release experiments, assuming that the total amount of added DOX was absorbed by the nanoparticles.

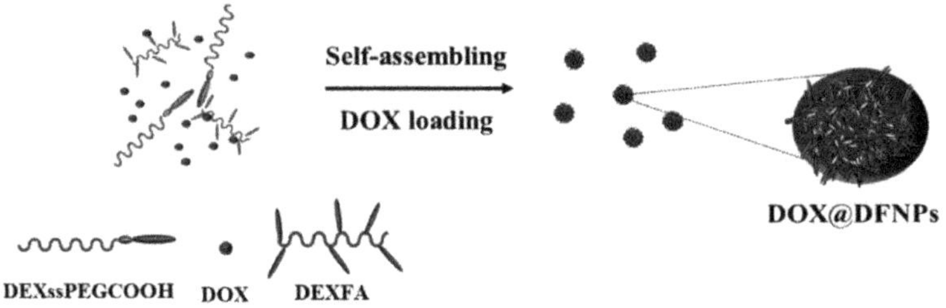

Figure 6. Schematization of DOX loading procedure.

Release experiments from DOX@DFNPs were performed in media mimicking the extracellular (phosphate buffer at pH 7.4) and intracellular (acetate buffer at pH 5.0) environments with or without GSH 10 mM (Figure 7).

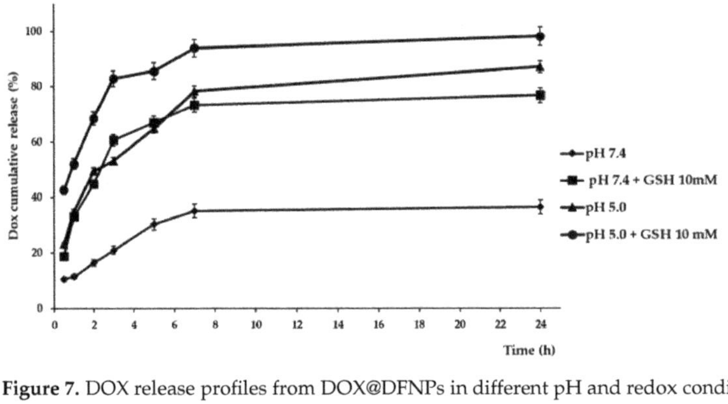

Figure 7. DOX release profiles from DOX@DFNPs in different pH and redox conditions.

At pH 7.4, simulating normal physiological conditions, a controlled release profile, not exceeding 36% in the first 24 h, was recorded, indicating that the nanoparticles were stable in the extracellular medium of healthy tissues and only a small amount of drug would be leaked during blood circulation.

In pH 7.4 medium containing 10 mM GSH and in acetate buffer at pH 5.0, similar release profiles were recorded, with percentages near to 35% and 80% in the first 60 min and 24 h, respectively. The data, in accordance with the stability experiments, are a consequence of the destabilization of the nanoparticle structure due, in the case of GSH in phosphate buffer, to the breakage of disulfide bonds, and in the case of acetate buffer at pH 5.0, to the modification of the ionization state of both DFNPs and drug. These phenomena modified the drug to polymer interactions, enhancing the drug release. Finally, when GSH was added to acetate buffer, the highest release percentages were observed at all the experimental times (52% and 98% after 60 min and 24 h, respectively), demonstrating the synergistic effect of pH and redox stimuli on the in vitro drug release and the suitability of the proposed system as redox/pH responsive drug delivery device. The stability of the loaded nanoparticles in plasma-simulating fluid was also performed using an FBS-containing phosphate buffer solution as the release medium. As expected, a controlled release profile very similar to that observed in the phosphate buffer in absence of FBS was recorded, with the nanoparticles maintaining more than 70% of the drug loaded after 24 h.

3.7. Biological Characterization

The effect of DOX, DFNPs, and DOX@DFNPs on cell proliferation was evaluated in non-tumorigenic epithelial cells (MCF-10A) and human breast adenocarcinoma cells (MDA-MB-231) after 72 h incubation. At first, the cytotoxicity of DOX or DFNPs in both cell lines was evaluated in the 0.08–1.28 μg mL^{-1} and 0.625–1.0 mg mL^{-1} concentration ranges, respectively. For free DOX, negligible cytotoxic effects were recorded on healthy cells at all the tested concentrations, while in cancer cells, a progressive decrease in cell survival in response to increasing concentrations of drug was observed, with a half-maximal inhibitory concentration (IC_{50}) value of 0.58 μg mL^{-1} (Figure 8a).

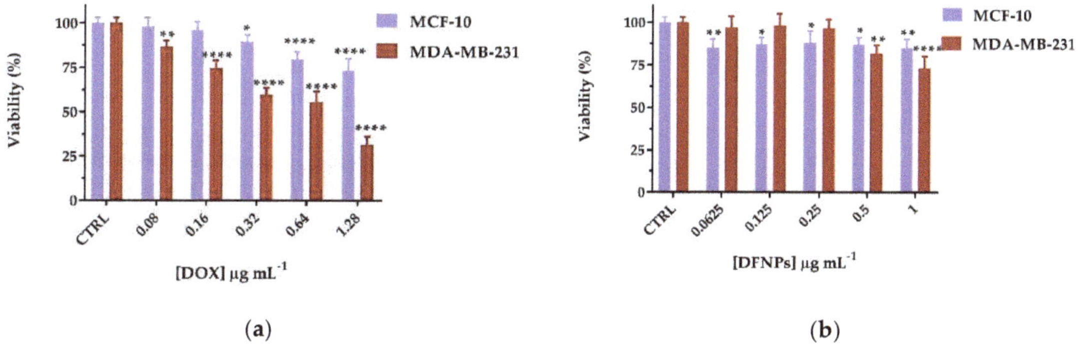

Figure 8. Viability percentages of (**a**) free DOX and (**b**) DFNPs on MCF-10A and MDA-MB-231 cell lines after 72 h. * $p < 0.05$, ** $p < 0.01$, and **** $p < 0.0001$ vs. each control (two-way ANOVA followed by Tukey's post-test; data are expressed as mean ± SD of three independent experiments).

Similarly, DFNPs did not exert any toxic effect on MCF-10A cells, while in MDA-MB-231 an inhibition of cell viability at a concentration up to 0.5 mg mL^{-1} was observed, as a confirmation of the safety of the proposed vehicle and the higher affinity towards tumor cells overexpressing the folate receptors.

Thus, the viability experiments with loaded nanoparticles (DOX@DFNPs) were performed fixing the nanoparticle concentration at 0.125 mg mL^{-1} to ensure both the effective formation of nanoparticles (as per CAC value of the conjugate) and negligible cytotoxic effects, and varying the DOX content according to the dose-response curve of free DOX. As depicted in Figure 9, nanoparticles retain their safety characteristics on healthy cells even when loaded with the cytotoxic drug, with cell death percentages not exceeding 20% at each tested concentration. However, when tested on MDA-MB-231, a dose-dependent decreasing of cell viability was recorded, with a significant increase of DOX activity at low drug concentration ($p < 0.001$ for 0.08 and 0.16 μg mL^{-1}), an IC_{50} value almost halved compared to free DOX (0.30), due to a more efficient cellular internalization in cancer cells overexpressing the folate receptors (Figure 9).

It is widely reported that free DOX is internalized via passive diffusion, while DOX@DFNPs are expected to enter the cells via endocytosis mediated by folate receptors, preventing the effluxion of pump-associated drug activity. The improved internalization of DOX after nanoparticle encapsulation was confirmed by cell uptake experiments. The exposure of the cancer cells to DOX@DFNPs results in a significantly enhanced fluorescent signal as compared to free DOX treatment (Figure 10), demonstrating the in vitro effectiveness of the proposed system. Furthermore, in order to assess the folate-targeting effect, both cell lines were pretreated with free FA before incubation with the nanoparticle formulation to block the folate receptors. As expected, a lower fluorescence signal was detected when the cancer cells were pretreated with free FA (Figure 11), as a demonstration of the receptor-mediated nanoparticles internalization. In addition, the lower expression of folate receptors in healthy cells was responsible for a reduced uptake of DOX@DFNPs compared to free DOX and to MDA-MB-231, with the FA pretreatment not affecting this behavior.

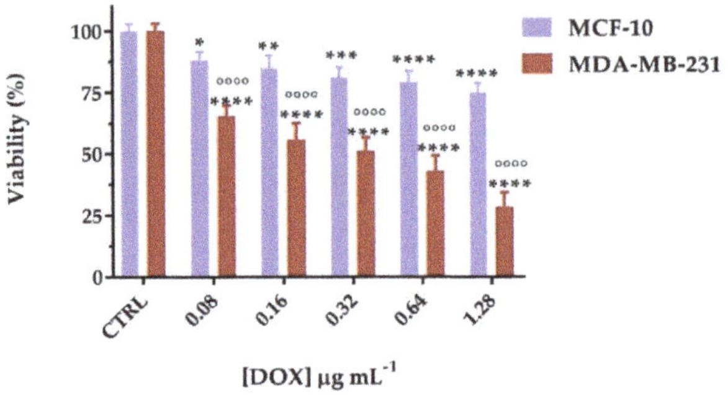

Figure 9. Cell viability of MCF-10A and MDA-MB-231 cells treated with DOX@DFNPs (drug concentration ranging from 0.08 to 1.28 µg mL^{-1}) after 72 h of culture. * $p < 0.05$, ** $p < 0.01$, *** $p < 0.001$, and **** $p < 0.0001$ vs. each control (CTRL); °°°° $p < 0.0001$ vs. the same drug equivalent concentration on MCF-10A (two-way ANOVA followed by Tukey's post-test; data are expressed as mean ± SD of three independent experiments).

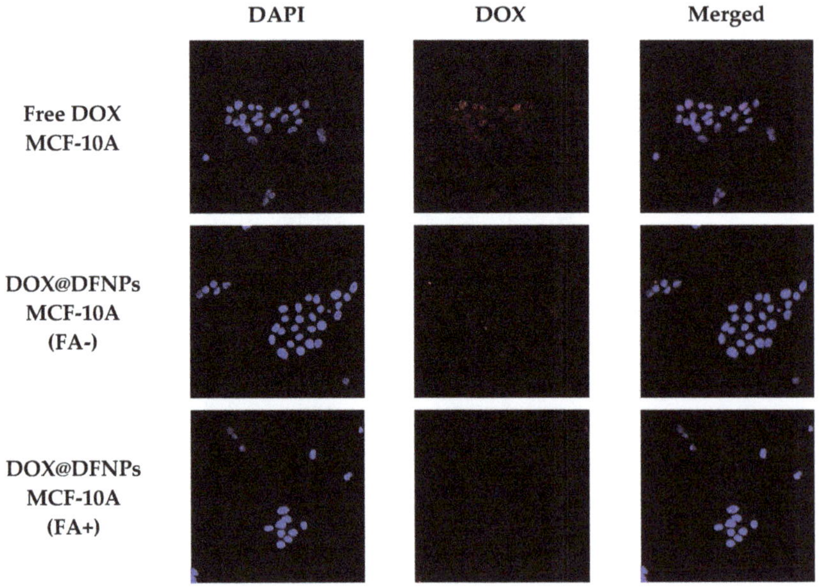

Figure 10. *Cont.*

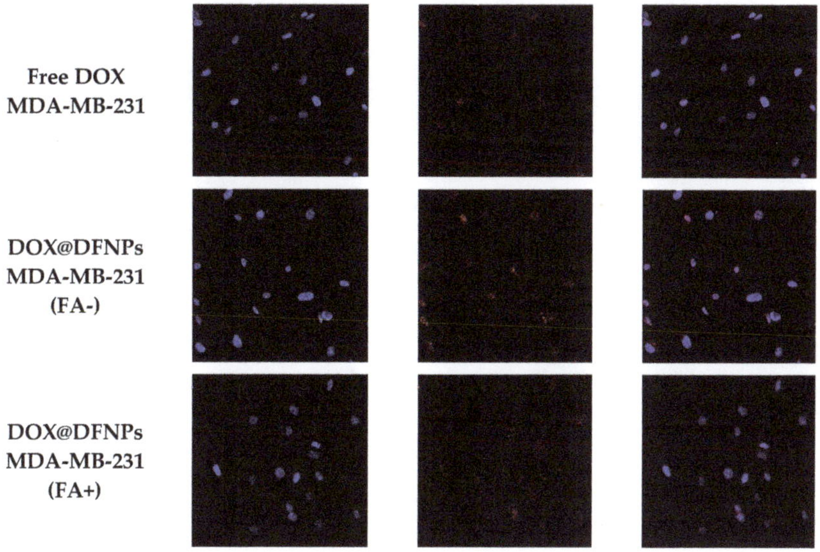

Figure 10. Confocal fluorescence images (40×) showing intracellular DOX or DOX@DFNPs (red fluorescence) and nuclear DAPI (blue signal) in MDA-MB-231 and MCF-10A cells. In (FA+) images, FA receptors were blocked by 1 h pretreatment with free FA, whereas (FA-) images were acquired in the absence of this receptor ligand.

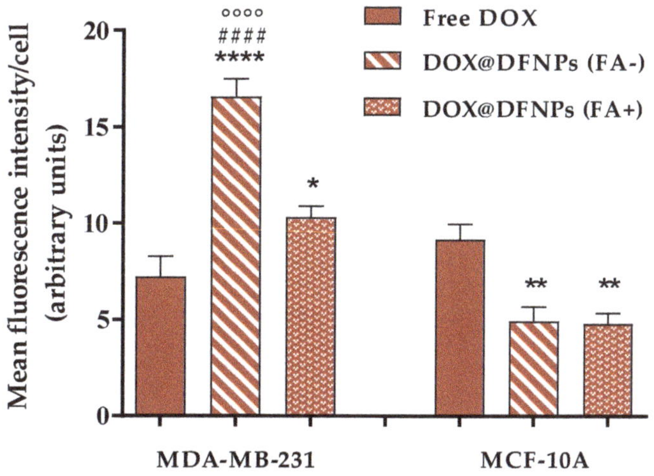

Figure 11. Quantitative analysis of red fluorescence intensity in MDA-MB-231 and MCF-10A cells (blocked-FA+ or unblocked-FA- receptors) exposed to DOX@DFNPs. Intensity of free DOX treatments was inserted as control. **** $p < 0.0001$ vs. free DOX; ** $p < 0.01$ vs. free DOX; * $p < 0.05$ vs. free DOX; #### $p < 0.0001$ vs. DOX@DFNPs (FA+); °°°° $p < 0.0001$ vs. MCF-10A cells (FA-). (One-way ANOVA followed by Tukey's post-test; data are expressed as mean ± SD of three independent experiments).

To better define the effect of free DOX and DOX@DFNPs in MDA-MB-231 cells, and to corroborate the above results, cell cycle analysis was performed. The data showed in Figure 12 demonstrated that free DOX decreased the G_0/G_1 phase and increased the G_2/M phase, compared to control cells [26].

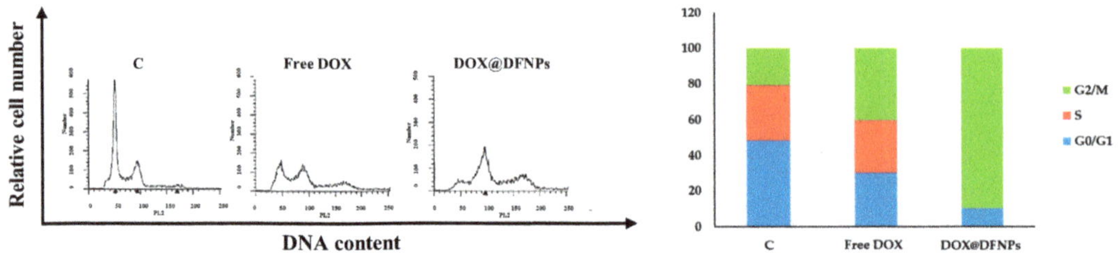

(a) (b)

Figure 12. Effects of free DOX and DOX@DFNPs on cell cycle distribution in breast cancer cells. (**a**) Flow cytometry analysis of the cycle profile of breast cancer cells. MDA-MB-231 treated with 0.32 μg mL^{-1} free DOX or DOX@DFNPs for 48 h, stained with propidium iodide (PI) and analyzed on a FACS flow cytometer. (**b**) Quantitative analysis of percentage gated cell at G_0/G_1, S, and G_2/M phases. C = control.

In cells treated with DOX@DFNPs, compared to free DOX, a more marked reduction of the G_0/G_1 phase, as well as a higher increase of G_2/M phase, was observed.

Thus, the DOX@DFNPs caused the block of cell cycle more efficiently than free DOX in the G_2/M phase.

4. Conclusions

DOX-loaded nanoparticles for targeted drug delivery were prepared by combining two self-assembling DEX conjugates with targeting (DEXFA) and pH/redox responsive (DEXssPEGCOOH) activity, respectively. The nanoparticles, with mean diameter of 50 nm and active targeting towards cancer cells via folate receptors, were found to be stable at physiological pH, while being destabilized at acidic pH and GSH concentration mimicking the intracellular environment, thus modulating the drug release profiles. Cell viability experiments on both healthy and cancer cells demonstrated the efficacy of the proposed system, with DOX@DFNPs showing a higher cytotoxic effect in MDA-MB-231 cells than in MCF-10A with respect to the free DOX. Moreover, cell uptake experiments confirmed the significant role of the nanoparticles in the vectorization of drug in cancer cells, demonstrating that this drug delivery system increased the effect on cell cycle arrest. Thus, this work proposed a simple and safe strategy to prepare new multifunctional DOX delivery systems with targeting and stimuli-responsive activity, with incoming in vivo studies being designed to further investigate the potential applications.

Supplementary Materials: The following Supplementary Materials are available online at https://www.mdpi.com/article/10.3390/nano11051108/s1: Figure S1: 2D-HSCQ spectrum of DEXcys, Figure S2: 2D-HSCQ spectrum of DEXssPEGCOOH, Figure S3: 2D-HSCQ spectrum of DEXFA, Figure S4: DOSY spectrum of DEX with fitting details, Figure S5: DOSY spectrum of DEXcys with fitting details, Figure S6: DOSY spectrum of DEXssPEGCOOH with fitting details.

Author Contributions: Conceptualization: M.C., F.I. and F.P.N.; methodology: M.C., A.P. and D.A.; validation: G.C., V.D.B., L.M., F.G. and D.A.; formal analysis: G.C. and L.M.; investigation: M.C., A.P., S.D.P. and M.F.; resources: F.I., L.M. and V.D.B.; data curation: F.P.N., L.M. and F.G.; writing—original draft preparation: M.C. and A.P.; writing—review and editing: G.C., L.M. and V.D.B.; visualization: M.C. and A.P.; supervision: F.I. and F.P.N. All authors have read and agreed to the published version of the manuscript.

Funding: This work was supported by MIUR Excellence Department Project funds (L.232/2016), awarded to the Department of Pharmacy, Health and Nutritional Sciences, University of Calabria, Italy, and by Università di Pisa under the "PRA—Progetti di Ricerca di Ateneo" (Institutional Research Grants)—PRA_2020-2021_58 "Agenti innovativi e nanosistemi per target molecolari nell'ambito dell'oncologia di precisione".

Institutional Review Board Statement: Not applicable.

Informed Consent Statement: Not applicable.

Data Availability Statement: Not applicable.

Conflicts of Interest: The authors declare no conflict of interest.

References

1. Aghebati-Maleki, A.; Dolati, S.; Ahmadi, M.; Baghbanzhadeh, A.; Asadi, M.; Fotouhi, A.; Yousefi, M.; Aghebati-Maleki, L. Nanoparticles and cancer therapy: Perspectives for application of nanoparticles in the treatment of cancers. *J. Cell Physiol.* **2020**, *235*, 1962–1972. [CrossRef]
2. Brigger, I.; Dubernet, C.; Couvreur, P. Nanoparticles in cancer therapy and diagnosis. *Adv. Drug Deliv. Rev.* **2012**, *64*, 24–36. [CrossRef]
3. Shi, Y.; Van der Meel, R.; Chen, X.Y.; Lammers, T. The EPR effect and beyond: Strategies to improve tumor targeting and cancer nanomedicine treatment efficacy. *Theranostics* **2020**, *10*, 7921–7924. [CrossRef]
4. Kang, H.; Rho, S.; Stiles, W.R.; Hu, S.; Baek, Y.; Hwang, D.W.; Kashiwagi, S.; Kim, M.S.; Choi, H.S. Size-Dependent EPR Effect of Polymeric Nanoparticles on Tumor Targeting. *Adv. Healthc. Mater.* **2020**, *9*, 1901223. [CrossRef] [PubMed]
5. Bi, Y.; Hao, F.; Yan, G.D.; Teng, L.S.; Lee, R.J.; Xie, J. Actively Targeted Nanoparticles for Drug Delivery to Tumor. *Curr. Drug Metab.* **2016**, *17*, 763–782. [CrossRef] [PubMed]
6. Biffi, S.; Voltan, R.; Bortot, B.; Zauli, G.; Secchiero, P. Actively targeted nanocarriers for drug delivery to cancer cells. *Expert Opin. Drug Del.* **2019**, *16*, 481–496. [CrossRef]
7. Du, J.Z.; Lane, L.A.; Nie, S.M. Stimuli-responsive nanoparticles for targeting the tumor microenvironment. *J. Control. Release* **2015**, *219*, 205–214. [CrossRef]
8. Fernandez, M.; Javaid, F.; Chudasama, V. Advances in targeting the folate receptor in the treatment/imaging of cancers. *Chem. Sci.* **2018**, *9*, 790–810. [CrossRef] [PubMed]
9. Scaranti, M.; Cojocaru, E.; Banerjee, S.; Banerji, U. Exploiting the folate receptor alpha in oncology. *Nat. Rev. Clin. Oncol.* **2020**, *17*, 349–359. [CrossRef]
10. Vaupel, P. Tumor microenvironmental physiology and its implications for radiation oncology. *Semin. Radiat. Oncol.* **2004**, *14*, 198–206. [CrossRef]
11. Kim, J.W.; Dang, C.V. Cancer's molecular sweet tooth and the Warburg effect. *Cancer Res.* **2006**, *66*, 8927–8930. [CrossRef] [PubMed]
12. Kanamala, M.; Wilson, W.R.; Yang, M.M.; Palmer, B.D.; Wu, Z.M. Mechanisms and biomaterials in pH-responsive tumour targeted drug delivery: A review. *Biomaterials* **2016**, *85*, 152–167. [CrossRef]
13. Xu, H.T.; Paxton, J.W.; Wu, Z.M. Enhanced pH-Responsiveness, Cellular Trafficking, Cytotoxicity and Long-circulation of PEGylated Liposomes with Post-insertion Technique Using Gemcitabine as a Model Drug. *Pharm. Res.* **2015**, *32*, 2428–2438. [CrossRef]
14. Schafer, F.Q.; Buettner, G.R. Redox environment of the cell as viewed through the redox state of the glutathione disulfide/glutathione couple. *Free Radic. Biol. Med.* **2001**, *30*, 1191–1212. [CrossRef]
15. Curcio, M.; Diaz-Gomez, L.; Cirillo, G.; Concheiro, A.; Iemma, F.; Alvarez-Lorenzo, C. pH/redox dual-sensitive dextran nanogels for enhanced intracellular drug delivery. *Eur. J. Pharm. Biopharm.* **2017**, *117*, 324–332. [CrossRef]
16. Fu, C.P.; Li, H.L.; Li, N.N.; Miao, X.W.; Xie, M.Q.; Du, W.J.; Zhang, L.M. Conjugating an anticancer drug onto thiolated hyaluronic acid by acid liable hydrazone linkage for its gelation and dual stimuli-response release. *Carbohyd. Polym.* **2015**, *128*, 163–170. [CrossRef] [PubMed]
17. Liu, Z.H.; Jiao, Y.P.; Wang, Y.F.; Zhou, C.R.; Zhang, Z.Y. Polysaccharides-based nanoparticles as drug delivery systems. *Adv. Drug Deliv. Rev.* **2008**, *60*, 1650–1662. [CrossRef] [PubMed]
18. Li, Y.L.; Zhu, L.; Liu, Z.Z.; Cheng, R.; Meng, F.H.; Cui, J.H.; Ji, S.J.; Zhong, Z.Y. Reversibly Stabilized Multifunctional Dextran Nanoparticles Efficiently Deliver Doxorubicin into the Nuclei of Cancer Cells. *Angew Chem. Int. Ed.* **2009**, *48*, 9914–9918. [CrossRef] [PubMed]
19. Liu, P.; Shi, B.H.; Yue, C.X.; Gao, G.H.; Li, P.; Yi, H.Q.; Li, M.X.; Wang, B.; Ma, Y.F.; Cai, L.T. Dextran-based redox-responsive doxorubicin prodrug micelles for overcoming multidrug resistance. *Polym. Chem.* **2013**, *4*, 5793–5799. [CrossRef]
20. Banerjee, A.; Bandopadhyay, R. Use of dextran nanoparticle: A paradigm shift in bacterial exopolysaccharide based biomedical applications. *Int. J. Biol. Macromol.* **2016**, *87*, 295–301. [CrossRef]
21. Curcio, M.; Cirillo, G.; Paoli, A.; Naimo, G.D.; Mauro, L.; Amantea, D.; Leggio, A.; Nicoletta, F.P.; Lemma, F. Self-assembling Dextran prodrug for redox- and pH-responsive co-delivery of therapeutics in cancer cells. *Colloid Surf. B* **2020**, *185*, 110537. [CrossRef] [PubMed]
22. Tang, Y.X.; Li, Y.H.; Xu, R.; Li, S.; Hu, H.; Xiao, C.; Wu, H.L.; Zhu, L.; Ming, J.X.; Chu, Z.; et al. Self-assembly of folic acid dextran conjugates for cancer chemotherapy. *Nanoscale* **2018**, *10*, 17265–17274. [CrossRef] [PubMed]
23. Butzbach, K.; Konhauser, M.; Fach, M.; Bamberger, D.N.; Breitenbach, B.; Epe, B.; Wich, P.R. Receptor-mediated Uptake of Folic Acid-functionalized Dextran Nanoparticles for Applications in Photodynamic Therapy. *Polymers* **2019**, *11*, 896. [CrossRef]

24. Wang, H.; Dai, T.T.; Zhou, S.Y.; Huang, X.X.; Li, S.Y.; Sun, K.; Zhou, G.D.; Dou, H.J. Self-Assembly Assisted Fabrication of Dextran-Based Nanohydrogels with Reduction-Cleavable Junctions for Applications as Efficient Drug Delivery Systems. *Sci. Rep.* **2017**, *7*, 40011. [CrossRef]
25. Zhang, Y.N.; Wang, H.L.; Mukerabigwi, J.F.; Liu, M.; Luo, S.Y.; Lei, S.J.; Cao, Y.; Huang, X.Y.; He, H.X. Self-organized nanoparticle drug delivery systems from a folate-targeted dextran-doxorubicin conjugate loaded with doxorubicin against multidrug resistance. *RSC Adv.* **2015**, *5*, 71164–71173. [CrossRef]
26. Bar-On, O.; Shapira, M.; Hershko, D.D. Differential effects of doxorubicin treatment on cell cycle arrest and Skp2 expression in breast cancer cells. *Anti-Cancer Drug* **2007**, *18*, 1113–1121. [CrossRef]
27. Volokhova, A.S.; Edgar, K.J.; Matson, J.B. Polysaccharide-containing block copolymers: Synthesis and applications. *Mater. Chem. Front.* **2020**, *4*, 99–112. [CrossRef]
28. Bosker, W.T.E.; Agoston, K.; Stuart, M.A.C.; Norde, W.; Timmermans, J.W.; Slaghek, T.M. Synthesis and interfacial behavior of polystyrene-polysaccharide diblock copolymers. *Macromolecules* **2003**, *36*, 1982–1987. [CrossRef]
29. Besheer, A.; Hause, G.; Kressler, J.; Mader, K. Hydrophobically modified hydroxyethyl starch: Synthesis, characterization and aqueous self-assembly into nano-sized polymeric micelles and vesicles. *Biomacromolecules* **2007**, *8*, 359–367. [CrossRef]
30. Curcio, M.; Mauro, L.; Naimo, G.D.; Amantea, D.; Cirillo, G.; Tavano, L.; Casaburi, I.; Nicoletta, F.P.; Alvarez-Lorenzo, C.; Iemma, F. Facile synthesis of pH-responsive polymersomes based on lipidized PEG for intracellular co-delivery of curcumin and methotrexate. *Colloid Surf. B* **2018**, *167*, 568–576. [CrossRef] [PubMed]
31. Liu, C.X.; Jiang, T.T.; Yuan, Z.X.; Lu, Y. Self-Assembled Casein Nanoparticles Loading Triptolide for the Enhancement of Oral Bioavailability. *Nat. Prod. Commun.* **2020**, *15*. [CrossRef]
32. Tavano, L.; Muzzalupo, R.; Mauro, L.; Pellegrino, M.; Ando, S.; Picci, N. Transferrin-Conjugated Pluronic Niosomes as a New Drug Delivery System for Anticancer Therapy. *Langmuir* **2013**, *29*, 12638–12646. [CrossRef]
33. Provencher, S.W. A constrained regularization method for inverting data represented by linear algebraic or integral equations. *Comput. Phys. Commun.* **1982**, *27*, 213–227. [CrossRef]
34. Zhuo, X.Z.; Lei, T.; Miao, L.L.; Chu, W.; Li, X.W.; Luo, L.F.; Gou, J.X.; Zhang, Y.; Yin, T.; He, H.B.; et al. Disulfiram-loaded mixed nanoparticles with high drug-loading and plasma stability by reducing the core crystallinity for intravenous delivery. *J. Colloid Interf. Sci.* **2018**, *529*, 34–43. [CrossRef] [PubMed]
35. Soe, Z.C.; Poudel, B.K.; Nguyen, H.T.; Thapa, R.K.; Ou, W.Q.; Gautam, M.; Poudel, K.; Jin, S.G.; Jeong, J.H.; Ku, S.K.; et al. Folate-targeted nanostructured chitosan/chondroitin sulfate complex carriers for enhanced delivery of bortezomib to colorectal cancer cells. *Asian J. Pharm. Sci.* **2019**, *14*, 40–51. [CrossRef]
36. Kuz'mina, N.E.; Moiseev, S.V.; Krylov, V.I.; Yashkir, V.A.; Merkulov, V.A. Quantitative determination of the average molecular weights of dextrans by diffusion ordered NMR spectroscopy. *J. Anal. Chem.* **2014**, *69*, 953–959. [CrossRef]
37. Spijker, H.J.; Dirks, A.J.; Van Hest, J.C.M. Synthesis and assembly behavior of nucleobase-functionalized block copolymers. *J. Polym. Sci. Pol. Chem.* **2006**, *44*, 4242–4250. [CrossRef]
38. Wang, H.B.; Li, Y.; Bai, H.S.; Shen, J.; Chen, X.; Ping, Y.; Tang, G.P. A Cooperative Dimensional Strategy for Enhanced Nucleus-Targeted Delivery of Anticancer Drugs. *Adv. Funct. Mater.* **2017**, *27*. [CrossRef]
39. Cheng, R.; Meng, F.H.; Deng, C.; Klok, H.A.; Zhong, Z.Y. Dual and multi-stimuli responsive polymeric nanoparticles for programmed site-specific drug delivery. *Biomaterials* **2013**, *34*, 3647–3657. [CrossRef]
40. Curcio, M.; Altimari, I.; Spizzirri, U.G.; Cirillo, G.; Vittorio, O.; Puoci, F.; Picci, N.; Iemma, F. Biodegradable gelatin-based nanospheres as pH-responsive drug delivery systems. *J. Nanopart. Res.* **2013**, *15*, 1581. [CrossRef]
41. Lin, C.; Zhong, Z.Y.; Lok, M.C.; Jiang, X.L.; Hennink, W.E.; Feijen, J.; Engbersen, J.F.J. Novel bioreducible poly(amido amine)s for highly efficient gene delivery. *Bioconjugate Chem.* **2007**, *18*, 138–145. [CrossRef] [PubMed]
42. Dong, W.Y.; Zhou, Y.F.; Yan, D.Y.; Li, H.Q.; Liu, Y. PH-responsive self-assembly of carboxyl-terminated hyperbranched polymers. *Phys. Chem. Chem. Phys.* **2007**, *9*, 1255–1262. [CrossRef] [PubMed]

MDPI
St. Alban-Anlage 66
4052 Basel
Switzerland
Tel. +41 61 683 77 34
Fax +41 61 302 89 18
www.mdpi.com

Nanomaterials Editorial Office
E-mail: nanomaterials@mdpi.com
www.mdpi.com/journal/nanomaterials

www.ingramcontent.com/pod-product-compliance
Lightning Source LLC
LaVergne TN
LVHW070425100526
838202LV00014B/1523
* 9 7 8 3 0 3 6 5 6 7 0 3 7 *